METEOROLOGICAL MONOGRAPHS

VOLUME 29　　　　　　　　JANUARY 2003　　　　　　　　NUMBER 51

CLOUD SYSTEMS, HURRICANES, AND THE TROPICAL RAINFALL MEASURING MISSION (TRMM)

—A TRIBUTE TO DR. JOANNE SIMPSON

Edited by

Dr. Wei-Kuo Tao
Dr. Robert Adler

American Meteorological Society
45 Beacon Street, Boston, Massachusetts 02108

ISBN 1-878220-54-3
ISSN 0065-9401

Support for this monograph has been provided by the Cooperative Research Centre for Southern Hemisphere Meteorology, Australia, and NASA Goddard Space Flight Center.

Published by the American Meteorological Society
45 Beacon St., Boston, MA 02108

Printed in the United States of America
by Allen Press, Inc., Lawrence, KS

TABLE OF CONTENTS

PREFACE

This monograph is dedicated to Dr. Joanne Simpson for her many pioneering research efforts in tropical meteorology during her 50-year career. Dr. Simpson's major areas of scientific research involved the "hot tower" hypothesis and its role in hurricanes, structure and maintenance of trade winds, air–sea interaction, and observations, and the mechanism for hurricanes and water spouts. She was also a pioneer in cloud modeling with the first one-dimensional model and had the first cumulus model on a computer. She also played a major role in planning and leading observational experiments on convective cloud systems.

The launch of the Tropical Rainfall Measuring Mission (TRMM) satellite, a joint United States–Japan project, in November of 1997 made it possible for quantitative measurements of tropical rainfall to be obtained on a continuous basis over the entire global Tropics. Dr. Simpson was the TRMM project scientist from 1986 until its launch in 1997. Her efforts during this crucial period ensured that the mission was both well planned scientifically and well engineered, as well as within budget. This achievement represents one of the high points in Dr. Simpson's 50 years in meteorology.

To honor Dr. Simpson's 50-year continuous contribution to improve our understanding of rain processes in the Tropics, a symposium took place at the National Aeronautics and Space Administration (NASA) Goddard Space Flight Center from 1 to 3 December 1999. This symposium consisted of presentations that focused on historical and personal points of view concerning Dr. Simpson's research career; her interactions with the American Meteorological Society; her leadership in TRMM; scientific interactions with Dr. Simpson that influenced personal research; research related to observations and modeling of clouds, cloud systems, and hurricanes; and research related to TRMM. There were a total of 36 presentations and 103 participants from the United States, Japan, and Australia.

The scientists who participated in the symposium contributed to this monograph, which is divided into five major parts. Part I consists of a summary paper that describes Dr. Simpson's major research on hurricanes, tropical clouds, and clouds systems and a paper that describes how she provides an ideal model of mentorship. Part II discusses Joanne's hot tower hypothesis and advances in tropical convection that have been observed during field campaigns and TRMM. Part III covers the development of cloud models and their application to the study of clouds and cloud systems. The development and application of the Goddard Cumulus Ensemble Model, Joanne's major project over the past two decades, are described. Hurricane and tropical cyclone genesis and organization are reviewed and discussed in part IV. The influence of mesoscale vortex dynamics and hot towers on hurricanes are included. Part V provides an overview of the TRMM project from historical and management perspectives. TRMM-related science, including measuring concepts, rain algorithm development, and the first precipitation radar in space, is described in part VI. Joanne's key role in the TRMM mission and design process is covered in these two latter sections.

The preparation of this monograph was supported by NASA's TRMM Program headed by Dr. Ramesh Kakar, TRMM Program Scientist.

Wei-Kuo Tao and Robert Adler, Editors

CONTRIBUTORS

Robert Adler
NASA Goddard Space Flight Center
Mail Code 912
Greenbelt, MD 20771
E-mail: adler@agnes.gsfc.nasa.gov

Isabella Angelini
462 Cedar Drive
Livermore, CA 94550
E-mail: ima4g@unix.mail.virginia.edu

Richard A. Anthes
University Corporation for Atmospheric Research
P.O. Box 3000
Boulder, CO 80307
E-mail: anthes@ucar.edu

David Bolvin
SSAI
NASA Goddard Space Flight Center
Mail Code 912
Greenbelt, MD 20771
E-mail: bolvin@agnes.gsfc.nasa.gov

William R. Cotton
Department of Atmospheric Science
Colorado State University
Fort Collins, CO 80523
E-mail: cotton@atmos.colostate.edu

Hillândia B. Cunha
Coordinator, Department of Geoscience Research
Instituto Nacional de Pesquisas da Amazonia
Manaus, AM, 69.083-000, Brazil
E-mail: hilandia@inpa.gov.br

Dr. Scott Curtis
Joint Center for Earth Systems Technology
University of Maryland, Baltimore County
NASA Goddard Space Flight Center
Mail Code 912
Greenbelt, MD 20771
E-mail: curtis@agnes.gsfc.nasa.gov

Dr. Michael Garstang
Department of Environmental Science
University of Virginia
Clark Hall
P.O. Box 400123
Charlottesville, VA 22904-4123
E-mail: mxg@swa.com

William M. Gray
Department of Atmospheric Science
Colorado State University
Fort Collins, CO 80523
E-mail: gray@atmos.colostate.edu

Dr. Jeffrey B. Halverson
JCET UMBC
NASA Goddard Space Flight Center
Mail Code 912
Greenbelt, MD 20771
E-mail: halverson@gilbert.gsfc.nasa.gov

Throy Hollis
4360 S. Phoenix St
Danis-Monthan AFB
Tucson, AZ 85707
E-mail: throy.hollis@dm.af.mil

Robert A. Houze Jr.
Department of Atmospheric Sciences
University of Washington
Box 351640
Seattle, WA 98195-1640
E-mail: houze@atmos.washington.edu

Thomas Keating
3915 Isbell Street
Silver Spring, MD 20906
E-mail: thomasjtkeating@aol.com

Dr. Chris Kidd
School of Geography
University of Birmingham
Edgbaston, Birmingham B15 2TT
United Kingdom
E-mail: c.kidd@bham.ac.uk

Christian D. Kummerow
Department of Atmospheric Science
Colorado State University
Fort Collins, CO 80523
E-mail: kummerow@atmos.colostate.edu

Thomas A. LaVigna
3913 York Lane
Bowie, MD 20715
E-mail: annandtomlavigna@erols.com

Margaret LeMone
National Center for Atmospheric Research
PO Box 3000
Boulder, CO 80307
E-mail: lemone@ucar.edu

Stephen Macko
Department of Environmental Sciences
University of Virginia
P.O. Box 400123
Charlottesville, VA 22903
E-mail: sam8f@virginia.edu

Gerald R. North
Department of Atmospheric Sciences
Texas A&M University
College Station, TX 77843-3150
E-mail: g-north@tamu.edu

Ken'ichi Okamoto
Department of Aerospace Engineering
Graduate School, Osaka Prefecture University
1-1 Gakuen-cho, Sakai, Osaka 599-8531, Japan
E-mail: Okamoto@aero.osakafu-u.ac.jp

Roger A. Pielke Sr.
Professor and State Climatologist
Department of Atmospheric Science
Colorado State University
Fort Collins, CO 80523
E-mail: pielke@atmos.colostate.edu

Elizabeth A. Ritchie
Department of Electrical Engineering
University of New Mexico
Room 125 EECE Bldg.
Albuquerque, NM 87131-1356
E-mail: ritchie@eece.unm.edu

Daniel Rosenfeld
Institute of Earth Sciences
The Hebrew University of Jerusalem
Jerusalem 91904, Israel
E-mail: daniel@vms.huji.ac.il

Dr. Eric A. Smith
NASA Goddard Space Flight Center
Code 912.1
Greenbelt, MD 20771
E-mail: easmith@pop900.gsfc.nasa.gov

Derek Stewart
Sandia National Laboratories
MS 9161, PO Box 969
7011 East Ave
Livermore, CA 94551
E-mail: dstewart@sandia.gov

Robert J. Swap
Department of Environmental Sciences
University of Virginia
P.O. Box 400123
Charlottesville, VA 22904-4123
E-mail: rjs8g@virginia.edu

Dr. Wei-Kuo Tao
Laboratory for Atmospheres
NASA Goddard Space Flight Center
Mail Code 912
Greenbelt, MD 20771
E-mail: tao@agnes.gsfc.nasa.gov

John Theon
6801 Lupine Lane
McLean, VA 22101
E-mail: jtheon@pop.erols.com

Dr. Thomas T. Wilheit
Department of Atmospheric Sciences
Texas A&M University
College Station, TX 77843-3150
E-mail: wilheit@tamu.edu

William L. Woodley
11 White Fir Court
Littleton, CO 80127
E-mail: williamlwoodley@cs.com

Edward J. Zipser
Department of Meteorology
University of Utah
135 S 1460 E, Room 819
Salt Lake City, UT 84112-0110
E-mail: ezipser@met.utah.edu

Chapter 1

The Research of Dr. Joanne Simpson: Fifty Years Investigating Hurricanes, Tropical Clouds, and Cloud Systems

W.-K. Tao,* J. Halverson,[+] M. LeMone,[#] R. Adler,* M. Garstang,[@] R. Houze Jr.,[&]
R. Pielke Sr.,** and W. Woodley[++]

*Laboratory for Atmospheres, NASA Goddard Space Flight Center, Greenbelt, Maryland
[+]Joint Center for Earth Systems Technology, University of Maryland, Baltimore County, Baltimore, Maryland
[#]NCAR, Boulder, Colorado
[@]Simpson Weather Associates, Inc., Charlottesville, Virginia
[&]Department of Atmospheric Sciences, University of Washington, Seattle, Washington
**Colorado State University, Department of Atmospheric Science, Fort Collins, Colorado
[++]Woodley Weather Consultants, Littleton, Colorado

ABSTRACT

Dr. Joanne Simpson's nine specific research contributions to the field of meteorology during her 50-year career—1) the hot tower hypothesis, 2) hurricanes, 3) airflow and clouds over heated islands, 4) cloud models, 5) trade winds and their role in cumulus development, 6) air–sea interaction, 7) cloud–cloud interactions and mergers, 8) waterspouts, and 9) TRMM science—will be described and discussed in this paper.

1. Introduction

Joanne Simpson's major areas of scientific research involved the "hot tower" hypothesis and its role in the global heat balance, hurricanes, the structure and maintenance of trade winds, air–sea interaction, and observations and the dynamics of hurricanes and waterspouts. She was also a pioneer in cloud modeling with the first one-dimensional model and created the first computer cumulus model. She led the work into multidimensional cloud modeling via observations of mergers and cloud interactions in convective lines. She played a major role in planning and leading observational experiments on convective cloud systems, such as the joint National Oceanic and Atmospheric Administration (NOAA)–Navy Project Stormfury, and the Florida Area Cumulus Experiment (FACE). She was a leading participant in the aircraft aspects of several Global Atmospheric Research Program (GARP) experiments, particularly the GARP Atlantic Tropical Experiment (GATE), Monsoon Experiment (MONEX) and Tropical Oceans Global Atmosphere Coupled Ocean–Atmosphere Response Experiment (TOGA COARE).

Her father's interest in aviation inspired Joanne's curiosity about weather and led her to a meeting with Carl Rossby at the University of Chicago. Joanne says, "Within ten minutes I was entrained into his orbit." Later her training qualified her to teach meteorology courses for aviation officers at both the University of Chicago and New York University (NYU) during World War II. She was one of seven women out of 200 students in that meteorology training course, and Joanne likes to note that all seven were in the top half of the class. Joanne became Herbert Riehl's student at the University of Chicago, focusing on the study of tropical convective clouds and their role in tropical wind systems. After receiving her Ph.D[1] Joanne started her experimental career at the Woods Hole Oceanographic Institution, moving on to the University of California at Los Angeles, NOAA, the University of Virginia, and in 1979 to the National Aeronautics and Space Administration/Goddard Space Flight Center (NASA/GSFC) as head of the Severe Storms Branch and later as the project scientist of the Tropical Rainfall Measuring Mission (TRMM).

Her many awards include the American Meteorological Society (AMS) Meisinger Award (1962), AMS Rossby Medal (1983), AMS Charles Franklin Brooks Award (1992), AMS Charles E. Anderson Award (2001), the Department of Commerce's Gold (1972) and Silver (1967) Medals, NASA's Exceptional Scientific Achievement (1982) and Outstanding Leadership (1998) Medals, and the first William Nordberg Memorial Award for Earth Sciences (1994). Dr. Simpson was elected an AMS Fellow (1969), a Fellow of the American Geophysical Union (1994), an Honorary Member of the AMS (1995), an Honorary Member of the Royal Meteorological Society (1999), and a member

---

[1] Dr. Simpson received her Ph.D. from the University of Chicago in 1949 and was the first female to obtain a Ph.D. in Meteorology.

of the National Academy of Engineering (1988). She has devoted herself to the AMS and served in many positions, culminating in the presidency of the organization. Her science partner and husband, Bob Simpson, has had a major influence on her career and provided her with unstinting moral support.

Joanne has made many major contributions to understanding the physical processes associated with hurricanes, tropical clouds, and cloud systems in the first 50 years of her career. In this paper, her research contributions in nine areas are reviewed and discussed. These include: 1) the hot tower hypothesis, 2) hurricanes, 3) airflow and clouds over heated islands, 4) cloud models, 5) trade winds and their role in cumulus development, 6) air–sea interaction, 7) cloud–cloud interactions and mergers, 8) waterspouts, and 9) TRMM science.

2. Major research areas and achievements

a. "Hot tower" hypothesis (1958 and 1979)

Riehl and Malkus (1958) analyzed synoptic upper-air sounding data for all stations between 30°S and 30°N. Specifically, they obtained temperature, humidity, and geopotential height at the 850-, 700-, 500-, 300-, 200-, 150-, and 100-mb levels from the U.S. Weather Bureau's publication "Monthly Climatic Data for the World." They used data "from 1952 onward," presumably about five years worth of data. The uppermost two levels were available only from 1956 onward. Riehl and Malkus obtained surface conditions from maps in the textbooks *Climatology* by Haurwitz and Austin (1944), and *Tropical Meteorology* by Riehl (1954). They considered two seasons (Dec–Feb and June–Aug) and constructed mean latitude–height cross sections of the synoptic variables.

From these cross sections, they determined the mean vertical profile of moist static energy Q at the location of the equatorial surface trough and at a distance of 20° of latitude from the trough (Fig. 1.1). The decrease of Q with height in the low levels, above the well-mixed layer to about 750 mb, was previously recognized. The increase of Q above 750 mb was the focus of the paper. They proposed a hypothesis to account for the high values of Q in the upper levels. They noted the following:

> Patently, a simple mass circulation will not be able to do this, for the horizontal export to high latitudes cannot be balanced above the minimum Q by a vertical advection. Radiation losses further compound the situation. We see that without some mechanism of transport other than a gradual mass circulation, the upper equatorial troposphere would be cooling at a rate of about 2°C per day!

It is worth noting that Riehl and Malkus's paper contained absolutely no data on clouds and/or precipitation. Nevertheless, they postulated that clouds explained the vertical profile of Q. They point out that small-to-mod-

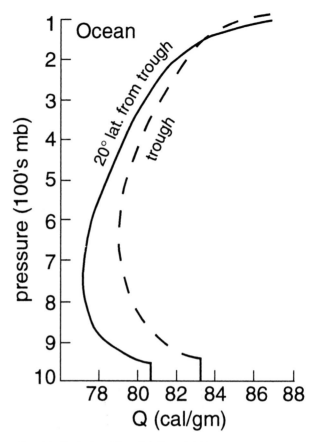

FIG. 1.1. Vertical profiles of Q (from Riehl and Malkus 1958).

erate entraining cumulus clouds can and probably do transport moist static energy upward to about the level of minimum Q. This fact is elegantly contained in later cumulus parameterization schemes such as those of Arakawa and Schubert (1974). Entraining cumulus do not penetrate much higher than the level of minimum Q. Somehow air parcels from the lowest levels of the atmosphere must be able to rise to the highest levels of the troposphere without becoming diluted by entrainment of dry environmental air. Riehl and Malkus recognized that this process must occur in ". . . embedded central cores in cumulonimbus clouds which are protected from mixing with the surroundings by the large cross section of the clouds." Their model for explaining the vertical profile of Q in the equatorial trough zone (Fig. 1.2) shows the lower layer of "mixing" by entraining cumulus with "convection" consisting of "undiluted chimneys penetrating to 40,000–50,000 feet in tropical disturbances" accounting for the heat transfer to upper levels. They estimated that 1500–5000 giant clouds would maintain the observed profile of Q. They postulated that in the core of the hot tower, fluxes of heat reached 2500 W m², a value confirmed by direct measurement in GATE (Emmitt 1978).

Riehl and Malkus's (1958) suggestion became known as the hot tower hypothesis. Because of the great amount

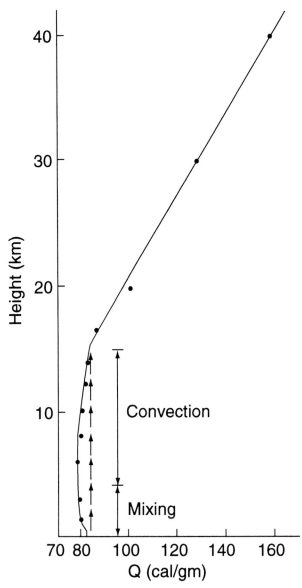

FIG. 1.2. Model of the mean vertical distribution of Q. Mechanisms of upward heat flow in the troposphere, and the limit of the upward penetration of heat gained by the atmosphere from the ground (from Riehl and Malkus 1958).

popular terminology in this second paper) assuming that a hot tower had an average area of 25 km^2 and an updraft velocity of 2–3 m s^{-1}. There was much better ocean transport data available for the 1979 paper. They found that the ocean transported heat poleward comparable to that transported by the atmosphere.

It should be noted that Riehl and Malkus (1958) and Riehl and Simpson (1979) recognized that the hot towers had to be embedded in mesoscale convective systems, which were an order of magnitude greater in area than the hot tower updrafts. This awareness was based on much personal visual observation of clouds on their part (e.g., Malkus and Riehl 1964), and it foreshadowed the work on the mesoscale organization of the circulation (Zipser 1969, 1977) and precipitation (Houze 1977) in the mesoscale convective systems containing the hot towers. The hot tower also explains why changes in the locations of tropical cumulunimbus cloud clusters, such as those due to an El Niño Southern Oscillation (ENSO) event or long-term land use change, can have such a large effect on the earth's global weather patterns (Hou 1998; Pielke 2001b; Chase et al. 2001).

b. Hurricanes

Beginning in the late 1940s and into the 1950s, reconnaissance flights into tropical cyclones for operational and research purposes were collecting the first datasets documenting the inner core and rainband structure. These observations laid the foundation for Joanne's theoretical investigations into the energetics of the hurricane eye and eyewall. The first of two landmark papers was published in 1958 under the title "On the structure and maintenance of the mature hurricane eye." In this paper, significant features of the mature hurricane eye were developed in terms of a simple model describing the physical processes undergone by the air composing it. These processes include detrainment of cloudy air from the tops of eyewall cumulonimbi, and the entrainment and mixing of eyewall and rain region air and angular momentum into hypothetical eye air parcels. The resulting theoretical thermodynamic vertical structure of the eye computed for a moderate and an intense hurricane agreed quite closely with observed data. These exercises led to the conclusion that much of the pronounced warmth in the eye occurs mainly above 700 mb, and the hydrostatic contribution to low surface pressures must derive chiefly from the eye column between 700 and 100 mb. In addition, it was hypothesized that the descent leading to eye warming, which was computed to be 16 cm s^{-1} at 850 mb, derives from low-level outflow caused by unbalanced centrifugal and Coriolis forces.

In her second significant paper, "On the dynamics and energy transformations in steady-state hurricanes" (Malkus and Riehl 1960), Joanne and coauthor H. Riehl presented a complex dynamical model of the inflow layer in a moderate, steady hurricane. Their goal was

of attention given to this idea, they revisited the hypothesis 21 years later (Riehl and Simpson 1979). In the second paper, the authors are not clear about how they upgraded the synoptic dataset, but evidently ~15 years of additional sounding data were used. They used surface maps from the revised textbook of Riehl (1979). They alluded to satellite data, which were by then widely available, but did not include visible or infrared imagery in the paper. Using a typical size for a mesoscale convective cloud system gleaned from satellite imagery, they made otherwise similar arguments to those in the earlier paper. These arguments led them to an estimate of 1600–2400 undiluted hot towers (they adopted the

to study the mechanisms by which energy release along a logarithmic spiral rainband is used to maintain the pressure field. This model relates wind speed, pressure gradient, surface shearing stress, mass flow, and convergence. The ambient atmosphere can only support a 10–20-hPa drop in surface pressure. To maintain the observed core pressure gradient, latent heat release must occur at a higher heat content than the mean tropical subcloud air, and derives from enhancement of oceanic sensible and latent heat fluxes driven by increased winds. From this work emerged the important relationship $-dp_s = 2.5 \ \theta_e$, which relates the surface pressure fall to the surface θ_e of air ascending in undilute eyewall hot towers. The importance of sensible heat fluxes is underscored by the observation that while the inflow air undergoes appreciable adiabatic expansion during its migration toward the eyewall, the temperature remains approximately constant, thus necessitating large fluxes from the ocean. Additional observations of the warm core in Hurricane Daisy reinforce the fact that about 75% of the surface pressure lowering derives hydrostatically from warming above 500 mb.

During the austral summer of 1993, a remarkable set of NASA instrumented aircraft flights into developing Tropical Cyclone Oliver captured the early formative stage of this storm, providing one of the very few research datasets describing the genesis of tropical cyclones (Simpson et al. 1997, 1998a). Given the multitude of cloud clusters over tropical oceans, it is perplexing why so few develop into tropical cyclones. The complexity of scale interactions, particularly those acting on the mesoscale, may play a crucial role. In the case of Oliver, cloud systems in the preformation environment contained mesoscale convective vortices, which were shown to interact and merge (Fig. 1.3). The interaction is essentially stochastic, but when these interactions occur in a vorticity-rich monsoon trough environment, it is shown (through the use of a baroclinic model) that the efficiency of mergers and the amplitude of the dominant emerging vortex is enhanced. It is hypothesized that once Oliver's vortex extended to the sea surface, the increase in winds enhanced the oceanic energy flux, thus fueling the hot towers necessary to initiate Oliver's rapid cyclogenesis. Building upon the earlier work of Ooyama (1982) and Schubert and Hack (1982), Joanne and colleagues speculate that by reducing the effective Rossby radius of deformation, the vortex merging process also facilitates confinement of latent heat to a warm core on the storm scale, rather than its diffusion away by gravity waves. Her 1958 hypothesis of the crucial role of undiluted hot towers was further developed in a second Oliver paper (Simpson et al. 1998b) and given observational support by the airborne measurements of Bonnie in 1998 by Heymfield et al. (2001).

c. Airflow and clouds over heated islands

Beginning with the observational study of airflow over Nantucket Island in Massachusetts (Malkus and Bunker 1952), Joanne Simpson introduced original concepts into understanding this flow. Joanne's two-part paper (Malkus and Stern 1953a,b) applied a theoretical model to describe the airflow observed over Nantucket Island, as well as to provide a more general description of this type of mesoscale system for any flat, heated island.

Malkus and Stern (1953b) introduced the concept of an "equivalent mountain," which corresponds to the influence of a heated island on the airflow. The equivalent mountain effect is shown to depend only upon the temperature distribution along the surface, the wind speed, the eddy turbulent conductivity in the surface layer, and the undisturbed thermodynamic stability. Figure 1.4 illustrates the form of this equivalent mountain for idealized flow, while Fig. 1.5 presents observational data for a case study day for Nantucket Island. This relation between air motion and the heating of a flat island has been confirmed repeatedly in more recent years, as illustrated, for example, in chapter 2 of this monograph.

In Malkus (1963), the concept of the equivalent mountain was used to propose that cumulus convective rainfall could be enhanced by asphalt ground coatings. Using the small flat island of Anegada in the West Indies, she documented the generation of a cumulus cloud street by the heated island. She concluded that an asphalt coating is sufficiently promising as a way to promote rainfall, that it should be investigated further. This work anticipates more recent studies that have shown the major role of land-surface properties, including the net radiation received at the ground surface and the partitioning of turbulence into sensible and latent heat forms on the occurrence, intensity, and patterning of cumulus convective rainfall (Pielke 2001).

d. The development of a cloud model in the context of cloud seeding experiments

Joanne pioneered the development of a computerized, one-dimensional, mathematical cloud model (Simpson et al. 1965) in the context of cloud seeding experiments (Simpson et al. 1967a; Simpson and Wiggert 1969, 1971). These experiments in cumulus dynamics were designed to evaluate the effect of silver iodide seeding upon individual tropical cumulus clouds. The cloud model predicts the rise rate, top heights, and in-cloud properties of both seeded and unseeded cumuli as a function of the ambient sounding, horizontal tower dimensions, and cloud-base conditions.

The initial randomized cloud seeding experiment was carried out on 23 tropical oceanic cumulus clouds on nine days in the summer of 1965 as part of the joint NOAA–Navy Project Stormfury. Following instructions in sealed envelopes, an aircraft seeded 14 of the clouds with 8–16 pyrotechnic silver iodide generators called Alecto units. Each unit released about 1.2 kg of silver iodide smoke. The nine remaining clouds were studied

FIG. 1.3. Tropical Cyclone Oliver color overlay of Willis Island radar echoes on a Geostationary Meteorological Satellite (GMS) satellite image. The red area represents cloud tops colder than 208 K, while the yellow is 184 K. The radar is not calibrated exactly; the stronger echoes are in lighter blue. The typhoon symbol denotes the pressure and wind vortex center, coincident within analysis uncertainty. The hot tower locations from the lidar are denoted as A–E. Tower D in the forming eyewall tops at 17 km while the tallest tower C is in the major rainband in the western mesoscale convective system (MCS). The solid lines are isobars. The W symbols denote warm core, with $W1_L$ apparently west of $W1_U$; W2 is a weaker warm core remaining with the weaker, sheared MCS. Asterisks denote the 10 locations of strong electric fields (from Simpson et al. 1998b).

in an identical manner as controls, using the same stack of four instrumented aircraft to penetrate the cloud before and after the seeding run. Cloud growth was documented by aircraft, radar, and photogrammetry. The seeded clouds grew vertically an average of 1.6 km more following the seeding run than did the control clouds; the difference is significant at the 0.01 level.

The cloud model was used to predict "seedability" and "seeding effect." Seedability is defined as the difference between the seeded and unseeded top of the same model cloud. Seeding effect is defined as the difference between the observed top and the predicted unseeded top of the same cloud. Both parameters were computed and graphed for all 23 clouds. The seeded and unseeded clouds separated into two distinct populations as shown in Fig. 1.6. Note that the seeded cases

(in circles) follow closely the dashed line of perfect prediction. The correlation between observed and predicted cloud growth is 0.973 ($P < 0.01$). The unseeded cases (in boxes) obviously constitute a different population since both the means and regression slopes of the two groups differ significantly ($P < 0.01$). The unseeded cases would fall along the horizontal line (zero effect) if the physical model made perfect predictions. That is, the unseeded clouds had growth potential (seedability) that was never realized (i.e., zero seeding effect) because they were never seeded.

This analysis shows clearly that seeding has an effect under the conditions specified and that the numerical model has considerable skill in specifying these conditions quantitatively and in predicting vertical growth for both seeded and unseeded clouds. It is the foundation

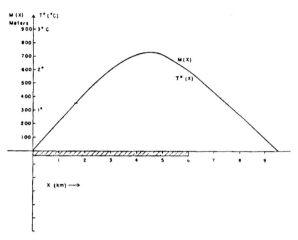

FIG. 1.4. Equivalent mountain corresponding to a rectangular-shaped temperature profile. Elevation of mountain $M(x)$ as a function of x is shown by the heavy line. The amount by which the island surface temperature exceeds that of the water determines the effective height of the equivalent mountain. The distance of exponential decay of the mountain in this example is 0.3 times the island width. This would represent a distance of 3 km if 2D were 10 km, $U = 2$ m s^{-1} and $K = 2.6 \times 10^5$ cm^2 s^{-1}. For a 2°C temperature perturbation of the island temperature, the maximum height reached by the mountain would be 667 m (adapted from Malkus and Stern 1953b).

upon which dynamic-mode seeding concepts have been laid, giving credibility to this approach to cloud seeding. Comparable model analyses were done for the Florida single cloud experiments (Simpson and Wiggert 1969, 1971).

Besides being of great value to dynamic-mode cloud seeding experimentation over the years, the Simpson cloud model is highly useful even today for the quantification of atmospheric properties in different large-scale environments and for the testing of various convective hypotheses. It can also be used to provide dynamic and thermodynamic datasets for remote sensor algorithms.

The dynamic-mode seeding experiments were moved from Puerto Rico and the Caribbean to Florida in 1967, and randomized seeding of individual supercooled convective clouds was conducted successfully in 1968 and 1971. Once again, statistically significant increases in cloud growth were noted following seeding along with strong indications for increased rainfall (Simpson and Woodley 1971). The Simpson 1D cloud model was refined further and applied to the new series of experiments (Simpson and Wiggert 1969, 1971). Enhanced cloud mergers appeared to be a primary effect of dynamic mode seeding with greatly increased rainfall from the merged convective entities over the rainfall that would have been produced had the convective clouds remained separate.

Although Joanne formally left the Florida cloud seeding program in 1974, she continued to make important contributions to the program, being the first to apply Bayesian statistical procedures for the analysis of seed-

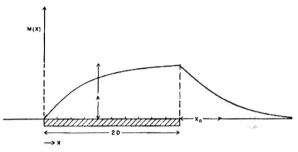

FIG. 1.5. Equivalent mountain for Nantucket Island on 14 Aug 1950 [case 4 studied by Malkus and Bunker (1952)]. Here T degrees (x) was the observed temperature profile along the island surface. The mountain function $M(x)$ is shown to have a similar shape to T degrees (x). The horizontal extent of the island is shown by the hatched region (adapted from Malkus and Stern 1953b).

ing effects (Simpson et al. 1975) and emphasizing the importance of downdrafts as linkages in dynamic cumulus seeding effects (Simpson 1980). In the latter paper, Simpson postulates that downdrafts provide the linkage between dynamically invigorated seeded cloud towers and those events below cloud base that cause enhanced inflow, new tower growth leading to cloud expansion, and frequent merger with neighboring clouds. She shows how acceleration of the cloud tops invigorated by seeding can lead to enhanced dynamic entrainment, increased evaporation, and hence to more rapidly formed and stronger downdrafts than would be

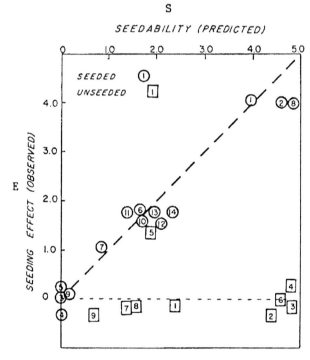

FIG. 1.6. Seeding effect vs seedability for the 23 clouds obtained during the randomized Caribbean cloud seeding experimentation in 1965. The seeded clouds are in circles and the control clouds are in boxes. See the text for details.

the case without seeding. Evidence taken from two series of experiments on relatively isolated cumuli in the Tropics and subtropics and combined with evidence derived from observational material on downdrafts collected since the late 1960s are cited in support of her arguments.

The current version of the dynamic-mode seeding experiments, as articulated by Rosenfeld and Woodley (1993) and refined further since then, shows the influence of Joanne over the years. In its present form the conceptual model involves a hypothesized series of meteorological events beginning initially on the scale of individual treated clouds or cells and cascading ultimately to the scale of clusters of clouds. This seeding is hypothesized to produce rapid glaciation of the supercooled cloud liquid water content (SLWC) in the updraft by freezing preferentially the largest drops so they can be rimed by the rest of the cloud water into graupel. This seeding-induced graupel is postulated to grow much faster than raindrops of the same mass so that a larger fraction of the cloud water is converted into precipitation before being lost to other processes. Ice multiplication is not viewed as a significant factor until most of the cloud water has been converted into precipitation. This faster conversion of cloud water into ice precipitation enhances the release of latent heat, increases cloud buoyancy, invigorates the updraft, and acts to spur additional cloud growth and/or support the growing ice hydrometeors produced by the seeding. These processes result in increased precipitation and stronger downdrafts from the seeded cloud and increased rainfall in the unit overall through downdraft interactions between groups of seeded and nonseeded clouds, which enhance their growth and merger. "Secondary seeding," whereby nonseeded clouds ingest ice nuclei and ice crystals produced by earlier seedings, is thought also to play a role in the precipitation enhancements.

Thus, one of the many research legacies of Joanne is the lead role that she played, both conceptually and with numerical models, in the development and testing of dynamic-mode seeding concepts that are still being refined and tested today.

e. Trade wind studies and the role of trade wind cumulus in their maintenance

Beginning in the mid-1950s (Malkus 1954, 1956, 1957, 1958b; Riehl and Malkus 1957; Brier and Simpson 1969) and strongly predicted upon the pioneering field studies in the Caribbean using an instrumented aircraft, Joanne began to grapple with the fundamental energetics of the trade wind system. Were the trades driving or driven? Are the trades simply a response to high-latitude forcing or are they being maintained by processes embedded within this limb of the general circulation? What was the basis for the observed remarkably steady-state nature of the trade winds? And how

does one account for the paradox that the atmosphere's most violent vortex, the hurricane, occurs intermittently within this steady-state system?

Remarkably, in these studies Joanne not only perceived that the system was highly scale dependent, but she was able to link these disparate scales into a coherent whole. While we are still grappling with the question of how fluxes of heat, moisture, and momentum across the tropical ocean interface are directly coupled to the convective and larger scales of the deeper atmosphere, Joanne quickly established the crucial role played by both shallow and deep cumulus clouds in the tropical atmosphere.

In her 1954 paper, Joanne established from aircraft penetrations of cumulus clouds the basic processes of entrainment (and detrainment). The trade wind cumuli act to elevate moisture along the trajectory, progressively raising the trade wind inversion. She also began, in this paper, formulating early thoughts on cloud models and cloud mergers.

By 1956, Joanne had produced a simplified model of the trade winds that incorporated cloud-produced diabatic heating. The model simulated most of the important features of the trades such as subsidence, downstream acceleration, increased vertical shear of the horizontal wind, and a downstream pressure drop (Fig. 1.7). Agreement between the motion and pressure fields depended upon the introduction of frictional stresses. The scale-dependent nature of the system is illustrated by the interplay between large-scale subsidence, which suppresses convection and convective growth, which, if enhanced, would trigger greater subsidence. The balance that is struck between these two scales results in the observed steady state of the trades.

Work completed in 1957 and 1958 (Riehl and Malkus 1957; Malkus 1958a,b) shows that the trades export through the circuitous route of deep convection in the equatorial trough, sensible and latent heat to higher latitudes. The sensible heat gain is related to the surface pressure distribution and the trades are found to be a self-driving system.

Joanne grappled throughout her career with how, in this steady-state system, regions of deep clouds and rainfall formed that were the necessary but not sufficient conditions for hurricane formation. In her work with Glenn Brier (Brier and Simpson 1969), she demonstrated the sensitivity of the tropical atmosphere to small but periodic changes in the fields of divergence. In an earlier study (Simpson et al. 1967b) she had shown that mesoscale to synoptic-scale systems in the Tropics can occur without any detectable perturbation of the horizontal velocity field. Classical wave or vortex structures in the velocity field were not a necessary condition for the development or organized deep convection.

In these and other studies Joanne has pointed to the need to adequately describe the three-or even two-dimensional velocity and moisture fields of the Tropics. These two variables, velocity and moisture, remain the

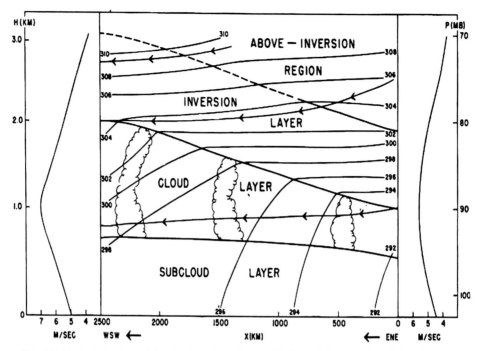

FIG. 1.7. Vertical structure of the air along the trajectory. Horizontal distance given in km downstream of entrance end. Vertical coordinate on the right in mb, and on the left in km (approximate). The heavy lines separate the layers. The lines with arrows are trajectories while the lighter solid lines are potential temperature isopleths labeled in degrees absolute. Trade cumulus clouds are entered schematically. To the right is the profile of wind speed (along the section) at the upstream end and to the left is the wind speed profile at the downstream end. Wind speeds are in m s⁻¹ (adapted from Malkus 1956).

keys to unlocking the remaining puzzles of the tropical atmosphere.

f. Air–sea interaction

Joanne's work has always portrayed air–sea interaction not as an isolated process, but as a piece of a complex, interacting system. Her understanding and enthusiasm were fueled by association with Woods Hole, which in the 1940s and 1950s did repeated measurements in the Caribbean using both ships and one of the first instrumented aircraft (Bunker 1955). This view has elucidated the role of sea–air interaction in the larger context of weather and climate, and has provided estimates of the flux interchange at the interface from both budgets and bulk aerodynamic techniques. Such crosschecks free the reader from excessive worry about the accuracy of the exchange coefficients in the bulk aerodynamic formulas used to predict air–sea fluxes, which continues to be an area of active research. Her work has spanned phenomena from the global circulations that serve as context to her work on the trade wind and equatorial trough zones, to the disturbances, hurricanes, and extratropical cyclones that modulate and feed off energy from the sea surface, to the smallest eddies transporting heat, moisture, and momentum from the sea surface and the clouds carrying the energy high into the

troposphere. As will be seen in this section, the role of the sea–air exchange varies with the context.

Joanne's paper on the role of sea–air interaction in the development of midlatitude cyclones (Simpson 1969) illustrates her holistic approach. Treatment of this subject had led to mixed conclusions; Simpson's synthesis of the known work at the time explored the reasons for these conclusions, drawing from her experience with convection in the Tropics. She began with the Petterssen et al. (1962) composite of a developing cyclone and showed the well-known concentration of strong fluxes in the unstable air (cold air over warm water) behind the cold front. Applying her experience with cloud development in the Tropics (Malkus and Ronne 1960), Joanne quickly arrived at the essential point, that it is not the fluxes themselves, but whether the fluxes occurred under conditions that favored deep penetrative convection that can heat the atmosphere through a deep layer, that determined whether or not the fluxes contributed to cyclone deepening. She used the well-known study of Winston (1955) as an example of rapid cyclone deepening from fluxes and penetrative convection under favorable synoptic conditions (positive vorticity and vertical velocity); the Winston case at later times and a storm documented by Bunker (1957) provided counterexamples, with strong surface fluxes having little effect because of strong low-level inversions. The Petterssen et al. (1962) composite provided an intermediate example, with some areas of deep convection over the strong fluxes, but only a small effect apparently because precipitation (latent heating) was small. Early numerical simulations of developing marine extratropical cyclones at the Geophysical Fluid Dynamics Laboratory (GFDL) indicated that inclusion of surface fluxes improved forecasts (e.g., Miyakoda et al. 1969), especially when sea surface temperature anomalies were included.

The 1969 synthesis has been refined but not changed in essence by high-resolution numerical simulations and focused field programs such as the Experiment on Rapidly Intensifying Cyclones over the Atlantic (ERICA) field experiment, which was held over the North Atlantic from 1 December 1988 to 28 February 1989 (Hadlock and Kreitzberg 1988). The importance of latent heat release in rapid cyclogenesis was demonstrated in a number of numerical experiments (e.g., Anthes 1983; Kuo and Reed 1988) and confirmed for a number of ERICA storms (e.g., Reed et al. 1993; Neiman et al. 1993; Rausch and Smith 1996). Air joining such cyclones may have already been moisture enriched well away from the storm, but well-placed locally enhanced fluxes have been shown to play a role in the intensification of some storms, particularly over warm water (Roebber 1989; Reed et al. 1993; Wakimoto et al. 1995). However, current papers focus on flux-enriched air being available for slantwise as well as vertical convection (e.g., Reuter and Yau 1993).

Joanne is well known for her important work done on the role of air–sea exchange in the trades, drawing from measurements over both the Atlantic [e.g., Woods Hole expeditions described in Woodcock and Wyman (1947); Bunker et al. (1949); Malkus (1958b); Colon (1960)] and Pacific (Riehl et al. 1951; Riehl and Malkus 1957). As in the review paper on extratropical cyclogenesis, Simpson's article in *The Sea* quickly emphasized that large surface energy fluxes do not lead to convection in the presence of a capping inversion—in this case, the trade wind inversion.[2] Rather, the trade inversion contains the moistening air until it reaches the equatorial trough zone, where more favorable thermal stratification and convergence allows the air to rise in deep convective hot towers that heat the middle and upper troposphere in the equatorial trough zone. Riehl et al. (1951) and Colon (1960) found the surface latent and sensible heat fluxes as residuals in the energy budgets that are consistent with bulk formulas and current estimates. Riehl and Malkus (1957) showed that the along-stream pressure gradient in the Pacific trades could be accounted for by the observed heating. From the budgets in Riehl et al. (1951), the streamwise pressure gradient is, to close approximation, also related to

——————
[2] While the modulation of deep convection by easterly waves in GATE is consistent with this finding, the modulation of convection by larger-scale motions in the Pacific warm pool during TOGA COARE was not so clear (e.g., Raymond 1995).

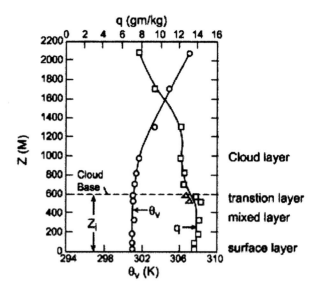

FIG. 1.8. Subdivision of the trade wind boundary layer, based on Malkus (1958b).

the vertical divergence of momentum transport. Combined, these two relationships link the sensible heat flux and streamwise surface stress, which to a good approximation is the total surface stress. By allowing the streamwise velocity to accelerate, and combining the kinematic equations, the gas law, and the first law of thermodynamics under the above approximations, Joanne (Malkus 1956) showed how heating, a streamwise pressure drop, and subsidence are related in the trade wind zone, and predicted the often-observed wind maximum at cloud base.

Joanne and colleagues pioneered studies of the heat and moisture flux through the subcloud and cloud layers of the trade wind zone. The work was mainly based on the previously mentioned series of Woods Hole expeditions to the Caribbean. Simpson's classic work (Malkus 1958b) describes the structure and processes in the subcloud (divided into a well-mixed layer and transition layer) and cloud layers (Fig. 1.8). The Woods Hole group correctly deduced that the sensible heating profile in the subcloud layer goes negative at a surprisingly low level, typically about one-third of the way to cloud base ($\sim$200 m) based on aircraft data (Bunker 1956, his Fig. 53; Malkus 1962). This deduction was supported by Colon's (1960) budget for the Caribbean basin (Malkus 1962, Fig. 30), although the ratio of downward heat flux near the top of the mixed layer to that at the surface is slightly larger than generally observed. Similarly, the ratio of latent heat flux near cloud top to that at the surface, about 0.8, is within the range of values from later studies (0.75–0.85) obtained by Riehl et al. (1951), Augstein et al. (1973), and Esbensen (1975). To determine whether the budget-based vertical humidity transport within the cloud layer was reasonable, Joanne (Malkus 1962) subdivided the trade cumulus layer at $\sim$1400 m into cloudy (saturation mixing ratio at aver-

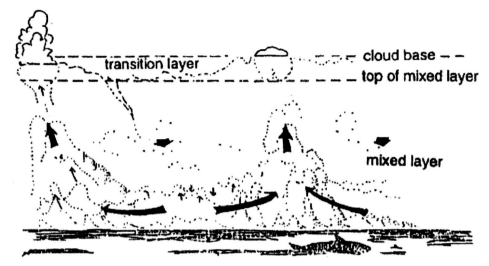

FIG. 1.9. Schematic of roll circulations and their relationship to cloud streets. The roll circulations (broad arrows) draw buoyant elements (outlined with stippling) into their upwelling regions and enable the buoyant elements to penetrate higher than otherwise, leading to cloud streets.

aged air temperature) and environmental air (measured mixing ratio), and combined the mass continuity and flux equations, to obtain an environmental subsidence of 1.2 cm s^{-1} for a typical cloud cover of 35%. For about 6% of the cloud being active and the remainder subsiding at a small speed (order 0.1 m s^{-1}), the observed water-vapor fluxes were consistent with active-cloud updrafts of 2 m s^{-1}, which is consistent with observations in Malkus (1954) and elsewhere.

Using smoke plumes, shipboard data, aircraft data, and reasoning based on budgets, Joanne (Malkus 1962) provided an early glimpse of how heat, vapor, and momentum are transported through the mixed layer. The ship data showed the expected association of low momentum with warm, moist air rising from the sea surface. She recognized that moisture, as well as temperature, contributes to the buoyancy of rising thermals (Malkus 1962, p. 207), a finding that even today catches the uninitiated by surprise. However, the lack of fast humidity measurements on the Woods Hole aircraft in 1958 (Malkus 1962, Fig. 53) precluded a detailed look at thermals. Once the thermals reached high enough altitudes, it was not possible to determine that their buoyancy was mainly due to water vapor. It was not until the 1960s that fast humidity measurements (e.g., Telford and Warner 1964) afforded the opportunity to trace water-vapor-induced buoyancy in plumes or bubbles through the mixed layer, and these two authors focused their attention on thermals over land, where temperature-induced buoyancy is relatively more important. Equipped with fast humidity measurements (Lenschow 1984) and inspired by the earlier Woods Hole efforts, LeMone and Pennell (1976) were able to connect cloud streets north of Puerto Rico to the sea surface by finding cloud source regions in the form of high virtual temperature upwelling regions of horizontal

roll vortices, whose buoyancy was entirely due to water vapor in the upper two-thirds of the subcloud layer. The roll upwelling regions were basically broad areas that favored the rise of thermals from the sea (Fig. 1.9). Directly linking clouds to the ocean surface via cloud-scale "roots" is more difficult. Lenschow and Stephens (1980) showed that convective thermals over the warm waters of the South China Sea,[3] identified by positive mixing ratio fluctuations, followed a coherent pattern through the subcloud layer, becoming steadily larger (diameter $d \sim z^{1/3}$) and widely spaced (number $N \sim z^{-1/3}$) with height z. Tying the thermals directly to clouds has proved more difficult. LeMone and Pennell (1976) were able to trace cloud-scale buoyant updrafts to about 100 m below cloud base in a regime with deeper trade wind cumulus; they argued the roots at lower altitudes, if present at all, could lie outside the aircraft track. Based on tethered balloon data, Garstang (Garstang and Fitzjarrald 1998) argued that the roots of growing cumulus extend downward as the clouds develop, presumably as a result of cloud-induced vertical pressure gradients (LeMone et al. 1988a,b).

Although Joanne's considerable contributions to the understanding of hurricanes are summarized elsewhere, we briefly summarize her role (Malkus and Riehl 1960; Riehl and Malkus 1961) in defining how heating and frictional drag worked in the inner core (less than $\sim$100 km from the center) of a moderate hurricane. By the time these papers were written, it was recognized (Byers 1944; Riehl 1954) that surface heating was necessary to lower pressure enough to make a hurricane. The mod-

---

[3] While it can be argued that this environment is fundamentally different from the trades, the temperature in the thermals behaves similarly to those in the trades, becoming lower than the environment about halfway up the subcloud layer.

el that grew out of the Malkus–Riehl collaboration is summarized in Malkus (1962). Assuming a symmetric pressure field, a constant drag coefficient and inflow depth and inflow winds independent of height, they obtained a wind field (and therefore stress field) that was reasonably consistent with observations and bulk aerodynamic estimates. The model results were then combined with thermodynamic data from Hurricane Daisy (in 1958) and an amazingly astute assumption of constant sea–air temperature contrast to derive a heat-energy budget that yielded surface latent and sensible heat fluxes as a residual. These fluxes agreed with bulk aerodynamic measurements to within 20%. A refined budget using actual wind measurements in Daisy and a smaller sea–air temperature contrast yielded flux values and a Bowen ratio (0.2) that matches values in similar hurricanes (Riehl and Malkus 1961; Cione et al. 2000).

This basic outline holds today, but enriched by details. Careful analysis of air–sea temperature contrast *on average* upholds the assumption of a near-constant air–sea temperature difference in the hurricane's inner core (Cione et al. 2000). More recent research focuses on what produces the air–sea temperature difference. Although it has been clear for several years that the sea surface temperature is cooled by wind-induced mixing with cooler waters below (e.g., Malkus 1962, p. 111), it is now recognized (Emanuel 1999) that the temperature stratification in the upper ocean in the hurricane's path can have a significant effect on the sea surface temperature, and thereby on surface fluxes and hurricane development. Similarly, hurricane researchers are looking more closely at the processes that affect the inflow temperature, such as intrusion of penetrative downdrafts from precipitating convection (e.g., Emanuel 1995; Cione et al. 2000), which bring low-θ_e air into the boundary layer, and entrainment (Barnes and Powell 1995), whose effect depends on differences between the inflow air and the air above. At larger distances (>150 km) from the hurricane center, Cione et al. (2000) found that θ_e actually drops as air converges inward. The values of the exchange coefficients relating air–sea temperature, specific humidity, and wind differences to the corresponding fluxes and the potential effect of sea spray remain tantalizing unknowns, but Simpson and Riehl's work suggests that the early surface flux estimates were not far off.

Joanne's work since this era (summarized elsewhere in this paper) has expanded to include numerical modeling but still involves the "expense and labor" needed to analyze the model results. A recent focus, along with colleagues (Wang et al. 1996, 2002) uses the Goddard Cumulus Ensemble (GCE) model to ascertain the role of air–sea fluxes in the development of mesoscale convective systems. As in the case of extratropical cyclones and the trade wind region, whether the fluxes have an impact has as much to do with whether they can feed clouds that penetrate into the upper troposphere. Results indicate that the significantly enhanced fluxes beneath the convection affect system development and precipitation amount far less than the transports in the environmental air that eventually forms the deep convection.

g. Cloud–cloud interactions and mergers

It has been recognized that the largest and most persistent convective clouds are often formed by the merging of two or more adjacent cells. Over the years, Joanne developed an original idea that showed how clouds merged into organized convective complexes that are the major producers of rainfall in the Tropics (Malkus 1954; Malkus and Riehl 1964; Simpson et al. 1980, 1993). Conventional digitized radar employed during FACE (1970–76) and GATE (1974) presented a unique, first opportunity to identify as well as quantify the importance of the growth of cumulus showers by cloud–cloud interaction and merging. In Simpson et al. (1980), a merger is defined as the consolidation of two previously separate echoes at the 1–mm h^{-1} isopleth of rain rate. A first-order merger is the result of the joining of two or more previously independent single echoes, and a second-order merger is formed by the juncture of two or more first- or second-order mergers. She and her colleagues found that the second-order mergers contributed about 68% of the rainfall and accounted for 53% of the area covered by echo. Single echoes and first-order mergers contributed similar properties of rainfall and total area. However, of all the echoes, nearly 90% were unmerged and only about 10% merged. In Florida, on relatively undisturbed days, she found that mergers of showers occurred predominantly in sea-breeze convergence zones, based on their predicted locations by a mesoscale model.

Simpson et al. (1993) studied a family of very tall (up to 20 km) cumulonimbus complexes (known locally as "Hectors") that developed almost daily over an adjacent pair of flat islands in the Maritime Continent region north of Darwin, Australia. About 90% of the total rainfall came from these merged systems, which comprised less than 10% of the convective systems. It was found that the behavior for the Hector and Florida storms was consistent and implied that total rainfall production comes from a multiplicative enhancement of production by cloud-scale interaction. Using rawinsonde and surface observations (including radiation and soil measurements), Joanne found that approximately 15% of the moisture source for the rainfall was provided by evapotranspiration, 45% by the sea-breeze convergence of warm, moist oceanic air, and 40% by "mining" of the original island boundary layer once the precipitation began.

Simpson et al. (1980) postulated that downdraft or gust front interaction is the primary mechanism of shower mergers. The approach or collision of gust fronts/downdrafts from adjacent clouds can force warm moist air upward, which in tropical air masses is both conditionally and convectively unstable. Her colleagues

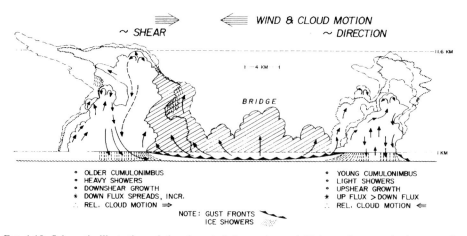

FIG. 1.10. Schematic illustration relating downdraft interaction to bridging and merger in the case of moderate shear opposite to the wind direction through most of the vertical extent of the cloud layer. The newer cumulonimbus on the right has predominately upward motion and moves faster than the ambient wind. The older cumulonimbus on the left has predominately downward motion and moves slower than the ambient wind so that the clouds move and propagate toward each other. The interaction of downdrafts enhances bridge development (adapted from Simpson et al. 1980).

Ulanski and Garstang (1978) found that stronger gust fronts were associated with moving, in contrast to stationary, showers. With wind shear, the merger processes should be different and more effective in joining and organizing cloud systems. Another important finding by Joanne was that the merging of convective showers is usually preceded by a "bridge" of visible smaller cumuli. The occurrence of the cloud bridge may simply manifest the importance of low-level convergence to the merging process. Joanne's proposed cloud–cloud interactions and merging mechanism is shown schematically in Fig. 1.10.

Using a three-dimensional cloud model to study cloud–cloud interactions and mergers, Tao and Simpson (1989) confirmed that the primary initiating mechanism for the occurrence of a second-order merger is the low-level convergence associated with cold cumulus outflows as suggested by observations. (See chapter 9 of this monograph for a review and discussion of other possible mechanisms associated with first-order mergers.)

h. Waterspouts

Starting back in the mid-1980s, Joanne and colleagues undertook research into the enigmatic origins of waterspouts. Two key published papers emerged (Simpson et al. 1986, 1991) that focused on waterspouts during GATE and over the Great Salt Lake. Building upon intensive observations published by Golden in the 1960s and 1970s, her approach was to combine conventional data (i.e., soundings, surface data, satellite imagery, and photography) with cloud resolving numerical models in order to elucidate the mechanisms for waterspout formation and dissipation.

In their GATE paper, Simpson et al. (1986) hypoth-

esize the necessity of parent vortices originating in the form of a cumulus-scale vortex pair. The vertical pair results from the tilting of ambient horizontal shear vorticity in the cloud environment. An active updraft intensifies the vortices below cloud base, with the relatively short lifetime of the waterspout coupled to the short duration of active updrafts within a growing congestus cloud. The three-dimensional numerical model of Schlesinger was used to simulate two cases of congestus growth based on observations collected from two days in GATE. In the one case with weak instability and low-level shear, the updraft was undercut early by a downdraft, and a vortex pair could not be extended to the surface. In the other, a more unstable subcloud layer was prescribed along with strong low-level shear, which resulted in strong parent vortices with maximum intensity at the surface. An unstable lower cloud layer appears to be a necessary but insufficient condition; given a favorable thermal environment, differences in the low-level wind profile are clearly crucial. It is noted that during both of the GATE case study days, waterspout formation occurred in the area where two gust fronts approached one another. Given a cumulus-scale parent vortex, the gust front interaction may provide the final convergence needed to stretch and amplify the subcloud vortex to waterspout strength. The mechanism is shown schematically in Fig. 1.11.

In the Great Salt Lake paper, Simpson et al. (1991) present detailed analyses of a waterspout that formed in a cumulus line over the lake. The spout was anticyclonic and formed in an environment of strong low-level instability and shear. A series of numerical experiments was conducted using the Goddard–Schlesinger cloud model. The model work capitalizes on several modifications made since the earlier (1986) waterspout work, including the introduction of ice-phase

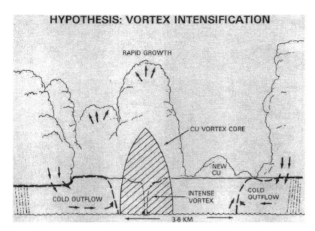

FIG. 1.11. Schematic illustration (not to scale) of cumulus–outflow interactions in relation to invigorated cumulus over a developing parent vortex. The shaded region approximately outlines the vortex core of one of the vortex pair organized by the model cloud F on day 186. The circle at the surface within the core is the sea surface dark spot; the dashed vertical lines show where the condensation funnel will appear as the central pressure drops. The gust front within and beneath cloud F is not shown on this diagram.

processes and an embedded high-resolution axisymmetric vortex model. The clouds were initiated using both a convergence line–type mechanism and a buoyant bubble. Both types resulted in formation of a cloud-scale vortex pair, but the line initiation developed stronger vortices, especially the anticyclonic member. An important finding is that the model cloud processes alone (i.e., preexisting anticyclonic rotation in the parent cloud) can produce a waterspout in the absence of external vorticity sources.

It should be noted that contrasting ideas have emerged regarding mechanisms responsible for formation of nonviolent tornadic vortices (i.e., waterspouts and landspouts), namely, Simpson's theory of preexisting organized vorticity contained within the cumulus cloud, and that of Wakimoto and Wilson (1989) as prescribed for nonsupercell tornado/landspout development. According to the Wakimoto and Wilson study, the rotation originates from low-level horizontal shear initially unrelated to (external to) overhead cloud processes, which then becomes titlted and stretched upward upon coincidence with a convective updraft. As noted by Simpson et al. (1991), in the case of waterspouts both mechanisms may be important, in varying degrees depending on the cloud environment.

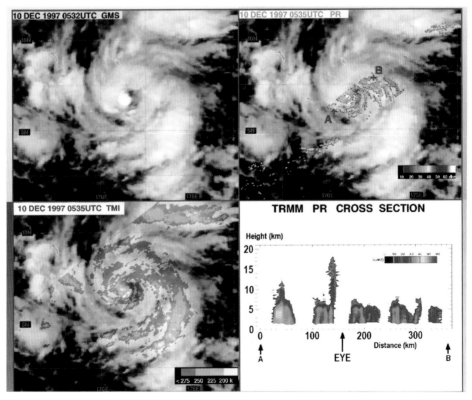

FIG. 1.12. (upper left) GMS IR image at 0532 UTC 10 Dec 1997. The coldest IR temperature (at this 4-km resolution) is 179 K. (upper right) TRMM PR image at 0535 UTC superimposed on the GMS image. Note the eye within the coldest portion of the GMS image, which showed no eye. (lower left) TRMM Microwave Imager (TMI) 85-Ghz channel superimposed on the GMS image. Wider swath on TMI shows the cloud pattern of a nearly mature tropical cyclone. (lower right) Cross section of TRMM PR from A to B shown in the upper-right PR image.

i. Tropical Rainfall Measuring Mission (TRMM) science

Joanne became project scientist for the fledgling Tropical Rainfall Measuring Mission (TRMM) in 1986. She brought to TRMM her 40 years or so of study and research in tropical meteorology and a particular knowledge of convective systems and storms in that region. She also had a very strong interest in using TRMM to answer questions related to tropical precipitating systems and their impact on climate processes. In this TRMM role, she led the development of the TRMM science goals and the observational requirements necessary to achieve them. This work was described in a 90-page TRMM science document authored by her and the initial TRMM Science Steering Group. Joanne also took the step at this early stage of TRMM to condense this scientific rationale for TRMM into an article for the "Bulletin of the American Meteorological Society" (AMS) (Simpson et al. 1988), complete with a cover image showing an artist's conception of the TRMM spacecraft. This early broad exposure of the TRMM concept, and the scientific basis for it, was accomplished more than nine years before the TRMM launch.

In the years leading up to the TRMM launch, Joanne continued her research focusing on convective systems using observations and models. In this pre-TRMM period, she resumed her use of field experiment data, including that from the airborne rain-mapping radar (AR-MAR), a simulator for the Precipitation Radar (PR) on TRMM. This field work culminated in a review of tropical cyclone formation emphasizing the role of hot towers in early intensification of systems (Simpson et al. 1998). By using the aircraft radar data and electrification and lightning information in these studies, she was preparing for the use of TRMM data in tropical cyclone studies. In addition, she contributed to algorithm research directly related to the TRMM mission in the area of satellite microwave observation of precipitation (Adler et al. 1991) and retrieval of latent heat profiles using models and satellite data (Tao et al. 1993).

With the successful launch of TRMM in late 1997, Joanne stepped down as TRMM project scientist and focused on TRMM-based tropical cyclone research, but also published a summary of early mission results (Simpson et al. 2000). Initial, exciting results of the TRMM views of Typhoon Paka (Simpson et al. 1998) gave an early indication of the importance of TRMM data in diagnosing convective bursts, showing a very tall TRMM radar echo (see Fig. 1.12). Joanne and colleagues are continuing to study TRMM data to understand the relations of these features, noted in numerous cases, to cyclone strength and intensification.

REFERENCES

Adler, R. F., H.-Y. M. Yeh, N. Prasad, W.-K. Tao, and J. Simpson, 1991: Microwave simulations of a tropical rainfall system with a three-dimensional cloud model. *J. Appl. Meteor.*, **30**, 924–953.

Anthes, R. A., 1983: Regional models of the atmosphere in middle latitudes. *Mon. Wea. Rev.*, **111**, 1306–1335.

Arakawa, A., and W. H. Schubert, 1974: Interaction of a cumulus ensemble with the large-scale environment: Part I. *J. Atmos. Sci.*, **31**, 674–701.

Augstein, E., H. Riehl, R. Ostapoff, and V. Wagner, 1973: Mass and energy transports in an undisturbed Atlantic trade-wind flow. *Mon. Wea. Rev.*, **101**, 101–111.

Barnes, G. M., and M. D. Powell, 1995: Evolution of the inflow boundary layer of hurricane Gilbert (1988). *Mon. Wea. Rev.*, **123**, 2348–2368.

Brier, G. W., and J. Simpson, 1969: Tropical cloudiness and rainfall related to pressure and tidal variations. *Quart. J. Roy. Meteor. Soc.*, **95**, 120–147.

Bunker, A. F., 1955: Turbulence and shearing stresses measured over the North Atlantic Ocean by an airplane acceleration technique. *J. Meteor.*, **12**, 445–455.

——, 1956: Measurements of counter-gradient heat flows in the atmosphere. *Austral. J. Phys.*, **9**, 133–143.

——, 1957: Turbulence measurements in a young cyclone over the ocean. *Bull. Amer. Meteor. Soc.*, **38**, 13–16.

——, B. Haurwitz, J. S. Malkus, and H. Stommel, 1949: *Vertical Distribution of Temperature and Humidity over the Caribbean Sea.* Papers in Physical Oceanography and Meteorology, No. 11, Massachusetts Institute of Technology and Woods Hole Oceanographic Institution, 82 pp.

Byers, H. R., 1944: *General Meteorology.* McGraw-Hill, 540 pp.

Chase, T. N., R. A. Pielke Sr., T. G. F. Kittel, R. R. Nemani, and S. W. Running, 2001: Simulated impacts of historical land cover changes on global climate. *Climate Dyn.*, **16**, 93–105.

Cione, J. J., P. G. Black, and S. H. Houston, 2000: Surface observations in the hurricane environment. *Mon. Wea. Rev.*, **128**, 1550–1561.

Colon, J., 1960: On the heat balance of the troposphere and water body of the Caribbean Sea. National Hurricane Res. Project Rep. 41, U.S. Dept. of Commerce, Washington, D.C., 65 pp.

Emanuel, K., 1995: Sensitivity of tropical cyclones to surface exchange coefficients and a revised steady-state model incorporating eye dynamics. *J. Atmos. Sci.*, **52**, 3969–3976.

Emmitt, G. D., 1978: Tropical cumulus interaction with and modification of the subcloud region. *J. Atmos. Sci.*, **35**, 1485–1502.

Esbensen, S., 1975: An analysis of subcloud-layer heat and moisture budgets in the western Atlantic trades. *J. Atmos. Sci.*, **32**, 1921–1933.

Garstang, M., and D. R. Fitzjarrald, 1998: *Observations of Surface to Atmosphere Interactions in the Tropics.* Oxford University Press, 384 pp.

Hadlock, R., and C. W. Kreitzberg, 1988: Experiment on rapidly intensifying cyclones over the Atlantic (ERICA): Field study, objectives, and plans. *Bull. Amer. Meteor. Soc.*, **69**, 1309–1320.

Haurwitz, B., and J. Austin, 1944: *Climatology.* McGraw Hill, 410 pp.

Heymsfield, G., J. Halverson, J. Simpson, L. Tian, and T. P. Bui, 2001: ER-2 Doppler radar investigations of the eyewall of hurricane Bonnie during the convection and moisture experiment-3. *J. Appl. Meteor.*, **40**, 1310–1330.

Hou, A. Y., 1998: Hadley circulation as a modulator of the extratropical climate. *J. Atmos. Sci.*, **55**, 2437–2457.

Houze, R. A., Jr., 1977: Structure and dynamics of a tropical squall-line system. *Mon. Wea. Rev.*, **105**, 1540–1567.

Kuo, Y.-H., and R. J. Reed, 1988: Numerical simulations of an explosively deepening cyclone in the eastern Pacific. *Mon. Wea. Rev.*, **116**, 2081–2105.

LeMone, M. A., and W. T. Pennell, 1976: The relationship of trade-wind cumulus distribution to subcloud-layer fluxes and structure. *Mon. Wea. Rev.*, **104**, 524–539.

——, G. M. Barnes, J. C. Fankhauser, and L. Tarleton, 1988a: Pressure fields measured by aircraft around the cloud-base updraft of deep convective clouds. *Mon. Wea. Rev.*, **116**, 313–337.

——, L. F. Tarleton, and G. M. Barnes, 1988b: Perturbation pressure at the cloud base of cumulus clouds in low shear. *Mon. Wea. Rev.,* **116,** 2062–2068.

Lenschow, D. H., 1984: Aircraft measurements in the boundary layer. *Probing the Atmospheric Boundary Layer,* D. H. Lenschow, Ed., Amer. Meteor. Soc., 39–55.

——, and P. L. Stephens, 1980: The role of thermals in the convective boundary layer. *Bound.-Layer Meteor.,* **19,** 509–532.

Malkus, J. S., 1954: Some results of a trade-cumulus cloud investigation. *J. Meteor.,* **11,** 220–237.

——, 1956: On the maintenance of the trade winds. *Tellus,* **8,** 335–350.

——, 1957: Trade cumulus cloud groups: Some observations suggesting a mechanism of their origin. *Tellus,* **9,** 33–44.

——, 1958a: On the structure and maintenance of the mature hurricane eye. *J. Meteor.,* **4,** 337–349.

——, 1958b: On the structure of the trade wind moist layer. *Pap. Phys. Oceanogr. Meteor.,* **13,** 3–47.

——, 1962: Large scale interactions. *The Sea: Ideas and Observations on Progress in the Study of the Seas.* Vol. 1, M. N. Hill, Ed., Interscience Publishers, 88–294.

——, 1963: Tropical rain induced by a small natural heat source. *J. Appl. Meteor.,* **2,** 547–556.

——, and A. Bunker, 1952: *Observational Studies of the Air Flow over Nantucket Island during the Summer of 1950.* Papers in Physical Oceanography and Meteorology, No. 12, Massachusetts Institute of Technology and Woods Hole Oceanographic Institution, 50 pp.

——, and H. Riehl, 1960: On the dynamics and energy transformations in steady-state hurricanes. *Tellus,* **12,** 1–20.

——, and ——, 1964: *Cloud Structure and Distributions over the Tropical Pacific Ocean.* University of California Press, 229 pp.

——, and C. Ronne, 1960: Cloud distributions over the tropical oceans in relation to large-scale flow patterns. *Physics of Precipitation,* Geopohys. Monogr., No. 746, Amer. Geophys. Union, 45–60.

——, and M. E. Stern, 1953a: The flow of a stable atmosphere over a heated island. Part I. *J. Meteor.,* **10,** 30–41.

——, and ——, 1953b: The flow of a stable atmosphere over a heated island. Part II. *J. Meteor.,* **10,** 105–120.

Miyakoda, K., J. Smagorinsky, R. F. Strickler, and G. D. Hembree, 1969: Experimental extended predictions with a nine-level hemispheric model. *Mon. Wea. Rev.,* **97,** 1–76.

Neiman, P. J., M. A. Shapiro, and L. S. Fedor, 1993: The life cycle of an extratropical marine cyclone. Part II: Mesoscale structure and diagnostics. *Mon. Wea. Rev.,* **121,** 2177–2199.

Ooyama, K. V., 1982: Conceptual evolution of the theory and modeling of the tropical cyclone. *J. Meteor. Soc. Japan,* **60,** 369–380.

Petterssen, S., D. L. Bradbury, and K. Pedersen, 1962: The Norwegian cyclone models in relation to heat and cold sources. *Geofys. Publ.,* **24,** 243–280.

Pielke, R. A., Sr., 2001: Influence of the spatial distribution of vegetation and soils on the prediction of cumulus convective rainfall. *Rev. Geophys.,* **39,** 151–177.

Rausch, R. L., and P. J. Smith, 1996: A diagnosis of a model-simulated explosively developing extratropical cyclone. *Mon. Wea. Rev.,* **124,** 875–904.

Raymond, D., 1995: Regulation of moist convection over the west Pacific warm pool. *J. Atmos. Sci.,* **52,** 3945–3959.

Reed, R. J., G. A. Grell, and Y.-H. Kuo, 1993: The ERICA IOP5 storm. Part II: Sensitivity tests and further diagnosis based on model output. *Mon. Wea. Rev.,* **121,** 1595–1612.

Reuter, G. W., and M. K. Yau, 1993: Assessment of slantwise convection in ERICA cyclones. *Mon. Wea. Rev.,* **121,** 375–386.

Riehl, H., 1954: *Tropical Meteorology.* McGraw-Hill, 392 pp.

——, 1979: *Climate and Weather in the Tropics.* Academic Press, 611 pp.

——, and J. S. Malkus, 1957: On the heat balance and maintenance of circulation in the trades. *Quart. J. Roy. Meteor. Soc.,* **83,** 21–29.

——, and ——, 1958: On the heat balance in the equatorial trough zone. *Geophysica,* **6,** 503–538.

——, and ——, 1961: Some aspects of Hurricane Daisy, 1958. *Tellus,* **13,** 181–213.

——, and J. S. Simpson, 1979: The heat balance of the equatorial trough zone, revisited. *Contrib. Atmos. Phys.,* **52,** 287–305.

——, T. C. Yeh, J. S. Malkus, and N. E. LaSeur, 1951: The northeast trade of the Pacific Ocean. *Quart. J. Roy. Meteor. Soc.,* **77,** 598–626.

Roebber, P. J., 1989: Role of the surface heat and moisture fluxes associated with large-scale ocean current meanders in maritime cyclogenesis. *Mon. Wea. Rev.,* **117,** 1676–1694.

Rosenfeld, D., and W. L. Woodley, 1993: Effects of cloud seeding in west Texas: Additional results and new insights. *J. Appl. Meteor.,* **32,** 1848–1866.

Schubert, W. H., and J. J. Hack, 1982: Inertial stability and tropical cyclone development. *J. Atmos. Sci.,* **39,** 1688–1697.

Simpson, J., 1969: On some aspects of sea–air interaction in middle latitudes. *Deep Sea Research and Oceanographic Abstracts,* Vol. 16, Pergamon Press, 233–261.

——, 1980: Downdrafts as linkages in dynamic cumulus seeding effects. *J. Appl. Meteor.,* **19,** 477–487.

——, and V. Wiggert, 1969: Models of precipitating cumulus towers. *Mon. Wea. Rev.,* **97,** 471–489.

——, and ——, 1971: 1968 Florida cumulus seeding experiment: Numerical model results. *Mon. Wea. Rev.,* **99,** 87–118.

——, and W. L. Woodley, 1971: Seeding cumulus in Florida: New 1970 results. *Science,* **173,** 117–126.

——, R. H. Simpson, D. A. Andrews, and M. A. Eaton, 1965: Experimental cumulus dynamics. *Rev. Geophys.,* **3,** 387–431.

——, G. W. Brier, and R. H. Simpson, 1967a: Stormfury cumulus seeding experiment 1965: Statistical analysis and main results. *J. Atmos. Sci.,* **24,** 508–521.

——, M. Garstang, E. J. Zipser, and G. A. Dean, 1967b: A study of a non-deepening tropical disturbance. *J. Appl. Meteor.,* **6,** 237–254.

——, A. R. Olsen, and J. C. Eden, 1975: A Bayesian analysis of a multiplicative treatment effect in weather modification. *Technometrics,* **17,** 161–166.

——, N. Westcott, R. Clerman, and R. A. Pielke, 1980: On cumulus mergers. *Archiv. Meteor. Geophys. Bioklim.,* **29A,** 1–40.

——, B. R. Morton, M. C. McCumber, and R. S. Penc, 1986: Observations and mechanisms of GATE waterspouts. *J. Atmos. Sci.,* **43,** 753–782.

——, R. Adler, and G. North, 1988: A proposed Tropical Rainfall Measuring Mission (TRMM) satellite. *Bull. Amer. Meteor. Soc.,* **69,** 278–295.

——, G. Roff, B. R. Morton, K. Labas, G. Dietachmayer, M. McCumber, and R. Penc, 1991: A Great Salt Lake waterspout. *Mon. Wea. Rev.,* **119,** 2741–2770.

——, T. D. Keenan, B. Ferrier, R. H. Simpson, and G. J. Holland, 1993: Cumulus mergers in the maritime continent region. *Meteor. Atmos. Phys.,* **51,** 73–99.

——, E. Ritchie, G. J. Holland, J. Halverson, and S. Stewart, 1997: Mesoscale interactions in tropical cyclone genesis. *Mon. Wea. Rev.,* **125,** 2643–2661.

——, J. Halverson, H. Pierce, C. Morales, and T. Iguchi, 1998a: Eyeing the eye: Exciting early stage science results from TRMM. *Bull. Amer. Meteor. Soc.,* **79,** 1711.

——, ——, B. S. Ferrier, W. A. Petersen, R. H. Simpson, R. Blakeslee, and S. L. Durden, 1998b: On the role of "hot towers" in tropical cyclone formation. *Meteor. Atmos. Phys.,* **67,** 15–35.

——, and Coauthors, 2000: The Tropical Rainfall Measuring Mission (TRMM). *Earth Obs. Remote Sens.,* **4,** 71–90.

Tao, W. K., and J. Simpson, 1989: A further study of cumulus interactions and mergers: Three-dimensional simulation with trajectory analyses. *J. Atmos. Sci.,* **46,** 2974–3004.

——, S. Lang, J. Simpson, and R. F. Adler, 1993: Retrieval algorithms for estimating the vertical profiles of latent heat release: Their applications for TRMM. *J. Meteor. Soc. Japan,* **71** (6), 685–700.

Telford, J. W., and J. Warner, 1964: Fluxes of heat and water vapor in the lower atmosphere derived from aircraft observations. *J. Atmos. Sci.,* **21,** 539–548.

Ulanski, S., and M. Garstang, 1978: The role of surface divergence and vorticity in the life cycle of convective rainfall. Part I: Observation and analysis. *J. Atmos. Sci.,* **35,** 1063–1069.

Wakimoto, R. M., and J. W. Wilson, 1989: Nonsupercell tornadoes. *Mon. Wea. Rev.,* **117,** 1113–1140.

——, N. A. Atkins, and C. Liu, 1995: Observations of the early evolution of an explosive oceanic cyclone during ERICA IOP5. Part II: Airborne Doppler analysis of the mesoscale circulation and frontal structure. *Mon. Wea. Rev.,* **123,** 1311–1327.

Wang, Y., W.-K. Tao, and J. Simpson, 1996: The impact of ocean surface fluxes on a TOGA COARE convective system. *Mon. Wea. Rev.,* **124,** 504–520.

——, ——, ——, and S. Lang, 2002: The sensitivity of tropical squall lines to surface fluxes: 3D Cloud resolving model simulations. *Quart. J. Roy. Meteor. Soc.,* in press.

Winston, J. S., 1955: Physical aspects of rapid cyclogenesis in the Gulf of Alaska. *Tellus,* **7,** 481–500.

Woodcock, A. H., and J. Wyman, 1947: Convective motion in air over the sea. *Ann. N. Y. Acad. Sci.,* **48,** 749–776.

Zipser, E. J., 1969: The role of organized unsaturated convective downdrafts in the structure and rapid decay of an equatorial disturbance. *J. Appl. Meteor.,* **8,** 799–814.

——, 1977: Mesoscale- and convective-scale downdrafts as distinct components of squall-line circulation. *Mon. Wea. Rev.,* **105,** 1568–1589.

Chapter 2

Joanne Simpson: An Ideal Model of Mentorship

ROGER A. PIELKE JR.

Department of Atmospheric Science, Colorado State University, Fort Collins, Colorado

ABSTRACT

The role of cumulus clouds in local, regional, and global weather and climate that is understood today is based to a large extent on the pioneering work of Joanne Simpson. Her involvement in this work is illustrated through the experiences as my career developed. She also was, and is, an ideal model of mentorship. This paper illustrates this model using my interactions during the 1970s and early 1980s, and how they have influenced research articles up to the present.

1. Experimental Meteorology Lab Years

In 1971, Dr. Simpson was director of the Experimental Meteorology Laboratory (EML) when I joined as a graduate student.[1] Her original papers (Stern and Malkus 1953, Malkus and Stern 1953) provided a framework to develop a sea-breeze numerical model and to apply this tool to better understand mesoscale and cumulus dynamics in south Florida. Figure 1 from her work demonstrates the concept of the "equivalent mountain," which results from surface heating of land (Fig. 2.1) Air is lifted, as a result, as it passes over the land. This provided an original explanation of why cumulus clouds preferentially develop over and downward of tropical islands even in the absence of topographical relief. Figure 2.2 illustrates this preference for rain showers downwind of such islands (in this case Grand Bahama, Eleuthera, and Andros Islands in the Bahamas, where the large-scale wind is southeasterly). When Joanne's work was performed, such direct observations of cloud fields were not available.

Since computers were available in the early 1970s, EML provided resources to construct a model of the sea breezes. Fortunately, the computer was sufficient to run a three-dimensional model of the sea breezes over south Florida. Using 11-km horizontal grid intervals over a 33×36 grid mesh with seven vertical levels, the model just fit on a CDC 6600 computer at the National Meteorological Center [NMC; now called the National Centers for Environmental Prediction (NCEP)]. Given the restraint on access to the computer and its comparative slowness, only one hour of simulation was possible each day. Ten hours of simulated time required ten days!

Dr. Simpson, however, felt strongly that students and research staff should not just be theoretical modelers. They need to also experience real weather. For this reason opportunities were provided for my participation in the Florida Area Cumulus Experiment (FACE; Simpson and Woodley 1975; Biondini et al. 1977; Woodley et al. 1977; see Fig. 2.3 for a caricature I drew during FACE). This experiment involved randomized cumulus cloud seeding (Simpson 1977). As part of Joanne's mentorship, I flew on several FACE cloud seeding experiments, both with the seeding aircraft and in a separate aircraft that often flew at just a few hundred feet above the ground! I was also permitted (and encouraged) to fly a hurricane Storm Fury flight (into Ginger in 1971).

This ability to participate in a variety of research programs was facilitated by the leadership of the government agencies that managed the weather research groups (the National Hurricane Research Lab and EML) and the National Weather Service [National Hurricane Center (NHC)]. These groups were collocated in a single multistory building (Fig. 2.4) with the Department of Atmospheric Sciences at the University of Miami in Coral Gables, Florida. This vertical linkage permitted a free exchange among academic, research, and operational weather communities.

This exposure to operational weather, which was permitted by the perceptive leadership of Joanne's husband, Dr. Robert H. Simpson, allowed researchers to experience the cold reality of making actual forecasts that affected lives and properties. We were permitted to watch NHC forecasters prepare hurricane watches and warnings, which allowed us to better relate our studies to actual public need.

Joanne is an observational meteorologist, as well as

---

[1] Dr. Bill Cotton, a graduate student when I first met him (and shared an office) at Penn State in 1968, had been previously hired by Joanne, and recommended me for a two-year tour of duty to complete my Ph.D. My personal and professional interactions with Bill continue to this day at Colorado State University.

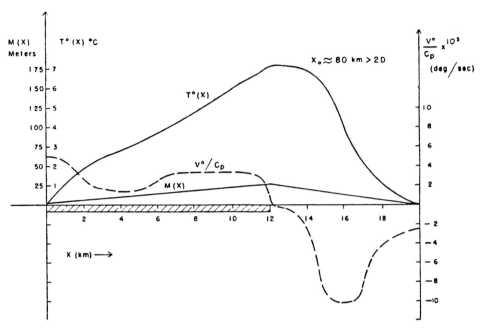

FIG. 2.1. Equivalent mountain for Nantucket Island on 9 Aug 1950 [case 3 studied by Malkus and Bunker (1952)]. Observed temperature profile along ground, $T°(x)$, is given by the heavy solid line. Heating function along the ground divided by specific heat at constant pressure, $V°/c_p$, was calculated from observed surface values of $U\partial T'/\partial x$ and is shown by the dashed line. Extent of island is indicated by the hatched region (from Stern and Malkus 1953).

a modeler. She provided resources to instrument 10 locations across south Florida with recording surface weather stations. Made by MRI Inc., these stations, which archived data on strip charts, were among the first applications of such portable automated weather stations. Such exposure of a student to real data helps establish a reality check when developing theoretical models. Figure 2.5 illustrates one of these observational platforms. The data from these mechanical weather sta-

tions, and from a set of volunteer observers that were recruited for FACE, resulted in several papers (Pielke and Cotton 1977; Pielke and Mahrer 1978). Among the results found was the prominent role of sea-breeze horizontal wind convergence in establishing favorable environments for thunderstorms and their merger. An example of observed and modeled winds across the east–west cross section that contained the MRI sites is displayed in Fig. 2.6. The wind convergence locally increased convective available potential energy (CAPE)

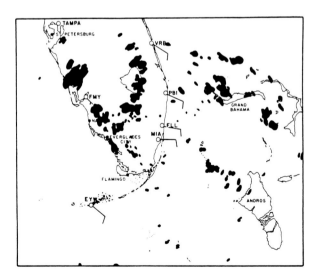

FIG. 2.2. Radar echo coverage at 1901 UTC 19 Aug 1971 as seen by the Miami WSR-57 10-cm radar (from Pielke 1974).

FIG. 2.3. Caricature drawn by R. A. Pielke during FACE of participants, including Joanne Simpson.

FIG. 2.4. Photograph of the building that housed the National Hurricane Research Lab, EML, and the National Weather Service at the University of Miami in Coral Gables, FL.

FIG. 2.5. Photograph of one of the MRI, Inc., weather stations (in the J. W. Corbett Wildlife Refuge west of the West Palm Beach, FL), which included a tipping bucket rain gauge, cup anemometer, wind vane, and a temperature sensor.

and physically moved cumulus clouds toward each other. As discussed in Simpson et al. (1980), Wescott (1994), and more recently in Simpson et al. (1993), merged cumulus clouds rain much more than the individual clouds would separately. This concept of cumulus cloud merger originated from Joanne's dynamic cloud seeding hypothesis (Cotton and Pielke 1995) and is an important reason for the high correlation between sea-breeze convergence zones and thunderstorms (Ulanski and Garstang 1978a,b; Cooper et al. 1982; Pielke et al. 1991; Garstang and Fitzgerald 1999; see also Fig. 2.7).

2. The University of Virginia years

As part of Joanne's position at the Center for Advanced Studies at the University of Virginia, I was hired as an assistant professor of environmental sciences in 1974. In this capacity, she continued to mentor my research and to aid in the development of an academic program. Among her major efforts was the recruitment of Dr. Ytzhaq Mahrer of the University of Jerusalem in Rehovot. This was a fruitful collaboration that permitted additional developments of the EML sea-breeze model to include topography (Mahrer and Pielke 1975, 1976) and improved computational solution techniques (Mahrer and Pielke 1978). Professor Mike Garstang and I also met during this time period, and developed a close research collaboration (e.g., see Pielke et al. 1987). Joanne was responsible for bringing Mike and I together, where, as an experimentalist and a numerical modeler, we were able to successfully pool our different expertise. This was another example of where Joanne recognized the need for combined observational and modeling studies.

In my role as an assistant professor and later associate professor, I adopted Joanne's philosophy of serving as a facilitator for student research, rather than a manager. Students who graduate from such a program benefit greatly from this combination of science exposure and collegial interaction. Among the major achievements was one of the first soil–vegetation–atmosphere transfer (SVAT) schemes that formed part of Mike McCumber's Ph.D. dissertation (McCumber 1980; McCumber and Pielke 1981).

An essential component of Joanne's leadership ethic was her treatment of support staff. Always requiring excellence, she still treated each person as an individual. They were as much a part of the research team as the scientific staff.

It was during these years that my professional association with Robert Simpson deepened. He was (and is) a source of strength for Joanne and is also an outstanding scientist on his own. I may be the only person who has separately published papers with both Simpsons.[2] This involvement with Bob also resulted in my involvement with both of them in the Typhoon Moderation (TYMOD) Project, which discussed the possibility of cloud seeding western Pacific typhoons (Simpson et al. 1978).

3. Subsequent years

After I left the University of Virginia, my research continued to be greatly influenced by Joanne's pioneering work. At Colorado State University, a new modeling system was developed that built on the EML sea-breeze model and the model developed independently by Bill Cotton. This new model was called the Regional

---

[2] My paper with Robert Simpson was Simpson and Pielke (1976).

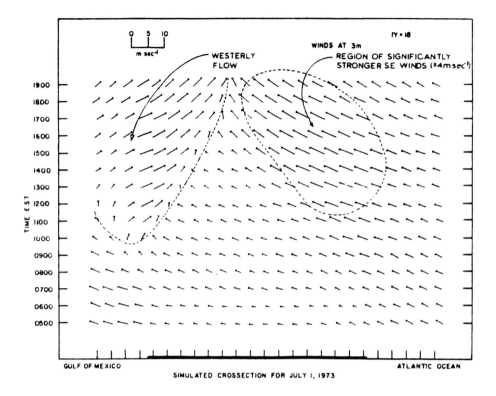

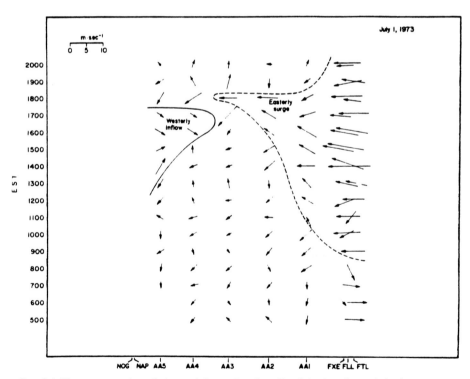

FIG. 2.6. Time cross section of observed (bottom) and predicted (top) surface winds along an east–west line from Fort Lauderdale, FL, to Naples, FL, along the 18th grid line from the southern edge of the model (from Pielke and Mahrer 1978).

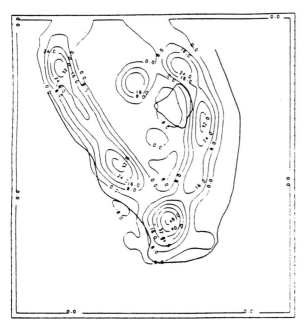
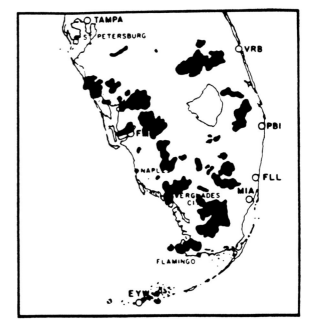

FIG. 2.7. (left) Vertical velocity prediction, 9.5 h after simulated sunrise, and (right) composite radar map at equivalent time for 29 Jun 1971 (from Pielke 1974).

Atmospheric Modeling System (RAMS; Pielke et al. 1992).[3]

My interest in hurricanes was intensified as a result of my collaboration with both Joanne and Bob. Thus it was natural that we would apply RAMS to simulate these tropical cyclones. We chose Hurricane Andrew (1992) since its intensification was underforecast and it made landfall over the same area in southern Dade

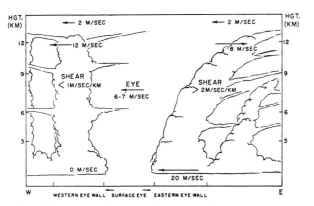

FIG. 2.8. Schematic illustration of asymmetry in eye structure when tropical easterly current has pronounced negative shear. The velocities given by the arrows illustrate a hypothetical set of winds (coordinates fixed to ground) resulting when a storm circulation with inflow at low levels and outflow aloft is superposed on a trade current of 10 m s^{-1} at the ground decreasing to 2 m s^{-1} in the stratosphere (from Malkus 1958).

_____

[3] Cotton summarizes the RAMS development and lists a spectrum of papers which resulted from the application of this modeling tool in chapter 8 of this monograph.

County, Florida, where I lived while working at EML. Our explicit modeling of deep thunderstorms in the eyewall and in the spiral bands of a hurricane (using horizontal grid increments of 5 km) was the key to properly simulating intensification of Andrew (Eastman et al. 1996). Such high spatial resolution is needed to accurately evacuate mass from the eyewall region. This was the first simulation of a specific hurricane with explicit microphysics at such fine spatial resolution.

The early observational papers by the Simpsons provided an essential assessment of the credibility of our hurricane model results. Malkus (1958), for example, provides an observed characterization of eyewall–eye dynamics, such as the influence of large-scale environmental shear (Fig. 2.8). In Nicholls and Pielke (1995), we investigated, using RAMS, the importance of wind shear associated with idealized hurricanes. Among our conclusions is that weak vertical wind shear can actually enhance the ability of a hurricane to intensify. Figure 2.9 illustrates the eye and eyewall structure as modeled by RAMS for a mature hurricane (Pielke and Pielke 1997, p. 78). Pielke (1990) used the Robert Simpson decision tree for hurricane development and intensification to illustrate the forecast procedure for these two aspects of tropical cyclones. In Pielke and Pielke (1997) we referred to several early papers by Bob (Simpson 1946; 1954; 1971; Simpson and Riehl 1981) that continue to provide a fundamental basis for prediction research.

On the global scale, the Riehl and Malkus (1958) paper is a major landmark in the study of the energy and water budget of the earth's atmosphere. The concept

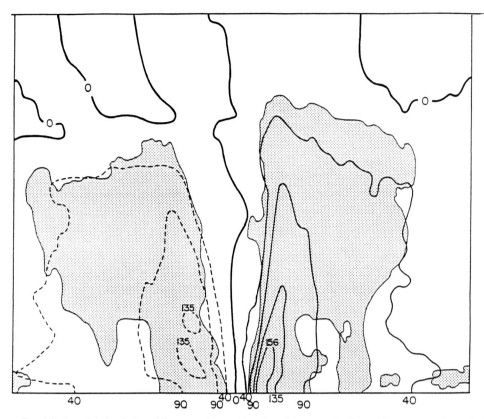

FIG. 2.9. A model simulation of the eye and eyewall region of a mature hurricane. The contours show wind speed in miles per hour. The solid contours represent winds moving into the page and dashed contours represent winds moving out of the page. The shaded region illustrates the locations of clouds. This figure was provided by Dr. Mel Nicholls of Colorado State University (from Pielke and Pielke 1997).

of undilute ascent in thunderstorms in the intertropical convergence zone, as one segment in the Hadley circulation, revolutionized the concept of how the atmosphere transfers mass and energy. The Riehl and Malkus study indicated that between 1500 and 5000 active undilute cumulus convective towers (i.e., "hot towers") are all that is needed for the equatorial side of the vertical transport associated with the Hadley circulation. Satellite images that subsequently became available confirmed this hypothesis and illustrate that mass and energy are transported poleward within high upper-tropospheric jet streams, which are referred to as subtropical jets. These jets are an effective mechanism for long-distance rapid communication of weather, a concept referred to as teleconnection.

This fundamental work is the basis for the conclusions of Chase et al. (1996, 2000) that tropical land-use change can alter the patterning of where these hot towers occur. Deforestation results in a drying and warming, such that cumulonimbus convection tends to occur more over nondeforested regions and over tropical ocean areas. This pattern shift results in alterations in the teleconnections with higher latitudes, such that weather patterns may be permanently changed. ENSO events, which also affect the patterning of thunderstorms, sim-

ilarly affect higher latitudes. Pitman and Zhao (2000) and Pitman et al. (1999) confirm the major role of land-use change affecting the Hadley circulation, and higher latitude weather. In one of my numerous fruitful collaborations with Bill Cotton, we published Cotton and Pielke (1995), where we connected our experiences associated with weather modification studies at EML with the current overselling of the ability to predict future climate. Land-use change was one of the issues that we used to illustrate that climate predictability inherently may be more limited than has been traditionally thought.

The role of the hot towers is critical to this human-caused climate sensitivity. Even though the location of the hot towers covers only a relatively small percentage of the earth's surface, large climate change effects can occur! This overlooked climate change issue is built on Joanne's early pioneering work.

4. Conclusions

The richness of Joanne Simpson's research accomplishments are best appreciated by tracking our current knowledge of the atmosphere to where these concepts were first discussed in the peer-reviewed literature. Her breadth of contribution is impressive and ranges from the cumulus cloud to global scale. Early in her career, she recognized the critical role of cumulus clouds in the earth's atmosphere, and now she continues to build on her innovative and broad expertise in such programs as the Tropical Rainfall Measuring Mission (TRMM; Kummerow et al. 1998) and the Tropical Ocean Global Atmosphere Coupled Ocean–Atmosphere Response Experiment (TOGA COARE; Halverson et al. 1999). When one uncovers the origin of many of our most basic concepts in atmospheric science, it is quite impressive how much of this knowledge is founded in her original work!

Finally, while completing pioneering research results, she also recognized the need to mentor the next generation of scientists. Her sacrifices are candidly summarized in Simpson (1973), a publication that was also extremely informative concerning the prejudices against women in science. Joanne's publication has sensitized many men to this obstacle.

Joanne's career has spanned all aspects of our profession, from research and mentorship to teaching and service. Her guidance as American Meteorological Society publication commissioner and scientific and technological activities commissioner and president has effectively led the society to its current very high level of achievement. She was also involved with Robert Simpson in setting up a private company (Simpson Weather Associates), which transfers research knowledge to help solve real-world problems. This responsibility of university professors is not widely recognized, yet this is a prime mechanism that can be used to interface with societal needs.

All the best to Joanne on an extremely valuable career that is still continuing full throttle!

Acknowledgments. This paper was prepared as part of an invited contribution to celebrate Joanne Simpson's career. The oral version was presented at the Symposium on Cloud Systems, Hurricanes and TRMM: Celebration of Dr. Joanne Simpson's Career—The First Fifty Years. Dallas Staley and Michelle McClaren ably handled the typing and the editing of the contribution.

REFERENCES

Biondini, R., J. S. Simpson, and W. Woodley, 1977: Empirical predictors for natural and seeded rainfall in the Florida Area Cumulus Experiment (FACE) 1970–1975. *J. Appl. Meteor.,* **16,** 12–97.

Chase, T. N., R. A. Pielke, T. G. F. Kittel, R. Nemani, and S. W. Running, 1996: The sensitivity of a general circulation model to global changes in leaf area index. *J. Geophys. Res.,* **101,** 7393–7408.

——, ——, ——, ——, and ——, 2000: Simulated impacts of historical land cover changes on global climate in northern winter. *Climate Dyn.,* **16,** 93–105.

Cooper, H. J., M. Garstang, and J. S. Simpson, 1982: The diurnal interaction between convection and peninsular-scale forcing over south Florida. *Mon. Wea. Rev.,* **110,** 486–503.

——, and R. A. Pielke, 1995: *Human Impacts on Weather and Climate.* Cambridge University Press, 288 pp.

Eastman, J. L., M. E. Nicholls, and R. A. Pielke, 1996: A numerical simulation of Hurricane Andrew. Presented at *Second International Symposium on Computational Wind Engineering,* Fort Collins, CO. Japan Association for Wind Engineering and the U.S. Wind Engineering Research Council.

Garstang, M., and D. R. Fitzjarrald, 1999: *Observations of Surface to Atmosphere Interactions in the Tropics.* Oxford University Press, 405 pp.

Halverson, J., B. S. Ferrier, T. M. Rickenbach, J. Simpson, and W.-K. Tao, 1999: An ensemble of convective systems on 11 February 1993 during TOGA COARE: Morphology, rainfall characteristics, and anvil cloud interactions. *Mon. Wea. Rev.,* **127,** 1208–1228.

Kummerow, C., W. Barnes, T. Kozu, J. Shuie, and J. S. Simpson, 1998: The Tropical Rainfall Measuring Mission (TRMM) sensor package. *J. Atmos. Oceanic Technol.,* **15,** 809–817.

Mahrer, Y., and R. A. Pielke, 1975: The numerical study of the air flow over mountains using the University of Virginia mesoscale model. *J. Atmos. Sci.,* **32,** 2144–2155.

——, and ——, 1976: The numerical simulation of the airflow over Barbados. *Mon. Wea. Rev.,* **104,** 1392–1402.

——, and ——, 1978: The meteorological effect of the changes in surface albedo and moisture. *Isr. Meteor. Res. Pap.,* **2,** 55–70.

Malkus, J. S., 1958: On the structure and maintenance of the mature hurricane eye. *J. Meteor.,* **15,** 337–349.

——, and A. Bunker, 1952: *Observational Studies of the Air Flow over Nantucket Island during the Summer of 1950.* Papers in Physical Oceanography and Meteorology, No. 12, Massachusetts Institute of Technology and Woods Hole Oceanographic Institution, 50 pp.

——, and M. Stern, 1953: The flow of a stable atmosphere over a heated island, Part I. *J. Meteor.,* **10,** 30–41.

McCumber, M., 1980: A numerical simulation of the influence of heat and moisture fluxes upon mesoscale circulations. Ph.D. dissertation, University of Virginia, 255 pp.

——, and R. A. Pielke, 1981: Simulation of the effects of surface fluxes of heat and moisture in a mesoscale numerical model—Part 1: Soil layer. *J. Geophys. Res.,* **86,** 9929–9938.

Nicholls, M. E., and R. A. Pielke, 1995: A numerical investigation of the effect of vertical wind shear on tropical cyclone intensification. Preprints, *21st Conf. on Hurricanes and Tropical Meteorology,* Miami, FL, Amer. Meteor. Soc., 339–341.

Pielke, R. A., 1974: A three-dimensional numerical model of the sea breezes over south Florida. *Mon. Wea. Rev.,* **102,** 115–139.

——, 1990: *The Hurricane.* Routledge Press, 228 pp.

——, and W. R. Cotton, 1977: A mesoscale analysis over south Florida for a high rainfall event. *Mon. Wea. Rev.,* **105,** 343–362.

——, and Y. Mahrer, 1978: Verification analysis of the University of Virginia three-dimensional mesoscale model prediction over south Florida for 1 July 1973. *Mon. Wea. Rev.,* **106,** 1568–1589.

——, and R. A. Pielke Sr., 1997: *Hurricanes: Their Nature and Impacts on Society.* John Wiley and Sons, 279 pp.

——, M. Garstang, C. Lindsey, and J. Gudsorf, 1987: Use of a synoptic classification scheme to define seasons. *Theor. Appl. Climatol.,* **38,** 57–68.

——, A. Song, P. J. Michaels, W. A. Lyons, and R. W. Arritt, 1991: The predictability of sea-breeze generated thunderstorms. *Atmosfera,* **4,** 65–78.

——, and Coauthors, 1992: A comprehensive meteorological modeling system—RAMS. *Meteor. Atmos. Phys.,* **49,** 69–91.

Pitman, A. J., and M. Zhao, 2000: The relative impact of observed change in land cover and carbon dioxide as simulated by a climate model. *Geophys. Res. Lett.,* **27,** 1267–1270.

——, R. Pielke, R. Avissar, M. Claussen, J. Gash, and H. Dolman, 1999: The role of the land surface in weather and climate: Does the land surface matter? *IGBP Newslett.,* **39,** 4–11.

Riehl, H., and J. S. Malkus, 1958: On the heat balance in the equatorial trough zone. *Geophysica,* **6,** 503–535.

Simpson, J. S., 1973: Meteorologist. *Ann. New York Acad. Sci.,* **208,** 41–46.

——, 1977: Empirical predictors for natural and seeded rainfall in the Florida Area Cumulus Experiment (FACE), 1970–1975. *J. Appl. Meteor.*, **16**, 585–594.

——, and W. L. Woodley, 1975: Florida Area Cumulus Experiments 1970–1975 rainfall results. *J. Appl. Meteor.*, **14**, 734–744.

——, N. E. Westcott, R. J. Clerman, and R. A. Pielke, 1979: On cumulus mergers. *Arch. Meteor. Geophys. Bioklim.*, **29A**, 1–40.

——, Th. D. Keenan, B. Ferrier, R. H. Simpson, and G. J. Holland, 1993: Cumulus mergers in the maritime continent region. *Meteor. Atmos. Phys.*, **51**, 73–99.

Simpson, R. G., and R. A. Pielke, 1976: Hurricane development and movement. *Appl. Mech. Rev.*, **29**, 601–609.

Simpson, R. H., 1946: On the movement of tropical cyclones. *Trans. Amer. Geophys. Union*, **27**, 641–655.

——, 1954: Hurricanes. *Sci. Amer.*, **190**, 32–37.

——, 1971: The decision process in hurricane forecasting. NOAA Tech. Memo. NWS SR-53, Fort Worth, TX, 35 pp.

——, and H. Riehl, 1981: *The Hurricane and Its Impact.* Louisiana State University Press, 398 pp.

——, and Coauthors, 1978: TYMOD: Typhoon Moderation. Virginia Technology, Inc. Final report for PAGASA, Republic of the Philippines.

Stern, M., and J. S. Malkus, 1953: The flow of a stable atmosphere over a heated island, Part II. *J. Meteor.*, **10**, 105–120.

Ulanski, S., and M. Garstang, 1978a: The role of surface divergence and vorticity in the lifecycle of convective rainfall. Part I: Observation and analysis. *J. Atmos. Sci.*, **35**, 1047–1062.

——, and ——, 1978b: The role of surface divergence and vorticity in the lifecycle of convective rainfall. Part II: Descriptive model. *J. Atmos. Sci.*, **35**, 1063–1069.

Wescott, N. E., 1994: Merging of convective clouds: Cloud initiation, bridging, and subsequent growth. *Mon. Wea. Rev.*, **122**, 780–790.

Woodley, W. L., and Cumulus Group, 1977: Rainfall results, 1970–1975: Florida Area Cumulus Experiment (FACE). *Science*, **195**, 735–742.

Chapter 3

What We Have Learned about Field Programs

MARGARET A. LEMONE

National Center for Atmospheric Research,* Boulder, Colorado

ABSTRACT

Based on personal experience and input from colleagues, the natural history of a field program is discussed, from conception through data analysis and synthesis of results. For convenience, the life cycle of a field program is divided into three phases: the prefield phase, the field phase, and the aftermath. As described here, the prefield phase involves conceiving the idea, developing the scientific objectives, naming the program, obtaining support, and arranging the logistics. The field phase discussion highlights the decision making process, balancing input from data and numerical models, and human interactions. The data are merged, analyzed, and synthesized into knowledge mainly after the field effort.

Three major conclusions are drawn. First, it is the people most of all who make a field program successful, and cooperation and collegial consensus building are vital during all phases; good health and a sense of humor both help make this possible. Second, although numerical models are now playing a central role in all phases of a field program, not paying adequate attention to the observations can lead to problems. And finally, it cannot be overemphasized that both funding agencies and participants must recognize that it takes several years to fully exploit the datasets collected, with the corollary that high-quality datasets should be available long term.

1. Introduction

Because my current work (on the diurnal cycle of the fair weather boundary layer over land) seemed a bit peripheral to the rest of the talks on the program, I asked Joanne what she would like to hear about—then at least *one* person would be interested. She suggested that I talk about field programs, and what we have learned. I decided to take a different tack on the subject (as she might have expected) and speak more about the field programs themselves than about the the science that we have learned.

2. The stages of a field program

I will describe the stages of a field program from its inception as an idea to the most important phase, the data analysis and synthesis, using cartoons and experiences drawn from several field programs.

a. Prefield phase

1) ORIGIN

Figure 3.1 idealizes two very different origins for a field program. In the first, an august committee of scientists of impeccable reputation and immense funding decree that a field program be conducted to answer a

Big Question, which will in turn guarantee them further immense funding and exposure on the Cable News Network (CNN). In this case, a group of scientists (identified here as "the usual suspects") are called in to fill in the details (Fig. 3.2). In the second example (Fig. 3.1b) a group of scientists (some august, some only January) get together and try to attack a question that has them stumped. They develop a hypothesis and design a modest field program to address the hypothesis. In this case, reviewers often point out that either (a) the dataset that the investigators want to collect is subcritical, so additional principal investigators (PIs) and/or instrumentation are needed, or (b) the funding agencies or facilities managers decide that the facilities needed (radars, aircraft, etc) are so expensive and the opportunities so large that additional PIs are needed. Once again, the usual suspects (Fig. 3.2) are brought in.

2) DETERMINING THE SCIENTIFIC OBJECTIVES

The field program that survives the origin/review stage is further designed in community planning meetings where someone constantly reminds everyone else that "We can't design the field program without a hypothesis!" During such meetings, the scientific objectives morph into something often somewhat different (read "bigger") from the original ones. This stage is illustrated in Fig. 3.3, drawn during a planning meeting sponsored by the Stormscale Operational and Research Meteorology (STORM) program in 1988; but the au-

* NCAR is sponsored by the National Science Foundation.

FIG. 3.1. (a) Top-down origin: August committee recommending new field program to address a Big Problem. (b) Bottom-up origin: Group of scientists develops hypothesis.

dience comments can be applied to most field programs. The expansion problem is compounded when the signal (real or imagined) that *funding* might be available gets out to the community. This leads to a snowball effect (even for tropical experiments), which increases the attendance to such planning meetings and the potential for broadened objectives.

In other cases, a field program is sufficiently similar to an earlier one that the PIs closely examine the objectives of earlier programs, to make sure everything is covered or to avoid being redundant. This is illustrated in Fig. 3.4, drawn during the early planning stages of the Monsoon Experiment (MONEX).

3) THE NAMING OF THE EXPERIMENT

Atmospheric scientists are masterful acronym creators, not only creating first-order acronyms, such as the Tropical Rainfall Measuring Mission (TRMM), but second-order acronyms, such as FIFE (First ISLSCP Field Experiment, where ISLSCP is the International Satellite Land Surface Climatology Project), and third-order ac-

FIG. 3.2. The Usual Suspects. Group of scientists called in to participate in experiment decreed by group in Fig. 3.1a, or group of scientists called in to make experiment proposed by group in Fig. 3.1b acceptable to scientists and administrators like those in Fig. 3.1a.

ronyms such as GAUC (GATE Aircraft Utilization Committee, where GATE is the GARP Atlantic Tropical Experiment, and GARP is the Global Atmospheric Research Program). The ideal acronym itself describes an aspect of the experiment [like STORM or Boreal Ecosystem Atmosphere Study (BOREAS)] and, even before this era of political correctness, care was taken to create acronyms that offend no one. GARP for example, was selected because it was not an offensive word in any language, at least until publication and filming of *The World According to Garp* years later. Thus some field program names changed (FOPS to FAPS and SCUM to

FIG. 3.3. Determining the objectives of a field program, as illustrated by Jul 1988 STORM meeting in Longmont, Colorado.

MONEX:

IN SEARCH FOR OBJECTIVES

LEMONS

FIG. 3.4. MONEX: "In Search of Objectives."

SCMS), but some genuinely humorous names have survived (BARFEX). BARFEX, by the way, stands for Boardman Atmospheric Radiation Flux Experiment.

4) OBTAINING THE NEEDED INSTRUMENTATION/ MONEY

This step is where reality sets in. The experiment organizers, with the help of more community planning meetings, rough out a design that describes objectives, location, timing, instrument platforms, and their use. Through this stage, the number of PIs entrained through interest in the subject matter, concentration of useful measurements, or a funding opportunity remains large, an advantage when going to the agencies ("347 scientists want to use the data from your new super airborne lidar"). Often, the needed facilities must come from more than one government agency, or even more than one government, each of which has its own goals, timetable, and application procedure. Potential participants are thus deeply involved in simultaneous applications for funding and facilities, and putting together an experiment plan despite the uncertainties. Once the applications and proposals are in and reviewed, the final deployment is outlined in an iterative procedure that *usually* fits the facilities to the experimental objectives, and that *usually* ensures that the PIs responsible for the needed platforms are participants in the experiment. Note: even though the number of PIs and instruments may be reduced in this phase, the objectives usually remain untouched.

Unexpected factors can make the planning exercise even more interesting at any time. Timing and even location can shift. Timing is usually shifted to give worthy scientists access to needed instruments. Location can be trickier. Veterans of GATE recall that it was originally planned for the tropical Pacific, until someone discovered that it was impossible to pronounce "GPTE."[1] Smaller experiments can also get shifted in location. Just months before the field phase of an experiment I was involved in, by which time we were well on our way to securing leased sites for surface instruments, a representative of the National Science Foundation (NSF) strongly suggested that we shift our experiment site, since a certain agency (to remain unnamed) wanted our instruments to beef up one of its field programs. We stayed where we were, because a third agency already had its instruments in place, and we weren't prepared to think up a new acronym!

5) LOGISTICS: PRACTICE, PRACTICE, PRACTICE, AND HEALTH ISSUES

Getting people and instruments coordinated is a big task. In GATE, the Preliminary Regional Experiment for STORM (PRESTORM), and Mesoscale Alpine Programme (MAP) there were actually rehearsals to teach scientists and air crews to work together in coordinating multiple aircraft flights. Smaller field programs have a few shakedown days at the beginning of the field phase. These shakedowns are valuable, because planners addressing the Big Questions sometimes forget how much trouble the average person (and average scientist) has with minutiae that can ruin a potentially perfect intensive operations period. When there are so may things to consider, even time and place can be difficult to straighten out. In this respect, the planners of GATE were brilliant in locating the experiment in the Greenwich time zone, so that we could live and collect data in UTC, and mercifully most activity was restricted to the Northern Hemisphere. On the other hand, participants in the Tropical Ocean Global Atmosphere Coupled Ocean–Atmosphere Response Experiment (TOGA COARE) had to simultaneously deal with three time zones (Townsville, Australia, time; Honiara, Solomon Islands, time; and UTC) and latitudes north and south of the equator.

One of the special attractions of tropical experiments is the possibility of getting malaria and a host of other interesting and exotic diseases. Before GATE, it wasn't certain whether the shots or the warnings were worse. Fig. 3.5 illustrates some of the potential dangers, based on a briefing before TOGA COARE. Unfortunately, there were even more dangers than we originally realized, resulting from the antimalarial medication Larium. As the experiment ensued, we became experts on the

---

[1] Actually, the decision was made for political reasons. For details, see online at http://uniblab.atmos.ucla.edu/tropic/newsletters/newsletter28.html.

FIG. 3.5. The health briefing prior to TOGA COARE. Do you *really* want to go to Honiara? Drawing from M. A. LeMone (1993, unpublished manuscript).

expected side effects ("vivid dreams,"[2] and "mental instability," or curious behavior even for scientists[3]. If someone was being particularly argumentative on a particular day, we would ask him/her whether they had just taken their Larium tablet, suspecting them of suffering from PMS (Post Mefloquine Syndrome, which lasted 24–48 h after ingestion). Unfortunately, a new form of (fortunately mild) malaria resistant to Larium infected several TOGA COARE participants.

b. The field phase

1) KEEPING TO THE SCIENTIFIC OBJECTIVES

Anyone familiar with large field programs know that the scientific objectives become so ambitious (or so diffuse) that almost no one can remember them. This inspired the cartoon drawn by Josh Holland during GATE (Fig. 3.6), probably after a debate about priorities.

2) RUNNING THE FIELD PROGRAM

Field programs, unlike armies, are run by consensus, which is sometimes difficult to obtain, particularly with

ambitious objectives and instrumentation that can interfere with other instruments or flight plans that can interfere with other instruments or other flight plans. Developing a consensus becomes particularly difficult for experiments run using more than one operations cen-

FIG. 3.6. GATE field phase. The scientific objectives are here somewhere! (MST = Mission Selection Team.) Sketch by Josh Holland, taken from M. A. LeMone (1974, unpublished manuscript).

---

[2] Some of the dreams were so entertaining, the name "hi-larium" might be more appropriate.

[3] Similar symptoms can originate from trying to keep track of the three time zones and northern vs southern latitudes while conducting aircraft missions.

FIG. 3.7. Coordination between multiple operations centers in the field. Tired Honiara participants taking advantage of the Townsville speaker's not seeing his audience: TOGA COARE, Feb 1993. From M. A. LeMone (1993, unpublished manuscript).

ter. This was done in the Taiwan Area Mesoscale Experiment (TAMEX) (1987; i.e., pre-internet), with communication by phone or fax between Taiwan and Okinawa, or by radio between Taipei and the P3 aircraft. TAMEX had fairly straightforward objectives, making multiple centers doable. When field programs have complex objectives and multiple centers [e.g., STORM Fronts Experiment Systems Test (STORM FEST) and TOGA COARE], the lack of face-to-face communications leads to misunderstandings and even resentment. Figure 3.7, which shows a difficult COARE conference call between Townsville and Honiara illustrates the "hearing but not listening" phenomenon that could not be solved even through the internet. The problem in GATE and COARE was compounded by the isolation of ships, which had different instruments and operational considerations from the aircraft. Television communication, used in STORM FEST, worked slightly better.

Lack of food, lack of sleep, and sometimes illness or simply the wrong kind of weather can also lead to conflict, as illustrated by Fig. 3.8, drawn during TOGA COARE. Figure 3.9 was prepared during STORM FEST, during which a stretch of adverse weather (i.e., "good weather" to anyone else) and lack of sleep led to such tensions that one student filled his journals with observations of human interaction instead of the pearls of wisdom he expected from map discussions.

In the Cooperative Atmosphere–Surface Exchange

Study (CASES-97),[4] we had a computer-generated sampler on the wall with words of wisdom on how to survive a field program. Things like "assume the other person is an idiot," and its companion "don't be offended if other people assume you are an idiot," and "redundancy is good." Watching out for one another and retaining our humility helped prevent forgetting the many details important to operating a field program.

Joanne Simpson taught me another good rule— "Don't sweat the small stuff"—during GATE. This is illustrated by the following story, which requires a little background. GATE was in 1974, before the "politically correct" era. The walls of the radio room in the GATE Operational Control Centre (GOCC, another third-order acronym) were covered with *Playboy* centerfolds. The analysis room had fewer but more select nude women, including Chesty Morgan, whose endowment matched her name (think "watermelons"). One day, a colleague returning from the States gave a secretary (female—this was 1974) and me a copy of *Playgirl*. During a quiet moment, we examined the male centerfold (one African-American football player) and showed him to the Senegalese women who worked in the analysis room. Having already noticed the imbalance in female (many) versus male (zero) nudes and European (many) versus African (zero) nudes, they decided to rectify the situation by posting the *Playgirl* centerfold on the wall. This

_____

[4] This program had an interesting evolution, even though it was conducted in Kansas.

Fig. 3.8. Decision making after two months in the field.

situation was most unacceptable to Colonel (Ret.) Barney (informally known as the GATE operations director) who immediately chewed out my supervisor (Ed Zipser), who was ordered to chew me out in turn for this flagrant violation of the rules. If the humor of this story is now obvious, it wasn't at the time, and when I told the story to Joanne while walking out to the airplanes, she just said to ignore the small stuff and focus on the real problems, which I have tried to do since. (I must admit, though, that the women in the centerfolds were always better posed than the men.)

3) MAKING DECISIONS BASED ON DATA

We have more data now than we did three decades ago, but we probably made better use of the data back then, because more of us understood the limitations of the instrumentation, and more of us knew how to use our eyes. Today, with a proliferation of instruments and

the use of computers, many are unaware or even uninterested in where the data come from. This led to a ludicrous situation, illustrated by Fig. 3.10, which really occurred in TAMEX.

TAMEX was conducted out of the Central Weather Bureau building in Taipei. TAMEX had the best ground–aircraft communication of any experiment I have participated in, truly remarkable given the mountains in Taiwan. But there was a bit too much focus on computers. We were on the sixth floor, which would have afforded a good view if the shades weren't always drawn to make it easier to read the computer screens.

Fig. 3.10. Computer output vs looking out the window. Which do you believe? (TAMEX 1987).

Fig. 3.9. Two scientists deciding on strategy and priorities for next intensive observing period. The reader is invited to fill in favorite field program, and two favorite scientists. From STORM FEST, 1992; based on figure generously provided by William Blumen.

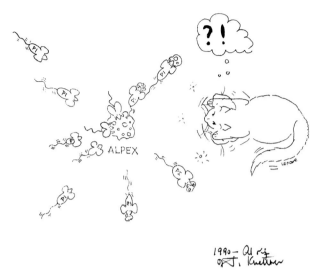

FIG. 3.11. Program coordinators for the GARP Alpine Experiment (ALPEX) (cat) gets dizzy trying to prevent PIs (mice) from running away with little bites of cheese (data) instead of pooling resources to answer the major questions. Cartoon suggested by Joachim Kuettner, 1990.

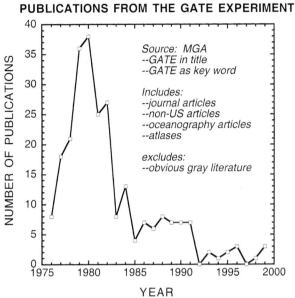

FIG. 3.12. Number of publications per year with "GATE" in the title or as a key word. Source: *Meteorological and Geophysical Abstracts.*

To see outside, we had to walk to the stairwell. The decision to have an intensive observing period in TA-MEX was based partially on the wind direction and speed—a good strong low-level jet slamming into the Island of Taiwan was ideal for setting up strong storms from the forced uplift or maintaining them through low-level shear as discussed by Rotunno, Klemp, and Weisman (1988). Weaker winds would result only in shorter-lived convectively generated storms. One day a colleague came into the TAMEX headquarters, very excited about the strong winds: cumulus were moving rapidly across the sky from the southwest and winds at the surface were gusty. However, the people inside denied the strong winds, saying that a radiosonde had just been sent up and the computed winds were light! I cannot recall whether eyeballs won out over the computer screen, but the eyeballs were right: it did rain. And rain.

All field programs are rich with examples of data misinterpretation, particularly when we are trying to use new types of data. In PRE-STORM (1985), we vectored the National Oceanic and Atmospheric Administration (NOAA) P3 to an apparent storm based on a cold cloud shield in the satellite infrared image. After all, several articles had been published showing strong precipitating convection under such cloud shields. A rather angry airborne scientist radioed us that there wasn't even any rain under the cloud. We had a similar experience in TOGA COARE with the National Center for Atmospheric Research (NCAR) Electra.

c. After the field program

The most important phase of the field program is the combination and synthesis of the resulting data, a process that takes continuing coordination and funding.

Figure 3.11 was designed by Joachim Kuettner to show how scientists tend to want to run off with little pieces of cheese (data) rather than working on the Big Problems, much to the frustration of the field program directors (the dizzy cat). This tendency to pick out small pieces enables quick writing of papers to please those providing the funding, even though those quick papers aren't necessarily addressing the Big Questions that the field program was designed to answer.

It takes several years to mine the data. LeMone (1983) found that the publications in American Meteorological Society (AMS) journals from GATE peaked in 1980, six years after the field program. For the Florida Area Cumulus Experiment (FACE), a smaller program, the lag time was four years (R. Holle 1985, personal communication), and for the Cooperative Convective Precipitation Experiment (CCOPE), a medium-sized program, the lag time was 5–6 years. The early papers were typically either from conventional instruments that could be checked quickly, from one instrument system, or were semiroutine descriptions of instruments and early results. It was the later papers that focused on and integrated data from more complex platforms (e.g., radar and aircraft) to gain physical insight. Figure 3.12 updates the earlier publication data and includes international journals. After a peak in 1979–80 (possibly slightly earlier than for AMS journal articles), the publications from GATE fell to 6–7 papers per year, and then 0–5, with most papers during the last decade related to the dynamical properties of precipitating convection, statistical studies of precipitation in preparation for TRMM, development of convection parameterization

FIG. 3.13. Blending data is not a linear process. Drawn after GATE, in 1974–75, clearly revealed by reference to the NOAA DC-6 (*Thirty-Nine Charlie*) and the adding machine the frustrated scientist is using. From M. A. LeMone (1974, unpublished manuscript).

schemes and radar algorithms, and most recently, comparisons with TOGA COARE.

Unfortunately, funding is often cut too quickly after field programs. TOGA COARE funding for turbulence and mesoscale data analysis and even tracking down radiosonde errors was cut prematurely, based on the decision that modeling efforts should receive preference in proposals submitted two-three years after the field effort. This philosophy resulted in a long delay in correction of severely flawed radiosonde data and in a group of scientists running out of funding before the aircraft turbulence data they proposed to analyze became available. However, the radiosonde story is also heartening: through persistence of a few, the radiosonde data problems were discovered and corrected. Zipser and Johnson (1998), Lucas and Zipser (1996, 2000), and Guichard et al. (2000) give parts of this story that one day should be told in its entirety. Similar situations have been repeated numerous times, with the result that many scientists are compelled to go out in the field yearly or even more often to keep funded, or reduced to analyzing datasets on the sly, with other sources of funding.

The arguments for expecting fast data processing and analysis (not unique to COARE) are typically that faster computers can make data analysis faster. This is false for many reasons. First, new and more complex instruments with greater data volume have increased the time required for data processing and analysis. Second, focus on new instruments sometimes leads to less attention to older, "proven" instruments, as in the case of the COARE radiosondes. Third, datasets from diverse sources must be understood and made compatible, a problem compounded by broader mix of instruments, objectives, and people. Blending datasets is not always straightforward, as illustrated by Fig. 3.13, drawn while validating GATE data. Fourth, the increased sophistication of numerical models has led to a new sort of study that integrates model and data in exercises with

the twin objectives of model verification and extending the data to study the physics. Finally, we aren't any smarter than we were 30 years ago. In order to tease new knowledge out of data, we need time to think.

Post-experiment funding failure is not new. *Ocean Enough and Time: Discovering the Waters around Antarctica,* by Gorman (1995), describes a six-ship expedition led by Charles Wilkes that headed to Antarctica in 1839 to survey the coast (they established that Antarctica was a continent) and collect specimens. Only two ships returned, laden with specimens that were delivered to Louis Agassiz at Harvard. Agassiz could not complete analysis of the specimens, because *there wasn't enough money.*

An additional problem, referred to earlier, is that the objectives of an experiment may have been frozen at their most optimistic maximum, before the funding, number of PIs, or instrumentation was finalized at a reduced level, the largest field programs probably have the most significant problem—objectives may have been determined in large, international meetings that are hard to repeat. Objectives outstripping resources leads to stretching resources in the field and in post-experiment analysis and synthesis. With smaller field programs, adjusting objectives is easier.

Finally, scientists share some of the blame for inadequate analysis of field data. Shortly after GATE, Joanne Simpson (Simpson 1976) noted the following.

> Unfortunately, my experience has been that where most field experiments fail is in their follow-up, namely analysis and publication. The fault in the past has lain with both management, who finds other pressures on funds, and scientists, who find more glamour in running out to the field again, than in gluing their bottoms to a desk chair to carry through the painful and laborious scrutiny, corrections analysis, [and] reanalysis of data.

NASA has bucked this trend, perhaps because long time lags associated with even getting data from space probes requires longer-term thinking. Funding for analysis of FIFE (1987, 1989) has been at such a sufficient level that a special FIFE edition of the "Journal of Atmospheric Sciences" was published in 1998. Improvement of the land-surface parameterization in the ECMWF model can be directly traced to what was learned from FIFE (Viterbo and Beljaars 1995). Much effort was made to make the data available in usable form, so that even nonparticipants could benefit from the data. The data management plan for TRMM reflects a similar philosophy. One hopes that sustained funding for TRMM PIs is maintained for a sufficient time so that the data can be used to its fullest.[5]

---

[5] Note that the author recommends this even though she is not a TRMM PI.

3. The use of models

I separate out the use of models because they are now used in all phases of a field program, from planning through execution and in postanalysis and synthesis. GATE was the first field program for which I saw participation of scientists involved in modeling and parameterization in the planning. While it was a struggle to bring these diverse groups together, GATE was designed to collect data for testing of both numerical models and convective parameterization schemes.[6] Today, numerical models are used in deciding where to put instruments, how many instruments are needed to achieve important objectives, and what measurements are needed. Models are tested with the results of field programs. Once verified using observational data, models can extend the domain and fill in data gaps, making possible detailed studies of important physical processes.

Models can also be misleading. Here, I use the term "model" to refer to conceptual or theoretical models as well as numerical models. One of my favorite examples is in the book *The Rejection of Continental Drift* by Naomi Oreskes (1999), which relates several reasons why scientists in the United States were so tardy in accepting continental drift. In this case, the problem was accepting as truth an assumption made by William Bowie [of American Geophysic Union (AGU) Bowie Medal fame] in a calculation. The calculation demonstrated to reasonable accuracy that land floating on a substrate could explain observed gravitational anomalies. The assumption was that the base of the earth's crust was flat. The success of the calculation led American scientists to begin thinking that continents did have flat bottoms, rather than "roots." It was hard to conceive of mantle currents dragging flat-bottomed continents around to produce continental drift! This was one of the factors that delayed the acceptance of plate tectonics in the United States.

The plate-tectonics episode is an example of an erroneous assumption taken too seriously. A more common problem occurs when models are taken too seriously, their assumptions and limitations ignored, misunderstood, or forgotten. My favorite example in meteorology relates to peoples' conception of the behavior of the wind profile in the convective boundary layer. When I was a student in the late 1960s and early 1970s, the wind in the planetary boundary layer (PBL) was characterized by an Ekman spiral (Fig. 3.14), epitomized by the famous Leipzig wind profile (Mildner 1932). Much work at that time was devoted to estimating the appropriate eddy-exchange coefficient (e.g., O'Brien 1970), studying instabilities for the Ekman spiral (Faller and Kaylor 1966; Lilly 1966; Brown 1970), or esti-

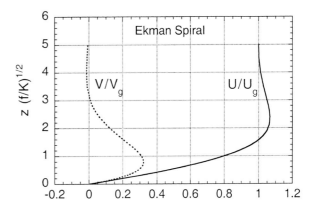

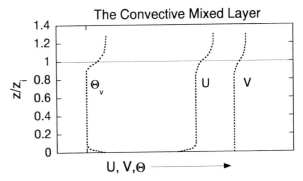

FIG. 3.14. Schematics of boundary layer wind characterized by (top) Ekman wind profile and (bottom) well-mixed momentum. My colleagues seemed to shift from Ekman thinking to mixed layer thinking in the early 1970s. In the figure, z_i is the top of the boundary layer, f is the coriolis acceleration, and K the (constant) eddy exchange coefficient.

mating the departures from the Ekman spiral created by thermal wind (Gray and Mendenhall 1973).[7]

Around 1970, perhaps partially inspired by the temperature and moisture "mixed layers" Malkus (1958) identified in the tropical oceanic subcloud layer, Geisler and Kraus (1969) found it useful to assume a mixed layer for wind as well. This idea was reinforced by Deardorff's (1972) early large eddy simulations (LESs), and such well-mixed layers are "observed" in current LESs of convective boundary layers (e.g., Moeng and Sullivan 1994). Thus boundary layer meteorologists changed from characterizing the daytime boundary layer wind in terms of the Ekman spiral to assuming that the wind is nearly constant with height except near the top and bottom (Fig. 3.14), as a result of the convectively driven mixing processes that similarly affect the temperature and mixing ratio. How accurate is this mixed-layer concept? Small vertical shears are reported by

---

[6] A one-dimensional numerical model was used in Joanne Simpson's 1968 cumulus seeding experiment, another example of her pioneering work (Simpson and Wiggert 1971).

[7] One of my favorite "informal" field programs was described to me by a professor in England. Convinced by the Ekman-instability work that Ekman spirals don't exist in the convective atmosphere, he launched and tracked a balloon for his dynamics class to show that the Leipzig wind profile was a rare occurrence. Unfortunately, for whatever reason, the data showed a perfect Ekman spiral!

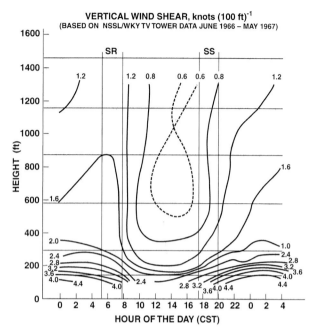

FIG. 3.15. Wind shear as a function of time, based on one year of data from the WKY-TV tower in Oklahoma. One kt $(100\ \text{ft})^{-1}$ corresponds to 1.7 m s^{-1} $(100\ \text{m})^{-1}$. Data were collected at 23, 246, 296, 581, 873, and 1458 ft above the ground (1 foot = 0.305 m). Figure from LeMone et al. (1999), adapted from Crawford and Hudson (1970). Reprinted with permission from Kluwer Academic.

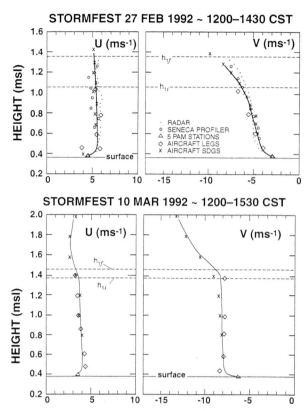

FIG. 3.16. Observed wind profiles for two days with similar pressure gradients and baroclinity, based on STORM FEST data. From LeMone et al. (1999), reprinted with permission from Kluwer Academic. Here, h_{1i} and h_{1f} are heights of the mixed layer, based on virtual potential temperature, at the beginning and ending of the averaging time.

Riehl et al. (1951) and Nicholls and LeMone (1980) in fair weather over the tropical oceans, where the diurnal variation is small. However, this was not the case for Pennell and LeMone (1974). Over land, the situation is more problematic, as illustrated in Fig. 3.15, which shows the wind shear vector magnitude as a function of height and time of day. Only between about 12:00 and 16:00 local time is the vertical shear small in the interior of the boundary layer. This also happens to be the time of day when many overland boundary layer experiments have been conducted to ensure that things are steady state enough for aircraft to collect a reasonable statistical sample for flux estimates. Similarly, LES modelers like to perform averages over a long enough period to ensure a good statistically significant ensemble average, leading to a preference for simulating near steady-state conditions.

A fortuitous combination of conditions led to documentation of significant boundary layer shear even during the early afternoon during STORM FEST and a reasonable hypothesis for its origin. (This is thanks to that string of "good" weather that led to a larger number of boundary layer studies than expected.) Figure 3.16 compares the wind profiles during the early afternoon for two days, 27 February and 10 March. While both U and V wind profiles on 10 March and the U profile for 27 February have the expected near-zero vertical shear, the V profile on 27 February has considerable vertical shear. Comparison of the two days revealed that

the synoptic and turbulence characteristics were virtually identical except for one very important feature: the temperature inversion was weaker on 27 February, enabling continued growth of the boundary layer during the afternoon, and with it, entrainment of air with higher northerly momentum. Thus, even during the hours when the boundary layer wind over land "should" be constant with height, it doesn't always happen.

4. Conclusions

I have followed the stages of field programs from their inceptions to the publication of results. A field program can be divided into three stages: the planning stage, the field stage, and most importantly, the data-analysis and synthesis stage. Although science and instruments are important, human interactions can make or break a field program. Humans determine the objectives, what is measured and when, and how much attention is given to the data afterward. To to this, we have to work as a team. I tried to reproduce some of the feelings and difficulties we experience through the three phases. Even though field programs can be frustrating, we keep going back again. The important thing is to keep one's sense of humor, remember that tired people (including experimentalists) are fallible, and to maintain the experiment team long enough after any experiment to reap the results and answer the Big Questions.

Acknowledgments. Many people contributed to the ideas in this manuscript. Above all, I acknowledge my colleagues in the many field programs in which I have participated, and thank them for mostly maintaining their sense of humor. I also acknowledge that I share the flaws attributed to anonymous colleagues. Many of the ideas for the cartoons were suggested to me by colleagues, who also suggested adding a few details. Among these are Ed Zipser, Julie Lundquist, Bob Grossman, Greg McFarquhar, Gary Barnes, Garpee Barleszi, David Jorgensen, and Brad Smull. Also, two anonymous reviewers provided or reminded me of additional insights, sometimes made while still under the influence of Larium in Honiara. I would also like to acknowledge the staff of UCAR's Joint Office of Scientific Support, who have helped me and a number of colleagues through tight spots on all three phases of many field programs, especially Jim Moore, Steve Williams, Dick Dirks, Niemal Gamage, and Greg Stossmeister. I also gratefully acknowledge the support and sacrifice of the staff of NCAR's Atmospheric Technology Division, and NOAA's Office of Aircraft Operations. Both groups could write volumes about field programs. Kyoko Ikeda proofread the manuscript.

Many of the cartoons are from two unpublished manuscripts prepared by the author after GATE and TOGA COARE. The GATE manuscript is titled "Senegal's chief import is nuts which (sic) are used to feed the mosquitoes." The COARE manuscript entitled "What REALLY happened in Honiara: An off-white paper on the TOGA COARE folks on the turboprops . . . and the mosquitoes who loved them." Both are available from the NCAR archives (P.O. Box 3000, Boulder, CO 80307).

REFERENCES

Brown, R. A., 1970: A secondary flow model for the planetary boundary layer. *J. Atmos. Sci.,* **27,** 742–757.

Crawford, K. C., and H. R. Hudson, 1970: Behavior of winds in the lowest 1500 feet in Central Oklahoma: June 1966–May 1967. ESSA Tech. Memo. ERTLM-NSSL 48, 57 pp.

Deardorff, J. W., 1972: Numerical investigations of neutral and unstable planetary boundary layers. *J. Atmos. Sci.,* **29,** 91–115.

Faller, A. J., and R. E. Kaylor, 1966: A numerical study of the laminar Ekman layer. *J. Atmos. Sci.,* **23,** 466–480.

Geisler, J. E., and E. B. Kraus, 1969: The well-mixed Ekman Boundary Layer. *Deep-Sea Res.,* **16** (Suppl.), 73–84.

Gorman, J., 1995: *Ocean Enough and Time.* HarperCollins, 190 pp.

Gray, W. M., and B. R. Mendenhall, 1973: A statistical analysis of factors influencing the wind veering in the planetary boundary layer. *Bonner Meteor. Abhandl.,* **17,** 167–194.

Guichard, F., D. Parsons, and E. Miller, 2000: Thermodynamic and radiative impact of the correction of sounding humidity bias in the Tropics. *J. Climate,* **13,** 3611–3624.

LeMone, M. A., 1983: The time between a field experiment and its published results. *Bull. Amer. Meteor. Soc.,* **64,** 614–615.

——, M. Zhou, C.-H. Moeng, D. H. Lenschow, L. J. Miller, and R. L. Grossman, 1999: An observational study of wind profiles in the baroclinic convective planetary boundary layer. *Bound.-Layer Meteor.,* **90,** 47–82.

Lilly, D. K., 1966: On the stability of Ekman boundary flow. *J. Atmos. Sci.,* **23,** 481–494.

Lucas, C., and E. J. Zipser, 1996: The variability of vertical profiles of wind, temperature, and moisture, during TOGA COARE. Preprints, *7th Conf. on Mesoscale Processes,* Reading, United Kingdom, Amer. Meteor. Soc., 125–127.

——, and ——, 2000: Environmental variability during TOGA COARE. *J. Atmos. Sci.,* **57,** 2333–2350.

Malkus, J. S., 1958: *On the Structure of the Trade Wind Moist Layer.* Papers in Physical Oceanography and Meteorology, No. 13, Massachusetts Institute of Technology and Woods Hole Oceanographic Institution, 47 pp.

Mildner, P., 1932: Uber die Reibung in Einer Speziellen Luftmasse in den Unterstehen der Atmosphare. *Beitr. Phys. Freien. Atmos,* **19,** 151.

Moeng, C.-H., and P. P. Sullivan, 1994: A comparison of shear- and buoyancy-driven planetary boundary layer flows. *J. Atmos. Sci.,* **51,** 999–1022.

Nicholls, S., and M. A. LeMone, 1980: The fair weather boundary layer in GATE: The relationship of subcloud fluxes and structure to the distribution and enhancement of cumulus clouds. *J. Atmos. Sci.,* **37,** 2051–2067.

O'Brien, J. J., 1970: A note on the vertical structure of the eddy exchange coefficient in the planetary boundary layer. *J. Atmos. Sci.,* **27,** 1213–1215.

Oreskes, N., 1999: The rejection of continental drift: Theory and method in American earth science. Oxford University Press, 420 pp.

Pennell, W. T., and M. A. LeMone, 1974: An experimental study of turbulence structure in the fair-weather trade wind boundary layer. *J. Atmos. Sci.,* **31,** 1308–1323.

Riehl, H., T. C. Yeh, J. S. Malkus, and N. E. LaSeur, 1951: The northeast trade of the Pacific Ocean. *Quart. J. Roy. Meteor. Soc.,* **77,** 598–626.

Rotunno, R., J. B. Klemp, and M. L. Weisman, 1988: A theory for strong, long-lived squall lines. *J. Atmos. Sci.,* **45,** 463–485.

Simpson, J., 1976: The GATE aircraft program: A personal view. *Bull. Amer. Meteor. Soc.,* **57,** 27–30.

——, and V. Wiggert, 1971: 1968 Florida cumulus seeing experiment: Numerical model results. *Mon. Wea. Rev.,* **99,** 87–118.

Viterbo, P., and A. C. M. Beljaars, 1995: An improved land surface parameterization scheme in the ECMWF model and its validation. ECMWF Res. Dept. Tech. Rep. 75, 52 pp.

Zipser, E. J., and R. H. Johnson, 1998: Systematic errors in radiosonde humidities: A global problem? Preprints, *10th Symp. on Measurements, Observations, and Instrumentation,* Phoenix, AZ, Amer. Meteor. Soc., 72–73.

Chapter 4

From Hot Towers to TRMM: Joanne Simpson and Advances in Tropical Convection Research

ROBERT A. HOUZE JR.

Department of Atmospheric Sciences, University of Washington, Seattle, Washington

ABSTRACT

Joanne Simpson began contributing to advances in tropical convection about half a century ago. The hot tower hypothesis jointly put forth by Joanne Simpson and Herbert Riehl postulated that deep convective clouds populating the "equatorial trough zone" were responsible for transporting heat from the boundary layer to the upper troposphere. This hypothesis was the beginning of a 50-year quest to describe and understand near-equatorial deep convection. Tropical field experiments in the 1970s [Global Atmospheric Research Program Atlantic Tropical Experiment (GATE) and the Monsoon Experiment (MONEX)] in which Joanne participated documented the mesoscale structure of the convective systems, in particular the deep, stratiform, dynamically active mesoscale clouds that are connected with the hot towers. In the 1980s these new data led to better understanding of how tropical mesoscale convective systems vertically transport heat and momentum. The role of the mesoscale stratiform circulation in this transport was quantified. Tropical field work in the 1990s [especially the Coupled Ocean–Atmosphere Response Experiment (COARE), in which Joanne again participated] showed the importance of a still larger scale of convective organization, the "supercluster." This larger scale of organization has a middle-level inflow circulation that appears to be an important transporter of momentum. The mesoscale and supercluster scale of organization in tropical convective systems are associated with the stratiform components of the cloud systems. Joint analysis of satellite and radar data from COARE show a complex, possibly chaotic relationship between cloud-top temperature and the size of a stratiform precipitation area. The Tropical Rainfall Measuring Mission (TRMM) satellite, for which Joanne served as project scientist for nearly a decade, is now providing a global census of mesoscale and supercluster-scale organization of tropical convection. The TRMM dataset should therefore provide some closure to the question of the nature of deep convection in the equatorial trough zone.

1. Introduction

Advances in understanding tropical atmospheric convection are interwoven with the career of Joanne (Malkus) Simpson. This chapter traces highlights of these advances and their connections with her career. By the early 1950s Joanne had contributed some of the first highly important observational studies of tropical convection. From early meteorological aircraft measurements she determined the basic entrainment characteristics of trade wind cumulus. In Malkus (1952) she related the tilted structure of trade cumulus to the entrainment of environmental momentum. In Malkus (1954) she examined the aircraft-measured thermodynamic properties of these clouds and concluded that while sometimes the clouds had protected inner cores with properties similar to steady-state entraining jets, the aircraft data argued primarily for the clouds being aggregates of entraining thermals in various stages of development. This latter result foreshadowed the modern view of entraining cumulus espoused by Raymond and Blyth (1986) and incorporated into the parameterization of convection in climate models by Emanuel et al.

(1994). Joanne then turned her attention to the deeper convection of the equatorial regions and joined Herbert Riehl to publish one of the landmark papers in meteorology, "On the heat balance in the equatorial trough zone" (Riehl and Malkus 1958). They postulated that undiluted updrafts in cumulonimbus maintained the mean thermodynamic stratification of the equatorial zone by transporting high moist static energy boundary layer air to the upper troposphere. Such "hot towers" would effectively bypass the middle troposphere, which is left with a generally low moist static energy. This hypothesis, which is largely accepted today, set the direction of observational research on tropical convection for the next four decades. That direction was to document and understand the phenomenology of the convective clouds containing hot towers. This paper will review briefly that research and note Joanne's contributions along the way.

2. The 1960s: Identification of the mesoscale

The hot tower paper was not a study of convection per se. It merely hypothesized the role of convective

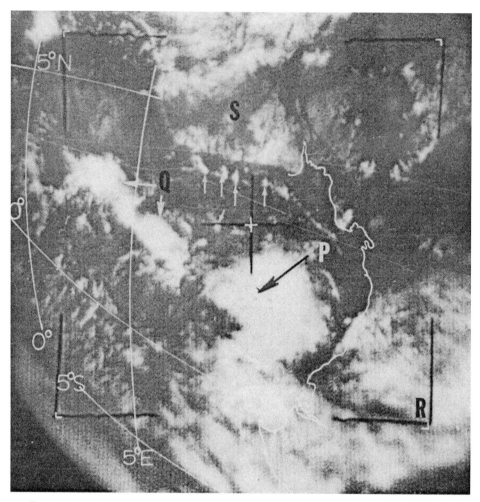

FIG. 4.1. Cumuliform clouds in the Gulf of Guinea, 0032 UTC 6 May 1964, Television Infrared
Observational Satellite *(TIROS)* 7, pass 4761T (from Anderson et al. 1966).

clouds in maintaining the large-scale mean thermodynamic structure of the equatorial atmosphere. After the
hot tower paper, Joanne continued her collaboration with
Herbert Riehl by immediately setting out to test the hot
tower hypothesis by making direct observations of the
detailed structure of convective clouds in the equatorial
zone via a series of aircraft flights over the tropical
Pacific. They mapped the clouds by photographic methods and visual observation. The results published in a
229-page monograph (Malkus and Riehl 1964) were
another groundbreaking contribution to tropical meteorology. This work, however, was limited by the absence
of modern aircraft instrumentation, little or no satellite
data, and sparse synoptic observations. Nonetheless, the
cloud mapping was effective and anticipated later studies by showing that the deep tropical convection, in
which hot towers occur, has a mesoscale organization.
On their flights, they noticed three levels of organization: "convective, meso-, and synoptic." The mesoscale
features had characteristic horizontal scales of 15–500
km. At the end of the monograph, they stated prescient-
ly, "So far as we know, the meso-scale regimes and
abrupt transitions between them have not been documented before. The main physical reasons for these transitions must be sought if attempts at theoretical modeling are to be fruitful."

Another important development during the 1960s was
the launch of the first weather satellites. These pictures
from space (e.g., Fig. 4.1) showed that the dominant
clouds of the equatorial trough zone were mesoscale,
that is, hundreds to thousands of kilometers in dimension. Since these pictures only showed the tops of the
clouds, it was easy to regard them as "cirrus canopies"
connecting skinny hot towers. This view was not really
consistent with the visual evidence of mesoscale organization in the aircraft studies of Malkus and Riehl
(1964). In an aircraft study aided by key sounding data
and early satellite data, Zipser (1969) was able to postulate the basic mesoscale structure of a tropical oceanic
mesoscale convective system (Fig. 4.2). The cirriform
cloud top was not a thin layer interconnecting individual
convective towers, as in Fig. 4.1, but rather was the top

of a deep precipitating stratiform cloud connected to a region of convective cells. Little has changed regarding Zipser's (1969) basic conceptual model. Later studies primarily have confirmed his picture, made it more quantitative, or added detail.

3. The 1970s: Documentation of the mesoscale

Despite the accumulating evidence of mesoscale organization of tropical convective systems in the 1960s, the oversimplified model of a convective population depicted in Fig. 4.3 remained popular. Ooyama (1971), Yanai et al. (1973), and Arakawa and Schubert (1974) built mathematical representations of tropical cloud ensembles on the notion that the convection was separated in scale from the larger synoptic-scale flow. A tractable mathematical approach to account for the mesoscale organization apparent in the earlier observational studies of Malkus and Riehl (1964) and Zipser (1969) awaited higher-resolution models, which became prevalent in the 1980s.

The conflict between the scale-separation theory and the potential importance of the mesoscale phenomena seen by satellite and aircraft was recognized in the planning of the Global Atmospheric Research Programme's (GARP) tropical field experiments in the 1970s. Joanne Simpson was a primary participant in the 1974 GARP Atlantic Tropical Experiment (GATE) and the 1978–9 Monsoon Experiment (MONEX). She made many aircraft flights in both of these projects and continued her efforts to map clouds photographically (Warner et al. 1980). However, GATE made use of a host of other types of measurements, which could better test the diverging conceptual models of tropical convection.

The basic design of GATE (Kuettner and Parker 1976) was to observe details of the clouds with a fleet of instrumented aircraft and simultaneously document the synoptic-scale environment and mesoscale wind pattern with supplemental soundings launched from an armada of ships, operating in a nested array off western Africa. Additionally, the boundary layer was sampled and radiation measured, as these were essential inputs to any diagnostic application of scale-separation theory. Fortunately, the ships were also equipped with three-dimensionally scanning, quantitative precipitation radars. The motivation for the radars was primarily to measure the rain, another required input, but the radars also provided detailed observations of the mesoscale organization of the precipitating disturbance. The prevailing acceptance of the scale-separation model sometimes led to the radar observations being thought to be only of marginal necessity. Participants in GATE planning recall the frequent suggestion to use the radars to track sounding balloons. This suggestion was fortunately not followed and the radar observations, combined with the mesoscale sounding array, turned out to be key in showing that the mesoscale nature of the con-

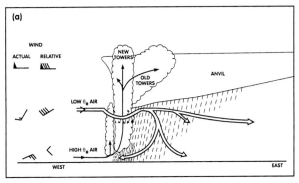

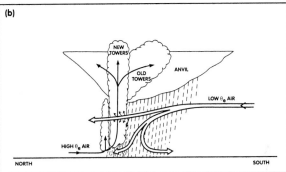

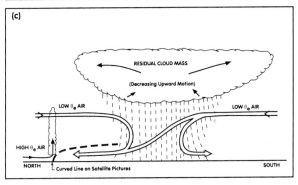

FIG. 4.2. (a) Schematic streamlines of airflow relative to a tropical oceanic convective cloud system in east–west section illustrating the mechanism of downdraft production. The low equivalent-potential-temperature (θ_e) air in the environment can pass under the anvil without necessarily intercepting convective towers, although such air that does intercept towers can be entrained by turbulent mixing into the towers and can also produce more intense and smaller-scale downdrafts than the direct large-scale production under the raining anvil. (b) Same as (a) except for north–south section. Both panels represent the intensification phase of the disturbance, with a large population of active cumulonimbus towers, either in individual clusters or in organized lines. (c) A north–south section similar to (b), but representing the dissipating phase of the disturbance, when maintenance of the downdraft is primarily by rain falling from the extensive cloud shield, although with considerable mesoscale variations in intensity not depicted in this diagram (from Zipser 1969).

vective systems was as anticipated by Malkus and Riehl (1964) and Zipser (1969).

GATE and MONEX ship and aircraft measurements made beneath the cirrus canopies showed that the rain from these clouds was dominated by cloud systems with rain areas ~100 km in dimension—manifestly meso-

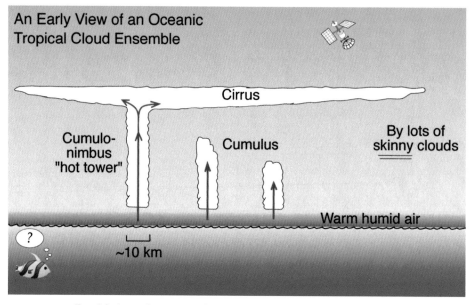

FIG. 4.3. A popular conceptual model of a tropical cloud ensemble, c. 1974.

scale. Radar observations showed that large portions of the precipitation were stratiform with pronounced melting layers (Shupiatsky et al. 1975, 1976; Houze 1977; Leary and Houze 1979), while other portions of the rain areas had embedded convective-scale cores of maximum intensity (hot towers). These radar observations substantiated the earlier visual impressions of Malkus and Riehl (1964) and Zipser (1969).

The new observations from GATE and MONEX led to a rethinking of how tropical convection vertically redistributes mass, heat, and momentum in the troposphere, and, as a result, the 1980s were a most productive period in tropical convection research.

4. The 1980s: Accounting for the mesoscale in transports of heat and momentum

Houze and Betts (1981) summarized the basic observations of convection in GATE, and Johnson and Houze (1987) summarized the MONEX observations. A major point of these summaries was the aforementioned ubiquitous occurrence of stratiform precipitation in the major rain-producing convective systems and the degree of mesoscale organization implied by this finding. Both rapidly propagating ("squall line") systems and slowly moving mesoscale convective systems exhibited stratiform precipitation regions in connection with the convection. From the Houze–Betts and Johnson–Houze summaries it was evident that a major new understanding of tropical convection would emerge from these rich datasets. A French field experiment, called Convection Tropicale Profonde (COPT-81) and conducted in the Ivory Coast (western equatorial Africa) in 1981 (Sommeria and Testud 1984) added a valuable continental dataset to the oceanic datasets of GATE and

MONEX. COPT-81 emphasized radar observations of squall-line mesoscale convective systems over land (Roux et al. 1984; Chong et al. 1987; Roux 1988; Chalon et al. 1988; Sun and Roux 1988; Chong and Hauser 1989).

One of the first important findings regarding the mesoscale organization of the convective cloud systems in GATE, MONEX, and COPT-81 was that the stratiform rain areas had their own dynamical structure, substantially different from the convective-scale dynamics of the hot towers populating the convective portion of the disturbance. The stratiform regions had convergence and vortex structure at midlevels and divergence at lower and upper levels (Figs. 4.4, 4.5; Gamache and Houze 1982, 1985; Houze and Rappaport 1984; Sun and Roux 1988; Lafore and Moncrieff 1989). As these circulation patterns became more widely known, investigators began to think about the implications of these findings for the vertical transports of momentum and heat.

LeMone (1983) analyzed the vertical redistribution of momentum by GATE mesoscale convective systems. By analyzing the GATE aircraft data, she found that a small-scale lower-tropospheric pressure minimum lay below the downshear-tilted updraft of a line of convective cells in a mesoscale convective system (Fig. 4.6). This pressure perturbation was a hydrostatic result of the warm, buoyant, sloping updraft lying overhead. LeMone further found that the pressure perturbation and associated circulation pattern of the convective system were consistent with the theoretical model of Moncrieff and Miller (1976; Fig. 4.7). Their model states that a two-dimensional steady-state convective system in an environment of specified shear similar to that of the large-scale environment in GATE must slope downshear with a low pressure perturbation underlying the sloping

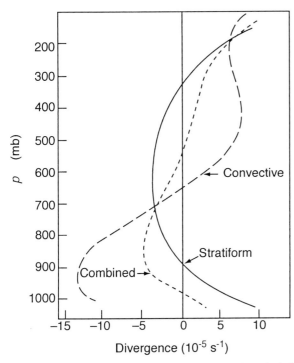

FIG. 4.4. Average divergence over the convective (long dashed), stratiform (solid), and combined (short dashed) regions of a tropical squall-line system (from Gamache and Houze 1982).

updraft. This flow configuration has a characteristic signature of vertical transport of horizontal momentum, in which cross-line momentum is removed from lower levels and added to upper levels (Fig. 4.8). The aircraft

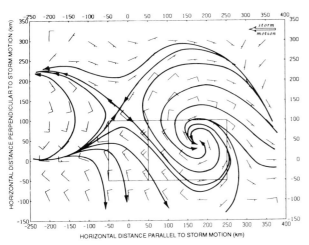

FIG. 4.5. Composite analysis of the mesoscale component of the wind at the 650-mb level of a tropical squall-line system. The smaller rectangle was the region containing the leading convective line seen on radar. The larger rectangle was the region of stratiform precipitation seen on radar. Winds were from ship-based soundings and aircraft measurements. The mesoscale component of the flow shown here was deduced by applying a mesoscale bandpass filter to the data to suppress larger- and smaller-scale wavelengths. A full wind barb represents 5 m s^{-1}. A vortex was evident toward the rear of the stratiform region (from Gamache and Houze 1985).

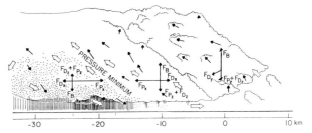

FIG. 4.6. Schematic of the processes leading to the momentum generation in the convective band. Here F_P, F_D, and F_B are forces on an air parcel owing to mesoscale pressure gradient, drag, and buoyancy, respectively. Horizontal and vertical components are indicated by subscripts x and z, respectively (from LeMone 1983).

data analyzed by LeMone (1983) implied such a transport (see her Figs. 13 and 14).

Houze (1982) analyzed the vertical redistribution of heating implied by the GATE mesoscale convective systems. Analysis of GATE, MONEX, and COPT-81 radar data showed that about 40% of the precipitation from mesoscale convective systems in these tropical regions was stratiform (Houze 1977; Cheng and Houze 1979; Gamache and Houze 1983; Leary 1984; Houze and Rappaport 1984; Churchill and Houze 1984; Houze and Wei 1987; Chong and Hauser 1989). This fact, coupled with the different dynamic structure of the stratiform region, suggested that a large proportion of the latent heat released in a convective system had a different vertical distribution than if the precipitation were of a purely convective nature. Using observations of the divergence profiles obtained by soundings and aircraft in convective and stratiform regions in GATE, Houze (1982, 1989) calculated the vertical distributions of heating in the convective and stratiform regions of a typical tropical

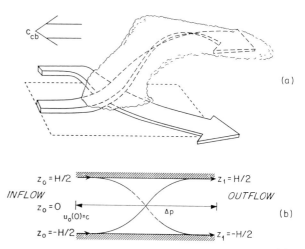

FIG. 4.7. (a) Schema of a propagating cumulonimbus; (b) limiting relative streamlines based on two-dimensional steady-state theory. On inflow, the velocity profile $U_o = u_o - c$, the static stability, and the parcel lapse rate are specified as a function of the inflow level z_o but c is calculated. Here Δp is the pressure difference across the system that must be consistent with the flow (from Moncrieff and Miller 1976).

Convective up and downdraft effects on the horizontal momentum field

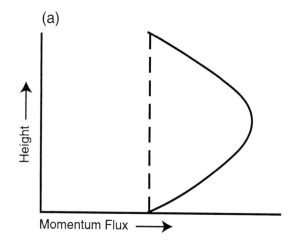

(a)

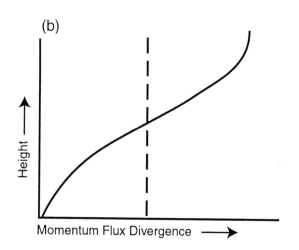

(b)

FIG. 4.8. (a) Idealized vertical flux of horizontal momentum by convective updrafts and downdrafts in a tropical oceanic organized convective system; (b) vertical divergence of the momentum flux in (a) [based on work of Moncrieff and Miller (1976) and LeMone (1983)].

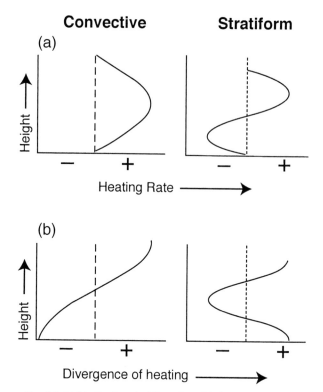

Convective   Stratiform

(a)

(b)

FIG. 4.9. (a) Idealized vertical flux of sensible heat in the convective and stratiform regions of a tropical mesoscale convective system; (b) vertical divergence of the fluxes in (a) [based on work of Houze (1982, 1989)].

mesoscale convective system, with precipitation 40% stratiform. Figure 4.9 sketches the shapes of profiles that he found to be characteristic of each type of region. The heating profile in the convective region forms a half sine wave, while that of the stratiform region forms a full sine wave. The tops of the profiles are at the tropopause. The lower panels show the profile of the vertical divergence of heating (which of course just mirrors the mass divergence profile) in each type of region. The convective region has a two-layer profile (convergence at low levels, divergence at upper levels), while the stratiform region has a three-layer profile, with convergence in middle levels sandwiched between divergence at low levels and aloft. The large-scale (i.e., balanced) flow responds directly to the profile of vertical divergence of heating, which is a source of large-scale potential vorticity (Haynes and MacIntyre 1987). To the extent that precipitating mesoscale convective systems constitute the large-scale heat source in the Tropics, the large-scale circulation must be largely a response to the convective and stratiform heating profiles. Mapes (1993) and Mapes and Houze (1995) later elaborated on this point. Hartmann et al. (1984) obtained a much more realistic Walker cell circulation over the tropical Pacific when they assumed a large-scale heating profile characteristic of mesoscale convective systems with strong stratiform components.

5. The 1990s: Superclusters and chaos

Early satellite pictures, like that in Fig. 4.1, indicated the mesoscale size of tropical convective systems dominating the equatorial regions, especially over the vast oceanic regions. As satellite sampling became more comprehensive, it became possible to characterize the cloud population of tropical regions statistically. One of the centers of tropical convection is the "warm pool" of the western Pacific and Indian Oceans. This region

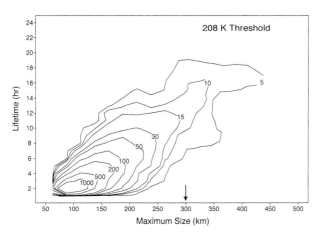

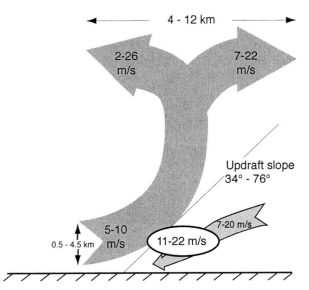

FIG. 4.10. Frequency of occurrence of convective systems tracked by an automatic tracking algorithm (per 25-km size interval per hour) as a function of the maximum size (abscissa) reached by a convective system during its lifetime (from the start to end of its life cycle) for all of the trackable clusters that occurred over the warm pool during COARE (from Chen and Houze 1997).

FIG. 4.11. Schematic of airflow in the convective regions of mesoscale convective systems observed by airborne Doppler radar in COARE. The numbers (from bottom to top) indicate the observed ranges of values for the depth of the inflow layer, horizontal relative velocity of inflow and outflow air currents, the slope of the updraft, and the width of the divergent region aloft. The horizontal directional differences of the low-level updraft inflow and midlevel downdraft inflow were often significantly different from 180° [based on figures and tables of Kingsmill and Houze (1999a)].

was the focus of COARE in 1992–93 (Webster and Lukas 1992; Godfrey et al. 1998). Joanne Simpson was a major participant in the aircraft program of COARE, especially with the National Aeronautics and Space Administration (NASA) DC-8.

Tracking methods applied to geosynchronous satellite data (e.g., Williams and Houze 1987) can indicate the lifetime and maximum size attained by mesoscale convective systems defined by a given threshold infrared temperature. The Williams–Houze method mimics the way an analyst would follow patterns in a series of satellite images by accounting for mergers and splits as the system goes through its life cycle. Chen et al. (1996) and Chen and Houze (1997) used the Williams–Houze method to determine statistics of the convection in COARE. Figure 4.10 shows their results for mesoscale convective systems defined by an infrared threshold of 208 K. The plot shows the frequency of occurrence by system lifetime and maximum size. Larger systems tend to last longer, but there is a lot of spread in the distribution. Systems reaching 300 km or more in maximum dimension (i.e., more than about 1 000 000 km² in area) last anywhere from 5 to 20 h. The longer lasting ones are what Nakazawa (1988) began to call "superclusters." Chen et al. (1996) called the systems > 300 km in dimension "superconvective systems," regardless of whether they were short or long lived. A lot of COARE research in the 1990s focused on these larger mesoscale convective systems.

Kingsmill and Houze (1999a,b) examined airborne dual-Doppler radar data collected on research flights on 25 different days of COARE. Many of these flights were in superconvective systems. Some flights were in convective regions and consistently showed deep updrafts consisting of sloping layers of air rising over advancing cold pools in circulations resembling those theorized by

Moncrieff and Miller (1976; Fig. 4.7). Figure 4.11 is a schematic composite of the structure seen by Kingsmill and Houze (1999a,b). These updrafts comprised the hot towers predicted by Riehl and Malkus (1958). However, they were not simply narrow vertical pipes carrying air from the boundary layer to the tropopause, as suggested by Fig. 4.3. The air entering the updraft came from a layer much deeper than the boundary layer. Typically, this layer was one to a few kilometers deep, and the air flowed rearward in a deep layer, extending from the middle through the upper troposphere.

Other flights were in the stratiform regions of superconvective systems, and the data from these flights invariably showed an extensive midlevel inflow, which lay just at the base of the precipitation anvil and sloped downward through the precipitation melting layer in the heart of the cloud system (Fig. 4.12). The midlevel inflow seen in these cases was similar to midlevel inflows in most mesoscale convective systems, except for the extreme horizontal extent. The direction of the midlevel inflow was related closely to the direction of large-scale environmental flow at midlevels.

Moncrieff and Klinker (1997) inferred the existence of superconvective systems with midlevel inflow entering from a direction determined by the large-scale environmental flow at midlevels (700–850 mb) when they used forecasts and analysis from an operational global numerical weather prediction model to study an especially active convective period in the COARE region (Fig. 4.13; note the horizontal scale of the disturbance

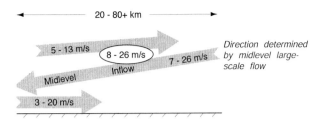

FIG. 4.12. Schematic of airflow in the stratiform regions of mesoscale convective systems observed by airborne Doppler radar in COARE. The numbers (from bottom to top) indicate the observed ranges of values for the horizontal relative wind velocity at low levels, the horizontal scale of the midlevel inflow, the horizontal relative velocity of the midlevel inflow and outflow air currents, the differential horizontal velocity between the middle and upper level, and the velocity at upper levels [based on figures and tables of Kingsmill and Houze (1999a)].

they describe, and the direction of the midlevel inflow). They found that these broad midlevel inflows of environmental air fed the mesoscale downdraft of the superconvective system. They further determined that the downward transport of momentum by these midlevel inflow mesoscale downdraft circulations played a strong role in balancing the large-scale momentum budget (see their Fig. 8b).

Using Doppler radar data from both ship and aircraft in superconvective systems in COARE, Houze et al. (2000) found evidence in the observations to confirm that the midlevel inflows of the superconvective systems transported momentum downward over large regions. They examined superconvective systems in different parts of the Kelvin–Rossby wave structure that dominated the wind pattern over the COARE region during December–February of the project (Fig. 4.14). They found that as the downdrafts associated with the extensive midlevel inflow circulations transported momentum downward, they enhanced the low-level westerly winds in the region of the westerly jet between the two large-scale gyres (a positive feedback). In the region of westerly onset (east of the gyres), the midlevel inflow circulations transported environmental easterly wind downward and thus reduced the incipient low-level westerlies (a negative feedback). This result, together with the Moncrieff and Klinker (1997) analysis, indicates the potentially large importance of the superconvective systems and of the mesoscale (as opposed to embedded convective scale) circulations in the large-scale momentum budget.

6. The year 2000 and beyond: Mapping convection in the equatorial trough zone with the TRMM satellite

The 1990s brought an increased awareness of the importance of circulations in superconvective systems, which adds to the knowledge of the importance of hot towers (Riehl and Malkus 1958) and mesoscale circulations in ordinary mesoscale convective systems (Zip-

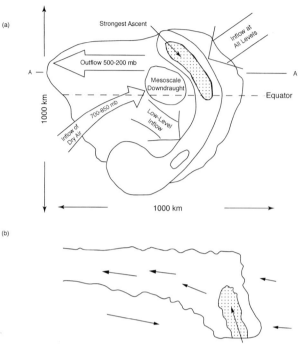

FIG. 4.13. Conceptual model of supercluster: (a) plan view, (b) zonal cross section through the line AA′ in (a) (from Moncrieff and Klinker 1997).

ser 1969). An important question remaining is the relative importance of the convective, mesoscale, and superconvective scales over the whole equatorial trough zone. Just how variable is convection in the Tropics? Is it predictable?

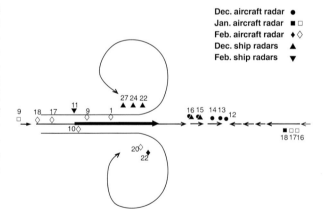

FIG. 4.14. Locations, in relation to the archetypical atmospheric Kelvin–Rossby wave, of ship and aircraft radar data obtained on different days during COARE. The idealized streamlines are drawn as they would appear on a map of the winds at the 850-mb level. The bold streamline indicates where the winds were strongest. Solid symbols indicate observations of superconvective systems. Open symbols indicate mesoscale convective systems <300 km in dimension. The solid symbols represent superconvective systems, which exceeded 300 km in horizontal scale. Numbers indicate the day of the month (from Houze et al. 2000).

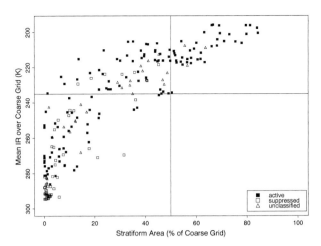

FIG. 4.15. Stratiform precipitation area determined by aircraft radar vs mean infrared temperature observed by satellite over the ~200-km scale region containing the radar observations. Data were from all the aircraft missions of COARE and thus represent convection over the western Pacific oceanic warm pool (from Yuter and Houze 1998).

The only instrument platform capable of making a census of convective phenomena throughout the equatorial regions is the meteorological satellite. Yuter and Houze (1998) analyzed COARE aircraft radar data taken concurrent with satellite infrared images. Since COARE was four months in duration, it provided an opportunity to analyze the joint variation of satellite and radar data over a specific region of the equatorial zone. Yuter and Houze (1998) applied an algorithm to aircraft radar data (lower fuselage data from the two P3 aircraft) to distinguish between the convective and stratiform components of the precipitation observed over a 200-km diameter area centered on the aircraft. The size of the stratiform precipitation area, observed by aircraft, is an

index of the mesoscale organization of the cloud system seen by the satellite. Figure 4.15 shows the relationship between the mean satellite-observed infrared temperature over the 200-km diameter region of radar observation. Each point represents one snapshot of concurrent satellite and radar data. The data showed a systematic relationship between the two variables, but not a strong correlation. Yuter and Houze (1998) likened this behavior to that of a phase space in a chaotic system (Lorenz 1993). Some combinations of variables just did not occur; specifically, large stratiform rain areas (>50% of the 200-km diameter sampled region around the aircraft) did not occur if the mean infrared temperature over the 200-km sampled region was greater than about 220 K. Smaller stratiform regions (<50% of the 200-km diameter sampled region around the aircraft) occurred with almost any mean cloud-top temperature over the sampled area, though the probability decreased with decreasing cloud-top temperature. A chaotic behavior in general would raise questions about the predictability of tropical mesoscale and superconvective systems.

As Tropical Rainfall Measuring Mission (TRMM) project scientist for a decade of her career, Joanne Simpson led the effort to put the satellite in space that may resolve such questions. The TRMM spacecraft has a meteorological radar onboard and thus provides the opportunity to determine the statistics of mesoscale and superconvective organization over the equatorial trough zone about which she and Herbert Riehl speculated in 1958. One of the TRMM products computed and archived every day is the spatial distribution of convective and stratiform precipitation shown by the TRMM Precipitation Radar (PR). The algorithm used to separate the echo into convective and stratiform components is that of Awaka et al. (1997). Figure 4.16 shows an ex-

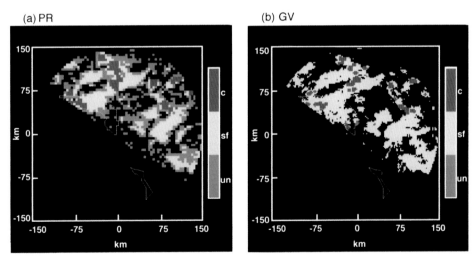

FIG. 4.16. Precipitation patterns indicated by (a) the PR on the TRMM satellite and (b) the Kwajalein TRMM ground-validation (GV) radar. Both radar echo patterns have been separated into convective (red) and stratiform (yellow) components. Undecided echo categories for the PR are indicated in green. The Kwajalein Atoll is indicated in white. Data are for 0251 UTC 25 Nov 1998, orbit 5712.

ample of the convective–stratiform pattern over the Kwajalein atoll. The Kwajalein ground-validation radar (www.atmos.washington.edu/gcg/MG/KWAJ/) pattern for the same time is shown for comparison. The results from the satellite and ground-based radars are clearly consistent.

The TRMM satellite appears to be capable of estimating close to the full amount of stratiform component of precipitation (Schumacher and Houze 2000). However, because the sensitivity of the PR is relatively low (minimum detectable reflectivity $\sim$17 dBZ) the total area covered by stratiform precipitation may not always be evident. However, the passive microwave data from TRMM can be helpful in filling out the pattern, so that it will be possible to determine statistics on the scales of stratiform precipitation areas in relation to cloud-top temperature (as in Fig. 4.15) for wide regions of the Tropics. It will be possible then to relate these statistics to terrestrial characteristics such as sea surface temperature, land versus ocean, and orography, as well as to features of the large-scale atmospheric circulation.

7. Final comments

Advances in tropical convection have pursued an understanding of the nature of the convective systems in which the "hot towers" hypothesized by Riehl and Malkus (1958) carry high moist static energy from the atmospheric boundary layer to the tropopause. In the first observational expedition to investigate these phenomena, Malkus and Riehl (1964) noticed that tropical convection was not just a matter of convective and synoptic scales, but that a "mesoscale" organization of the convection existed with horizontal dimensions $\sim$15–500 km. Over the next 30 years ever more sophisticated field experiments (in most of which Joanne participated) have elaborated on this mesoscale theme. Zipser (1969) described the mesoscale circulation in and below the massive, deep nimbostratus layer that identifies tropical mesoscale convective systems in satellite cloud-top images. GATE showed that $\sim$40% of rain in the oceanic equatorial zone fell from these nimbostratus "anvils" connected with the hot towers and that the anvil cloud layer was dynamically active (Houze 1977; Gamache and Houze 1982). Further analysis of data from GATE and MONEX showed how the mesoscale organization affects transports of momentum (LeMone 1983) and heat (Houze 1982). COARE focused attention on larger mesoscale phenomena, namely, superclusters (Nakazawa 1988) or more generically "superconvective systems" (Chen et al. 1996). Moncrieff and Klinker (1997) have used large-scale modeling and Houze et al. (2000) have used COARE ship and aircraft radar data to show that the superconvective systems are significant transporters of momentum through the action of organized midlevel inflows connected with large stratiform precipitation regions.

It remains to be determined the global impact of these mesoscale aspects of tropical convection. Joint variability of satellite-observed cloud-top temperature and the occurrence of stratiform precipitation (an index of mesoscale organization) suggest a complex relation between satellite-observed cloud-top imagery and radar-observed stratiform precipitation coverage (Fig. 4.15). The TRMM satellite offers the best way to explore this relationship over the next several years. The precipitation radar onboard the satellite is able to identify stratiform precipitation areas, within the limits of the radar's sensitivity. Outside these limits the passive microwave radiometers on the spacecraft can fill out the picture. Thus, Joanne Simpson's career in tropical convection comes full circle. In 1958 she helped to identify the active convection in the equatorial trough zone as the phenomenon on which to concentrate observational attention. Exactly four decades later, the TRMM satellite was flying and providing the means to make a global census of tropical convection and its mesoscale organization.

Acknowledgments. This work was sponsored by NASA Grant NAG5-4795.

REFERENCES

Anderson, R. K., E. W. Ferguson, and V. J. Oliver, 1966: The use of satellite pictures in weather analysis and forecasting. WMO Tech. Note 75, 184 pp.

Arakawa, A., and W. H. Schubert, 1974: Interaction of a cumulus ensemble with the large-scale environment: part I. *J. Atmos. Sci.,* **31,** 674–701.

Awaka, J., T. Iguchi, H. Kumagai, and K. Okamoto, 1997: Rain type classification algorithm for TRMM Precipitation Radar. *Proc. Int. Geoscience and Remote Sensing Symp.,* Singapore, IEEE, 1633–1635.

Chalon, J.-P., G. Jaubert, F. Roux, and J.-P. Lafore, 1988: The west African squall line observed on 23 June 1981: Mesoscale structure and transports. *J. Atmos. Sci.,* **45,** 2744–2763.

Chen, S. S., and R. A. Houze Jr., 1997: Diurnal variation and life-cycle of deep convective systems over the tropical Pacific warm pool. *Quart. J. Roy. Meteor. Soc.,* **123,** 357–388.

——, ——, and B. E. Mapes, 1996: Multiscale variability of deep convection in relation to large-scale circulation in TOGA COARE. *J. Atmos. Sci.,* **53,** 1380–1409.

Cheng, C.-P., and R. A. Houze Jr., 1979: The distribution of convective and mesoscale precipitation in GATE radar echo patterns. *Mon. Wea. Rev.,* **107,** 1370–1381.

Chong, M., and D. Hauser, 1989: A tropical squall line observed during the COPT 81 experiment in West Africa. Part II: Water budget. *Mon. Wea. Rev.,* **117,** 728–744.

——, P. Amayenc, G. Scialom, and J. Testud, 1987: A tropical squall line observed during the COPT 81 experiment in West Africa. Part I: Kinematic structure inferred from dual-Doppler radar data. *Mon. Wea. Rev.,* **115,** 670–694.

Churchill, D. D., and R. A. Houze Jr., 1984: Development and structure of winter monsoon cloud clusters on 10 December 1978. *J. Atmos. Sci.,* **41,** 933–960.

Emanuel, K. A., J. D. Neelin, and C. S. Bretherton, 1994: On large-scale circulations in convecting atmospheres. *Quart. J. Roy. Meteor. Soc.,* **120,** 1111–1143.

Gamache, J. F., and R. A. Houze Jr., 1982: Mesoscale air motions associated with a tropical squall line. *Mon. Wea. Rev.,* **110,** 118–135.

——, and ——, 1983: Water budget of a mesoscale convective system in the tropics. *J. Atmos. Sci., 40,* 1835–1850.

——, and ——, 1985: Further analysis of the composite wind and thermodynamic structure of the 12 September GATE squall line. *Mon. Wea. Rev., 113,* 1241–1259.

Godfrey, J. S., R. A. Houze Jr., R. H. Johnson, R. Lukas, J.-L. Redelsperger, A. Sumi, and R. Weller, 1998: COARE: An interim report. *J. Geophys. Res., 103,* 14 395–14 450.

Hartmann, D. L., H. H. Hendon, and R. A. Houze Jr., 1984: Some implications of the mesoscale circulations in tropical cloud clusters for large-scale dynamics and climate. *J. Atmos. Sci., 41,* 113–121.

Haynes, P. H., and M. E. MacIntyre, 1987: On the evolution of vorticity and potential vorticity in the presence of diabatic heating and frictional or other forces. *J. Atmos. Sci., 44,* 828–841.

Houze, R. A., Jr., 1977: Structure and dynamics of a tropical squall-line system. *Mon. Wea. Rev., 105,* 1540–1567.

——, 1982: Cloud clusters and large-scale vertical motions in the tropics. *J. Meteor. Soc. Japan, 60,* 396–410.

——, 1989: Observed structure of mesoscale convective systems and implications for large-scale heating. *Quart. J. Roy. Meteor. Soc., 115,* 425–461.

——, and A. K. Betts, 1981: Convection in GATE. *Rev. Geophys. Space Phys., 19,* 541–576.

——, and E. N. Rappaport, 1984: Air motions and precipitation structure of an early summer squall line over the eastern tropical Atlantic. *J. Atmos. Sci., 41,* 553–574.

——, and T. Wei, 1987: The GATE squall line of 9–10 August 1974. *Adv. Atmos. Sci., 4,* 85–92.

——, S. S. Chen, D. E. Kingsmill, Y. Serra, and S. E. Yuter, 2000: Convection over the Pacific warm pool in relation to the atmospheric Kelvin–Rossby wave. *J. Atmos. Sci., 57,* 3058–3089.

Johnson, R. H., and R. A. Houze Jr., 1987: Precipitating cloud systems of the Asian monsoon. *Monsoon Meteorology,* C. P. Chang and T. N. Krishnamurti, Eds., Clarendon Press, 298–353.

Kingsmill, D. E., and R. A. Houze Jr., 1999a: Kinematic characteristics of air flowing into and out of precipitating convection over the west Pacific warm pool: An airborne Doppler radar survey. *Quart. J. Roy. Meteor. Soc., 125,* 1165–1207.

——, and ——, 1999b: Thermodynamic characteristics of precipitating convection over the west Pacific warm pool. *Quart. J. Roy. Meteor. Soc., 125,* 1209–1229.

Kuettner, J. P., and D. E. Parker, 1976: GATE: Report on the field phase. *Bull. Amer. Meteor. Soc., 57,* 11–27.

Lafore, J. P., and M. W. Moncrieff, 1989: A numerical investigation of the organization and interaction of the convective and stratiform regions of tropical squall lines. *J. Atmos. Sci., 46,* 521–544.

Leary, C. A., 1984: Precipitation structure of the cloud clusters in a tropical easterly wave. *Mon. Wea. Rev., 112,* 313–325.

——, and R. A. Houze Jr., 1979: The structure and evolution of convection in a tropical cloud cluster. *J. Atmos. Sci., 36,* 437–457.

LeMone, M. A., 1983: Momentum transport by a line of cumulonimbus. *J. Atmos. Sci., 40,* 1815–1834.

Lorenz, E. N., 1993: *The Essence of Chaos.* University of Washington Press, 227 pp.

Malkus, J. S., 1952: The slopes of cumulus clouds in relation to external wind shear. *Quart. J. Roy. Meteor. Soc., 78,* 530–542.

——, 1954: Some results of a trade-cumulus cloud investigation. *J. Meteor., 11,* 222–237.

——, and H. Riehl, 1964: *Cloud Structure and Distributions over the Tropical Pacific Ocean.* University of California Press, Berkeley, 229 pp.

Mapes, B. E., 1993: Gregarious tropical convection. *J. Atmos. Sci., 50,* 2026–2037.

——, and R. A. Houze Jr., 1995: Diabatic divergence profiles in western Pacific mesoscale convective systems. *J. Atmos. Sci., 52,* 1807–1828.

Moncrieff, M. W., and M. J. Miller, 1976: The dynamics and simulation of tropical cumulonimbus and squall lines. *Quart. J. Roy. Met. Soc., 102,* 373–394.

——, and E. Klinker, 1997: Organized convective systems in the tropical western Pacific as a process in general circulation models: A TOGA COARE case study. *Quart. J. Roy. Met. Soc., 123,* 805–827.

Nakazawa, T., 1988: Tropical super clusters within intraseasonal variations over the western Pacific. *J. Meteor. Soc. Japan, 66,* 823–839.

Ooyama, K. 1971: A theory on parameterization of cumulus convection. *J. Meteor. Soc. Japan, 49,* 744–756.

Raymond, D. J., and A. M. Blyth, 1986: A stochastic mixing model for nonprecipitating cumulus clouds. *J. Atmos. Sci., 43,* 2708–2718.

Riehl, H., and J. S. Malkus, 1958: On the heat balance in the equatorial trough zone. *Geophysica, 6,* 503–538.

Roux, F., 1988: The west African squall line observed on 23 June 1981 during COPT 81: Kinematics and thermodynamics of the convective region. *J. Atmos. Sci., 45,* 406–426.

——, J. Testud, M. Payen, and B. Pinty, 1984: West African squall-line thermodynamic structure retrieved from dual-Doppler radar observations. *J. Atmos. Sci., 41,* 3104–3121.

Schumacher, C., and R. A. Houze Jr., 2000: Comparison of radar data from the TRMM satellite and Kwajalein oceanic validation site. *J. Appl. Meteor., 39,* 2151–2164.

Shupiatsky, A. B., A. I. Korotov, V. D. Menshenin, R. S. Pastushkov, and M. Jovasevic, 1975: Radar investigations of evolution of clouds in the eastern Atlantic. GATE Rep. 14, vol. 2, International Council of Scientific Unions/World Meteorological Organization.

——, ——, and R. S. Pastushkov, 1976: Radar investigations of the evolution of clouds in the East Atlantic. *TROPEX-74,* XXX, Ed., Vol. 1, *Atmosphere* (in Russian), Gidrometeoizdat, Leningrad, USSR, 508–514.

Sommeria, G., and J. Testud, 1984: COPT81: A field experiment designed for the study of dynamics and electrical activity of deep convection in continental tropical regions. *Bull. Amer. Meteor. Soc., 65,* 4–10.

Sun, J., and F. Roux, 1988: Thermodynamic structure of the trailing-stratiform regions of two west African squall lines. *Ann. Geophys., 6,* 659–670.

Warner, C., J. Simpson, G. van Helvoirt, D. W. Martin, D. Suchman, and G. L. Austin, 1980: Deep convection on day 261 of GATE. *Mon. Wea. Rev., 108,* 169–194.

Webster, P. J., and R. Lukas, 1992: TOGA COARE: The coupled ocean–atmosphere response experiment. *Bull. Amer. Meteor. Soc., 73,* 1377–1416.

Williams, M., and R. A. Houze Jr., 1987: Satellite–observed characteristics of winter monsoon cloud clusters. *Mon. Wea. Rev., 115,* 505–519.

Yanai, M., S. Esbensen, and J. H. Chu, 1973: Determination of bulk properties of tropical cloud clusters from large-scale heat and moisture budgets. *J. Atmos. Sci., 30,* 611–627.

Yuter, S. E., and R. A. Houze Jr., 1998: The natural variability of precipitating clouds over the western Pacific warm pool. *Quart. J. Roy. Met. Soc., 124,* 53–99.

Zipser, E. J., 1969: The role of organized unsaturated convective downdrafts in the structure and rapid decay of an equatorial disturbance. *J. Appl. Meteor., 8,* 799–814.

Chapter 5

Some Views On "Hot Towers" after 50 Years of Tropical Field Programs and Two Years of TRMM Data

EDWARD J. ZIPSER

Department of Meteorology, University of Utah, Salt Lake City, Utah

ABSTRACT

The "hot tower" hypothesis requires the existence of deep cumulonimbus clouds in the deep Tropics as essential agents, which accomplish the mass and energy transport essential for the maintenance of the general circulation. As the role of the deep convective clouds has been generally accepted, the popularity of referring to these deep "hot" towers as undilute towers also has gained acceptance. This paper examines the consequences of assuming that the deep convective clouds over tropical oceans consist of undilute ascent from the subcloud layer.

Using simple applications of parcel theory, it is concluded that observed properties of typical cumulonimbus updrafts in low- to midtroposphere over tropical oceans are inconsistent with the presence of undilute updrafts. Such undilute updrafts are far more consistent with observations in severe storms of midlatitudes. The observations over tropical oceans can be hypothetically explained by assuming large dilution of updrafts by entrainment below about 500 hPa, followed by freezing of condensate. This freezing and subsequent ascent along an ice adiabat reinvigorates the updrafts and permits them to reach the tropical tropopause with the necessary high values of moist static energy, as the hot tower hypothesis requires. The large difference observed between ocean and land clouds can be explained by assuming slightly smaller entrainment rates for clouds over land. These small entrainment differences have a very large effect on updrafts in the middle and upper troposphere and can presumably account for the large differences in convective vigor, ice scattering, and lightning flash rates that are observed. It follows that convective available potential energy (CAPE) is not a particularly good predictor of the behavior of deep convection.

Using the Tropical Rainfall Measuring Mission (TRMM) to map a proxy for the most intense storms on earth between 36°S and 36°N, they are found mostly outside the deep Tropics, with the notable exception of tropical Africa.

1. Introduction

Riehl and Malkus (1958) is one of the most quoted and influential papers in atmospheric science. The "hot tower" hypothesis is usually attributed to this paper, although those words are not used by Riehl or Malkus in any paper until several years later. What they actually wrote is that deep cumulonimbus clouds are essential to the general circulation and to the global energy balance, and that those clouds in the equatorial trough must necessarily transport moist static energy against its mean gradient above about 600 hPa. This is a conclusion that has influenced three generations of scientists and will surely stand unchallenged. Riehl and Malkus did illustrate the thermodynamic updraft path in such clouds, which was clearly intended to be pseudoadiabatic, and they indeed referred to them as "undilute" cores, later popularized as hot towers.

Why should such a universally accepted idea require renewed attention at this time? It has been known for 40 years that actual liquid water content in tropical clouds rarely exceeds 0.4 of adiabatic, although rapid coalescence growth and fallout of precipitation could perhaps have explained this fact. Recent decades of observations over tropical oceans—from 1974 in the tropical Atlantic (LeMone and Zipser 1980; Zipser and LeMone 1980), hurricanes[1] (Jorgensen et al. 1985), offshore of Taiwan (Jorgensen and LeMone 1989), offshore of tropical Australia (Lucas et al. 1994), and the warm pool of the equatorial Pacific (Wei et al. 1998; Igau et al. 1999)—have yielded data from aircraft penetrations of thousands of cumulonimbus clouds. The results reported in these papers show great consistency with one another. *Undilute updraft cores are not found.* The observed cores are of small diameter and have liquid water content and updraft velocities that are generally far smaller than would be consistent with adiabatic ascent.

---

[1] Although this database included hurricanes, convection in hurricane eyewalls is an obvious special case. The cited statistics of hurricane updrafts are similar to those of typical oceanic cumulonimbi, but there is ample evidence that the hurricane boundary layer can ascend undilute in eyewalls, and this paper does not consider them further.

The more recent papers also report that measured virtual temperature excesses, although a difficult measurement, also are far less than adiabatic. Wei et al. (1998) conclude that, on average, entrainment reduced buoyancy from undilute values by 2 K while the reduction due to the observed water loading is only 0.5 K. In a truly undilute core, reduction by water loading at 700 hPa would be 2 K.

Is it possible that out of thousands of opportunities to penetrate the strongest cores available, all of them could have missed all undilute cores over a 25-yr period? Proving a negative is difficult. The purpose of this paper is to offer an alternative explanation for how vigorous oceanic cumulonimbus manage to reach the tropical tropopause, in spite of the observations that apparently indicate that undilute cores are rare to nonexistent over tropical oceans. The proposed explanation is that the dilution of the cores by entrainment in the low-to-middle troposphere can be counteracted by the combined effects of freezing of condensate, and by shifting from the condensation to the sublimation latent heating rate.

This is hardly an original or revolutionary idea. Braham's (1952) classic paper summarizing results from the Thunderstorm Project observations includes a schematic illustration (his Fig. 9) of nearly compensating effects of entrainment and ice processes. Many thermodynamic texts represent an isothermal expansion along a parcel path during the freezing process, not entirely correctly labeled as the "hail stage." This term fell into disuse, mainly because it is not possible to state with generality how much condensate freezes at what temperature. However, the fact that such freezing can reinvigorate an updraft is universally known (e.g., Riehl 1979, p. 377; Johnson and Kriete 1982; Cotton and Anthes 1989, p. 470; Ooyama 1990; Johnson et al. 1999). The basis for the idea of "dynamic seeding" (e.g., Simpson and Wiggert 1969) was to artificially initiate or hasten the freezing process. It is not necessary to reopen the debates on how often Florida or Caribbean clouds are seedable (e.g., Sax 1969). The issue here is whether cumulonimbus in the equatorial trough glaciate naturally and at what temperature. The fact is, as postulated by Riehl and Malkus (1958), that these clouds are generally embedded in extensive regions of disturbed weather, and the writer's experience is that updrafts are most often embedded in extensive upper-tropospheric clouds, with precipitating ice. Allowing for some rare exceptions, we assume herein that cumulonimbus clouds over equatorial oceans glaciate naturally between −5° and −15°C. It will become obvious that the conclusions will not change by altering these temperatures within a reasonable range.

There is a common supposition that condensation loading can account for the departure of vertical velocity from undilute values. Xu and Emanuel (1989) argue that soundings are within observational error of neutrality over tropical oceans if the reference process is the reversible adiabatic rather than the pseudoadiabatic. These arguments are elaborated in Emanuel (1994, 463–467) in which he concludes that buoyancy in a radiative–convective atmosphere should be in the range of 1–2°C. However, it will be shown that the fully loaded undilute updraft applies not to the equatorial trough but rather to midlatitude severe storms, with surprising consequences. It will also be shown that realistic cumulonimbus cores in equatorial regions can indeed have buoyancies in the predicted range of 1–2°C but that they do not begin their ascent along reversible adiabats.

2. Approach to the issue

First, the properties of truly undilute cores as they are found in real midlatitude severe storms are considered. Their large diameter is consistent with minimal entrainment, and simple parcel theory assuming adiabatic water loading is in good agreement with observations. The same assumptions for clouds over tropical oceans are shown to be inconsistent with observations. Reconciliation is proposed by exceedingly crude assumptions of the fractional entrainment rate, and by consideration of ice processes. Only by very large departures from undilute ascent are these assumed conditions brought into agreement with common observations in the deep Tropics.

The examples presented herein will unapologetically neglect the virtual temperature correction, the specific heat of condensate, and differential fall speeds of condensate. When buoyancy is integrated to calculate convective available potential energy (CAPE) between two levels, accuracy better than 10%–20% is not attempted. When calculating vertical velocity from CAPE by the parcel method, the collective effects of nonhydrostatic pressure forces, form drag, and turbulence are considered simply by taking the square root of CAPE and not the square root of 2 × CAPE (in effect, assuming that the efficiency of conversion of CAPE into vertical kinetic energy is 50%).

It is fully recognized that the hypothesis presented herein requires validation by far more sophisticated models. That is why the arguments are made for the two strongly contrasting examples of a large and powerful hailstorm, and *ordinary* oceanic equatorial trough convection. The differences between the two are so great that they can survive the neglect of many complicating factors. Later, some qualitative discussion is made in an attempt to propose differences between typical, strong tropical ocean and continental convection, but in a more speculative mode.

3. Supercell examples

Some well-documented cases can easily illustrate that large undilute cores are common in severe storms. Davies-Jones (1974) and Bluestein et al. (1988) show a number of examples of balloon ascents in storm cores, with core updraft velocities as great as 50 m s^{-1} in the

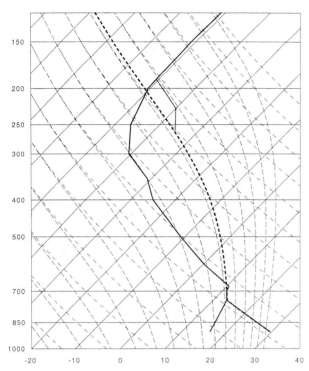

FIG. 5.1. Sounding adapted from Miller et al. (1988) representing the properties of the environment of a severe hailstorm (solid) and hypothetical pseudoadiabatic ascent from cloud base (dashed). The thin solid line near storm top shows the temperature increase from freezing of adiabatic water content between $-35°$ and $-40°C$. See Fig. 5.2 and text.

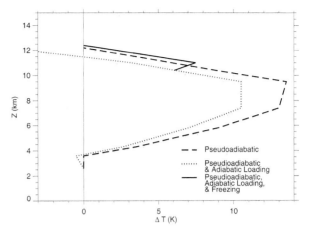

FIG. 5.2. Effective buoyancy of updraft from Fig. 5.1 according to various assumptions: standard pseudoadiabatic ascent (dashes), pseudoadiabatic ascent but adding drag from adiabatic water loading (dots), and the effect on the dotted curve of freezing the adiabatic water load (solid).

midtroposphere. Bluestein et al. calculate close agreement between CAPE and updraft velocity.

The hailstorm of 2 August 1981 passed through a dense observing network, including multiple Doppler radars and a penetration by an armored, instrumented aircraft (Musil et al. 1986). The environmental conditions were derived from multiple rawinsonde ascents and cloud-base updraft properties by aircraft. Figure 5.1 shows the environmental sounding and pseudoadiabatic updraft path (Miller et al. 1988). The radar data in both papers show an updraft greater than 10 km across, and the characteristic weak-echo region. Taken together, the case for an undilute updraft is overwhelming.

Miller et al. use the multiple Doppler data and a cloud model to simulate hail trajectories and hail growth, making the characteristic assumption of adiabatic cloud water content in the updraft until homogeneous freezing is assumed between $-35°$ and $-40°C$. At updraft speeds averaging $20–25$ m s^{-1}, air goes from cloud base to $-35°C$ in 6–8 min, which easily explains the lack of time for particle growth by any method except accretion. The very existence of a weak-echo region validates the assumption that large graupel and hail are not entering the core of the updraft. Nelson (1983), Miller and Fankhauser (1983), and many others have justified the use of the adiabatic undilute updraft core for estimating vertical velocity and liquid water content in the weak-echo

region. In the 2 August 1981 case, the penetrating aircraft measured updraft speeds as great as 50 m s^{-1} at 7 km, while the speed calculated from multiple Doppler data was $30–35$ m s^{-1}.

How much reduction in buoyancy is there from the adiabatic water loading? Figure 5.2 shows the buoyancy profile from the pseudoadiabatic and reversible adiabatic calculation. The condensate loading increases with height to 10 g kg^{-1} near 7.5 km, reducing the buoyancy from 13 to 10.5 K. The CAPE from the level of free convection to 7 km is reduced from 819 to 652 J kg^{-1}.

Assuming 100% efficiency of conversion from CAPE to vertical kinetic energy, the vertical velocity at 7 km would be reduced from 40 to 36 m s^{-1} by integrated water loading. Assuming 50% efficiency, the vertical velocity at 7 km would be reduced from 29 to 26 m s^{-1}. In supercell updrafts, nonhydrostatic effects could be a significant omission and their neglect could cause errors in estimated updraft speeds in either direction. Nonetheless, the range of calculated updraft speeds is in good agreement with the Doppler estimates but not quite as strong as the aircraft measurement. The overall conclusion is that the updraft core has nearly adiabatic water content (how could it not?!) and that parcel theory may somewhat underestimate peak observed velocities. When buoyancies are as great as they typically are, even adiabatic water loading, remarkably, is of secondary importance. Similar conclusions can be drawn from the Oklahoma examples of Nelson (1983) and Bluestein et al. (1988).

How much boost in buoyancy is there from freezing all condensate between $-35°$ and $-40°C$? More than 4 K. This may seem large, until one views the shape of the altered buoyancy profile as a function of height (Fig. 5.2). In this particular example, the equilibrium level is raised about 800 m, and of course the large updraft would be maintained or enhanced near storm top. But the CAPE is increased relatively little, because the increased buoyancy is integrated over a small vertical distance. There is an increase of some 7 K in the potential temperature (and correspondingly to the moist static energy and equivalent potential temperature) of the outflow from the storm, from the case without freezing. Thus, the outflow is potentially warmer than either the pseudoadiabatic or reversible adiabatic value (Fig. 5.1).

Bosart and Nielsen (1993) analyzed the interesting case of a rawinsonde balloon intercepting an outflow from a severe storm system in Louisiana. They argue that because the outflow was close to the same equivalent potential temperature as the inflow, they sampled outflow from an undilute updraft. If it was truly undilute, the argument herein would imply that the outflow should have been about 2 K higher in wet-bulb equivalent potential temperature, rather than the same value as the low-level inflow. A possible rationalization is that their inflow air may have been a degree or so lower in dewpoint, and/or that by the time the balloon sampled the outflow, some mixing had taken place that would have slightly reduced the temperature from totally undilute values.

4. Equatorial trough examples

The atmospheric structure hypothesized by Riehl and Malkus (1958) as the environment of the undilute towers is reproduced in Fig. 5.3. Its total pseudoadiabatic CAPE is 2045 J kg^{-1} and maximum buoyancy is 6 K at 400 hPa (7.56 km). The CAPE and the shape of the buoyancy profile are typical. Such soundings have been observed in the warm pool of the Pacific Ocean and shown to be representative of air entering deep convective systems (Lucas et al. 1994; LeMone et al. 1998). It does not have the slight stability near the 0°C isotherm, which is common in the warm pool region (Johnson et al. 1999); but as they have noted, this enhanced stability is more relevant to slowing ascent of cumulus congestus than of deep cumulonimbus. The assumed sounding is not quite as unstable as some Darwin, Australia soundings during the break period (Keenan and Carbone 1992) or those preceding intense thunderstorms in that region (Rutledge et al. 1992; Williams et al. 1992; Simpson et al. 1993). Later, some of the subtle but important distinctions between land and ocean convection in the equatorial trough zone will be discussed. First, the oceanic example is considered.

The pseudoadiabatic CAPE to 500 hPa is 447 J kg^{-1}. Cloud base mixing ratio is 19 g kg^{-1}, while the saturation value of the undilute parcel at 500 hPa and $T = -1°C$ is 7.4 g kg^{-1}. Therefore, the fully loaded CAPE with adiabatic liquid water content, essentially the reversible adiabatic CAPE, is 179 J kg^{-1}. Making the 50% efficiency assumption, the vertical velocity of an undilute parcel at 500 hPa is 21 m s^{-1} for pseudoadiabatic and 13 m s^{-1} for reversible adiabatic ascent, respectively. Assuming clean oceanic air, fully adiabatic water

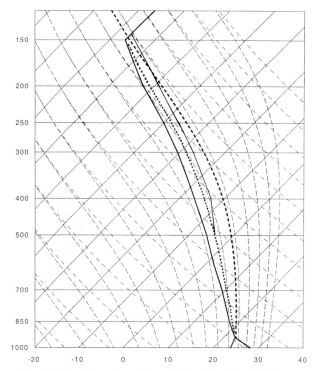

Fig. 5.3. Sounding adapted from Riehl and Malkus (1958) representing the properties of the environment in the equatorial trough zone (solid) and hypothetical pseudoadiabatic ascent from cloud base (dashed). The dotted curve represents the actual temperature of the ascent at one-third pseudoadiabatic buoyancy. The thin solid curve shows the temperature increase from that of the dotted curve from freezing one-third the adiabatic water load between 500 and 400 hPa, and subsequent ascent along the ice adiabat. See Fig. 5.4 and text.

loading would be very unlikely, as the average ascent rate of 6–7 m s^{-1} implies 13 min for the ascent, during which time some coalescence and fallout would occur. But any reduction in loading would increase vertical velocity still farther outside the observed range. Even the undilute reversible updraft of 13 m s^{-1} is outside the range of all known observations from the literature cited in the introduction. The reversible liquid water contents of 11 g kg^{-1} are larger than almost any known observations by a factor of 3. The thermal buoyancies of 3 K at 700 hPa and 5 K at 500 hPa for an undilute core are larger than most known observations by a factor of 2 or more (Lucas et al. 1994; Wei et al. 1998).

Therefore, it is concluded that undilute ascent in equatorial oceanic cumulonimbus is extremely rare, and if found, would constitute a special case requiring some special explanation. The conclusions of Wei et al. (1998) and many others, based upon careful analysis of penetration data, are that large dilution by entrainment is the norm, and that large conversion of cloud water to rainwater and consequent fallout from weak updrafts greatly reduces the loading. It will now be shown that despite large dilution, subcloud air can nevertheless easily ascend within cumulonimbi to the tropopause, with simple assumptions about freezing of condensate. It

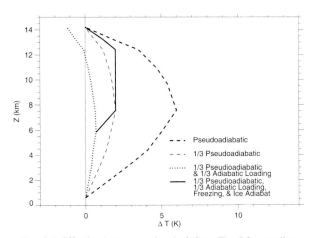

FIG. 5.4. Effective buoyancy of updraft from Fig. 5.3 according to various assumptions: standard pseudoadiabatic ascent (heavy dashes), one-third pseudoadiabatic buoyancy (light dashes), one-third pseudoadiabatic buoyancy but adding drag from one-third adiabatic water loading (dots), and the effect on the dotted curve of freezing one-third adiabatic water load and subsequently following the ice adiabat (solid).

must be emphasized that the explanation herein, while plausible, is offered as a simple hypothesis, which should not be accepted as fact without additional research.

Riehl (1979, p. 377) makes a compelling argument for the reasonableness of freezing of large amounts of condensate, and for use of the ice adiabat above 400 hPa. However, neither Riehl and Simpson (1979) nor Riehl (1979, p. 157–197) seem to question the assumption that undilute ascent is a necessity for cumulonimbus to reach the tropopause and perform their acknowledged function of energy transport.

First, consider a large (factor of 3) dilution of buoyancy from its pseudoadiabatic value by entrainment, illustrated in Fig. 5.4. By itself, this reduced buoyancy is in closer agreement with observations. Deciding on an appropriate water loading value is more difficult. The specific conceptual model of the updraft will matter. The updraft speed is now low enough to be near the fall speed of raindrops, so the drop size spectrum will matter. The simplicity of the arguments herein argues for extremely simple assumptions, so accordingly, the water loading is assumed to be a factor of 3 less than adiabatic throughout the depth of the updraft. This is on the high side of most measurements but is chosen intentionally to give a reasonably large buoyancy reduction.

With thermal buoyancy and water loading both reduced by factors of 3, effective CAPE between cloud base and 500 hPa is about 60 J kg^{-1}, reducing the updraft to under 8 m s^{-1}, again assuming 50% efficiency of conversion. This is still on the high side of the range of observations, about halfway between the highest 10th percentile for tropical oceans and the highest 10th percentile for the Thunderstorm Project observations over land (Lucas et al. 1994). Repeating the calculation for

the 700-hPa level, the updraft is about 4 m s^{-1}, about the same as the highest 10th percentile over oceans.

Next, consider this hypothetical 8 m s^{-1} updraft at 500 hPa, containing 3.9 g kg^{-1} condensate. The environment temperature is $-6°C$, and the assumed updraft temperature has been reduced to $-4.3°C$ from its pseudoadiabatic value of $-1°C$. Common research aircraft experience flying in disturbances over tropical oceans is that neither cloud water nor rainwater in updrafts of modest strength will remain unfrozen for more than a few minutes, even at temperatures as warm as $-5°C$, and certainly not by $-10°$ or $-15°C$ (e.g., Black and Hallett 1986; Willis and Hallett 1991). It is probably a conservative assumption that updrafts such as this one will freeze condensate between $-4°$ and $-15°C$ which, in this case, is between 500 and 400 hPa. The latent heat of sublimation rather than condensation is assumed from $-15°C$ to cloud top. The solid line in Fig. 5.4 gives the resulting parcel path.

The details of the events in the lower part of the troposphere are complex and surely require cloud resolving models for understanding. However, Fig. 5.4 is clear in demonstrating the following: if the updraft can survive to about $-5°C$ to loft several grams per Kilograms of condensate at more than a few meters per second, the buoyancy gain from freezing and subsequent shift to the ice adiabat is considerable. A large increase in updraft velocity results from integrating this enhanced buoyancy over a large vertical distance. In the example shown, water-loaded CAPE between 500 and 150 hPa is about 610 J kg^{-1}, which would increase the vertical velocity (again at 50% efficiency) by almost 25 m s^{-1} between 500 and 150 hPa.

Thus, one arrives at a schematic model of the vertical distribution of buoyancy in strong tropical oceanic cumulonimbus (Fig. 5.4), consistent with weak updrafts in the lower half of the troposphere and quite strong updrafts above midtroposphere. Such a model explains many of the observations. For example, vertically pointing wind profiler observations (e.g., Balsley et al. 1988) and observations from the airborne ER-2 Doppler radar looking downward from 20 km (Heymsfield et al. 1996; G. M. Heymsfield 1998 and 1999, personal communication) occasionally show updrafts in the 10–20-m s^{-1} range in the upper troposphere, but hardly ever in the lowtroposphere to midtroposphere. Ebert and Holland (1992) illustrate a case of extremely cold temperatures above a tropical cyclone that imply upper-tropospheric updrafts in the 20–m s^{-1} range with soundings similar to Fig. 5.3. While direct observations are scarce, the author's experience in the field includes encounters with a few updrafts of about 20 m s^{-1} in the upper troposphere over tropical oceans, for example, during the birth of Oliver (Simpson et al. 1997), but never in lowtroposphere to midtroposphere.

The hypothetical parcel paths are also shown on the thermodynamic diagram (Fig. 5.3). The essential point to note is that the equivalent potential temperature (and therefore also the moist static energy) of the outflow is nearly identical to that of the totally unrealistic pseudoadiabatic ascent. The moist static energy lost by entrainment and mixing below 500 hPa has been gained through the ice processes. This near equality is probably coincidental. Unfortunately, it has often been used (misused) to imply that undilute ascent is a common occurrence.

5. Differences between oceans and continents near the equator

The example above applies mainly to the cumulonimbus clouds of the oceanic equatorial trough zone. The observational database is much more solid over the ocean than over continents. More importantly, evidence is increasing that storms over tropical continents are much stronger than those over tropical oceans. The increased hazard of penetration is an important reason for the near absence of in-cloud data in strong continental storms.

Orville and Henderson (1986), Goodman and Christian (1993), and Goodman et al. (2000) used satellite observations of lightning to show that "lightning loves land." Rutledge et al. (1992) and Williams et al. (1992) showed that the continental regimes near Darwin, Australia, produced more intense storms than did the oceanic regimes. Zipser (1994), Zipser and Lutz (1994), and Petersen and Rutledge (1998) showed that the rain-to-lightning ratio is far greater over oceans, and that the reasons are related to the weaker updrafts over oceans, in agreement with findings in Lucas et al. (1994). Mohr and Zipser (1996) and Mohr et al. (1999) used passive microwave satellite data to show that the ice scattering signatures over land were consistently stronger than those over oceans. (The justification for using low brightness temperatures in the 37- and 85-GHz passive microwave channels is outlined in the next section.)

Joanne Simpson's leadership role in making the Tropical Rainfall Measuring Mission (TRMM) satellite a reality is well known (Simpson et al. 1988, 1996; Kummerow et al. 1998). It is fitting that we are now able to use the data from TRMM to address some of the quantitative issues about the characteristics of tropical convection in different regions directly. One step in that direction is to use not just passive microwave data, an indirect indicator of convective vigor, but the TRMM radar, which gives a direct measure of the height by large hydrometeors.

Nesbitt et al. (2000) made a systematic comparison of TRMM radar and passive microwave estimators of convective strength for two tropical Pacific Ocean regions and for two continental regions (tropical Africa and the Amazon basin) for a 3-month period. Not surprisingly, the results, especially their Figs. 5, 8, and 9, confirm earlier indications that by each measure, convection over the two oceanic regions is weaker than over either of the two land regions, but with storms over

equatorial Africa much stronger than those over the Amazon (see also McCollum et al. 2000). They also used the lightning imaging sensor on TRMM to demonstrate that the flash rate over the two land regions was greater than that over the two ocean regions by two orders of magnitude.

The question arises whether these admittedly stronger storms over land might have true undilute cores. That is always a possibility for the very strongest storms (more on this soon). However, it is argued here that a modest change in the assumed updraft strength in the midtroposphere can account for a major change in the properties of that storm. Zipser and Lutz (1994) point out that most raindrops can fall through a 5–m s⁻¹ updraft but not through a 10–m s⁻¹ updraft. The faster the updraft, other things being equal, the greater the supercooled liquid water content should be. Thus, a 10–m s⁻¹ updraft would have more condensate freezing, greater graupel riming rates, greater heights attained by strong radar echoes, and greater ice scattering signatures at 37 and 85 GHz. Many authors have pointed out that the noninductive ice–ice collision process requires substantial supercooled liquid water content, graupel, and small ice particles. Modeling studies by Baker et al. (1995, 1999) find great sensitivity of flash rate to updraft velocity, such that the difference between 5 and 10 m s⁻¹ may translate into a huge difference in flash rate. Following this reasoning, we explore how small changes in the assumed parcel path in Figs. 5.3 and 5.4 might have major effects.

Consider first a slightly greater entrainment rate below 500 hPa than assumed for Figs. 5.3 and 5.4, that might be more characteristic of oceanic clouds. As already shown, the parcel buoyancy in Fig. 5.4 is consistent with an 8–m s⁻¹ updraft at 500 hPa. A slightly greater dilution would reduce this toward 5 m s⁻¹. The time for the updraft to reach 500 hPa would increase, the fraction of condensate falling out would increase, and the updraft boost from condensate freezing would decrease. A slight decrease in vigor would, in fact, be more consistent with the typical properties of tropical oceanic cumulonimbus clouds without materially changing their ability to reach the upper troposphere and contribute their still ample moist static energy as required (Riehl and Malkus 1958).

Consider now a slightly smaller entrainment rate in the broader updraft cores that appear to be more characteristic of continental clouds. This slightly smaller dilution would easily increase updraft strength to 10 or 12 m s⁻¹. The time for the updraft to reach 500 hPa would decrease, the fraction of condensate falling out would decrease, and the supercooled cloud liquid water content would increase. The updraft boost from freezing of condensate would increase and the updraft speed in the upper troposphere (already nearly 25 m s⁻¹) could reach very large values. Thus, it is not necessary to assume undilute ascent from cloud base to explain extremely vigorous cumulonimbus clouds, with high lightning flash rates, large optical depth of graupel, and correspondingly low brightness temperatures at 37 and 85 GHz, as observed over tropical continents.

If it is true that highly diluted updrafts with small differences in entrainment rates result in updrafts of 5 m s^{-1} versus 10–12 m s^{-1} at 500 hPa, this corresponds to the difference observed between oceanic convection and "ordinary" continental convection summarized by Lucas et al. (1994). A sensitivity study using a one-dimensional model with entrainment and microphysics that supports this speculation was carried out by Ferrier and Houze (1989). They show that small changes in either entrainment or in low-level forcing result in large changes in updraft velocity. These and many other studies summarized in Blyth (1993) and Houze (1993) tend to show that the specific mechanism of entrainment is less important than the fractional dilution with outside air, which is a strong inverse function of updraft radius.

Why should updrafts over land have smaller entrainment? Lucas et al. (1994) have already speculated that these differences could be caused by larger diameter updrafts over land, related to the deeper mixed layers. See also Lucas et al. (1996), offering additional possibilities that the slightly greater convective inhibition often observed over land may result in larger updrafts from cloud base as would be required to overcome the inhibition (i.e., larger forcing). The oft-repeated contention that CAPE is greater over land than over water does not survive close scrutiny (Lucas et al. 1994), but the "shape of the CAPE" is very important, often fatter over land than over tropical oceans. More importantly, the main purpose of this paper is to demonstrate that CAPE *per se* is overrated as a predictor of the behavior of convection when entrainment, water loading, forcing, and the vertical distribution of buoyancy are taken into account. This is a point well made by Blanchard (1998) in the context of forecasting severe weather.

Following the reasoning in this section, it is concluded that strong continental convection often begins as moderately entraining updrafts in the low troposphere to midtroposphere, but that updraft speeds increase greatly in the upper troposphere, not infrequently exceeding 25–35 m s^{-1}. This behavior has been observed and modeled in the famous "Hectors," the cumulonimbus found on many days during November and December over the Tiwi Islands north of Darwin, Australia (Simpson et al. 1993).

6. Equatorial trough versus subtropical convection according to TRMM

There is already ample evidence that the convective systems of the equatorial trough zone, although heavily raining and unquestionably vital because of their collective role in the general circulation, are not as strong as many of the severe storms of higher latitudes. Mohr and Zipser (1996) defined mesoscale convective systems as "intense" according to numbers of pixels with 85-GHz polarization corrected temperature (PCT; see Spencer et al. 1989) less than threshold values, signifying large optical depth of ice scattering by dense ice particles (i.e., graupel or hail). The equatorial trough zone had few of these with the sole exception of continental Africa. Notably, there were few intense systems in tropical South America or the Indonesian (maritime continent) region. With some differences, the distribution of intense systems in Mohr and Zipser resembled those of large mesoscale convective complexes mapped by Laing and Fritsch (1997).

The physical significance of the ice scattering signature is that it is directly related to the ice water path with few approximations. As Vivekanandan et al. (1991) stated, ". . . the ice layer remains relatively unobscured from a spaceborne radiometer at 37 GHz and 85 GHz, perhaps presenting an ice-phase characterization as a more inherently retrievable property than the rain phase." They show that low brightness temperatures (T_b) at these frequencies are quantitatively related to ice density and size. While relationships are not unique, they provide strong justification for considering that very low T_b means large depth of high-density ice particles. The implication is that strong updrafts are required for large depths of graupel or hail, and therefore that low T_b can be used as a proxy for convective intensity. (This would seem to be a more direct proxy than cold IR cloud-top temperatures.)

This paper argues that true undilute updrafts from below cloud base are likely in certain severe storms but rarely in equatorial latitudes. Undilute updrafts starting with large water vapor content (such as the tropical ocean sounding) would certainly contain very large graupel or hail. Rather than use the 85-GHz channel, the 37-GHz (8-mm wavelength) channel is even more sensitive to large graupel and hail. We examined all orbits for a full year of TRMM data (March 1998–February 1999) and searched for the minimum PCT at 37 GHz in each and every precipitation feature that was observed. Then we simply plotted the locations of all systems containing at least one pixel colder than an arbitrary T_b, selected such that this T_b is only achieved on average of once per day. That 37-GHz T_b is 187 K, and the resulting distribution is shown in Fig. 5.5. With the notable exception of equatorial Africa (again!), the equatorial trough zone is hardly represented at all, especially the oceans.

Figure 5.5 is interpretable as a map of the most intense convective storms viewed by TRMM (36°N–36°S), and it includes a few surprising features. The total dominance of Africa over the maritime continent and tropical South America is surprising, although the lightning data and Nesbitt et al.'s (2000) data presage these results. The extension of subtropical maxima offshore of the U.S. east coast, the South African east coast, and offshore of southeast Australia is also surprising. The storms in the Mediterranean Sea are surprising; they appear only in autumn and may be related to cold air

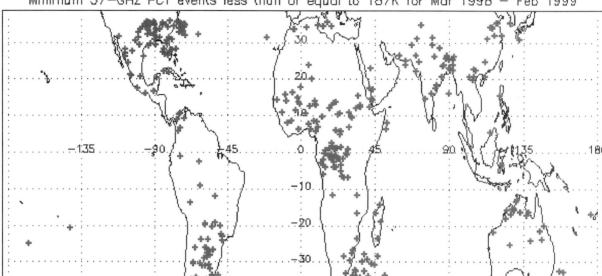

FIG. 5.5. The precipitation features observed by TRMM during one year with the coldest 37-GHz PCT (strongest ice scattering signature; see text). At 8-mm wavelength, colder PCT is proposed as a measure for largest optical depth of large graupel and/or hail, and as a crude proxy for updraft strength.

over warm water. However, the severe storm areas of the southern United States, subtropical South America, and India–Bangladesh appear in the locations and seasons expected.

If nearly undilute towers existed in the Tropics, it is concluded based on the above arguments that they would have updrafts from 10 to 20 m s^{-1} at $-5°$C with 7–12 g kg^{-1} of condensate. Before this condensate load could significantly slow the updraft, freezing would generate an additional 2–3-K buoyancy and invigorate the updrafts to 30–40 m s^{-1}. Assuming only that such clouds would have regions, perhaps between updrafts and downdrafts, where the combination of large supercooled liquid water content, large graupel content, and large numbers of small ice particles coexist, vigorous charge separation would be likely. The large depth of graupel and small hail would give very low brightness temperatures at 85 and 37 GHz.

The TRMM observations, of which Fig. 5.5 is but one small example, strongly suggest that this rarely, if ever, happens over tropical oceans, but elsewhere, especially in subtropical regions well known for severe weather. Such powerful storms appear mainly over Africa in the deep Tropics. McCollum et al. (2000) note that remote sensing techniques are overestimating rain in Africa, and speculate on reasons for the low precipitation efficiency of African storms. That paper, Mohr et al. (1999), and others are pointing the way for using TRMM and other satellite data to survey and compare the properties of storms over the globe. Among the most important of these properties is the degree of dilution of updrafts, their vertical velocity, and the intensity of

storms. This paper hypothesizes that the most intense storms (and undilute updrafts) are rarely found over tropical oceans, proposes some reasons for this, and suggests that careful exploitation of data from TRMM can broaden our understanding of the nature and distribution of intense convection around the world.

7. Conclusions

Observations of the properties of extreme midlatitude severe storms are consistent with the presence of undilute updrafts.

Observations of the properties of typical cumulonimbus clouds over tropical oceans are inconsistent with the presence of undilute updrafts.

The common observations over tropical oceans can hypothetically be explained by assuming large dilution of updrafts by entrainment below about 500 hPa, followed by freezing of condensate. This freezing and subsequent ascent along an ice adiabat reinvigorates the updrafts and permits them to reach the tropical tropopause with the necessarily high values of moist static energy, as the hot tower hypothesis requires. It is recognized that the explanation offered herein is speculative and probably controversial.

More speculatively, the large difference observed between typical ocean and land clouds can be explained by assuming slightly smaller entrainment rates for clouds over land. These small entrainment differences have a very large effect on updrafts in middle and upper troposphere and can presumably account for the large differences in convective vigor and ice scattering that are observed.

Also speculatively, small differences in entrainment rate can lead to differences in updraft strength in midtroposphere (5 m s^{-1} vs 10–12 m s^{-1}) that are typically observed between land and ocean, and in turn can lead to the very large differences in lightning flash rates that are observed.

CAPE is a measure of maximum *potential* vertical velocity. Without taking account of the vertical distribution of buoyancy, entrainment, water loading, and the degree of forcing, it is *not* by itself a good predictor of the intensity or vertical velocity in convective clouds.

Using TRMM to map a proxy for the most intense storms on earth between 36°S and 36°N, they are found mostly outside the deep Tropics, with the notable exception of tropical Africa.

Acknowledgments. Joanne Simpson has been an inspiration to this writer and countless numbers of his colleagues since his graduate school days. Together with Herbert Riehl, her writings on the importance of deep convection and the importance of the balance between theory, modeling, and observations have had a profound influence. The need for observations in the right time and place have stimulated the writer's addiction to field work. Her development of cloud models, more sophisticated than those needed for this paper, through colleagues such as Wei-Kuo Tao and Brad Ferrier, have put these valuable tools in our hands. Without her leadership and perseverance, of course, we would not have the unprecedented opportunities presented by TRMM, nor the field campaigns that have led to so much valuable data, which we shall be using for a decade or more.

Field programs do not succeed without immense effort and cooperation of many. At the risk of omitting very many, I thank Michael Garstang, Peggy LeMone, Robert Houze, Sandra Yuter, Steve Rutledge, Gerry Heymsfield, and Robbie Hood for their contributions and insights, without which many of the achievements of and resulting from our field work would not have been possible. I also cite a generation of pilots of research aircraft that were willing to fly wherever and whenever necessary to obtain data that could have been gathered nowhere else. For this paper, I thank Karen Mohr, Daniel Cecil, and Steven Nesbitt for devising the methodology for viewing data from SSM/I and TRMM in the form of mesoscale convective systems and precipitation features, and Steven Nesbitt and David Yorty for their help producing the figures.

This research was supported by several NASA TRMM grants, most recently #NAG5-8563.

REFERENCES

Baker, M. B., H. J. Christian, and J. Latham, 1995: A computational study of the relationships linking lightning frequency and other thundercloud parameters. *Quart. J. Roy. Meteor. Soc.,* **121,** 1525–1548.

——, A. M. Blyth, H. J. Christian, J. Latham, K. L. Miller, and A. M. Gadian, 1999: Relationships between lightning activity and various thundercloud parameters: Satellite and modeling studies. *Atmos. Res.,* **51** (3–4), 221–236.

Balsley, B. B., L. W. Ecklund, D. A. Carter, A. C. Riddle, and K. S. Gage, 1988: Average vertical motions in the tropical atmosphere observed by a radar wind profiler on Ponape (7°N latitude, 157°E longitude). *J. Atmos. Sci.,* **45,** 396–405.

Black, R. A., and J. Hallett, 1986: Observations of the distribution of ice in hurricanes. *J. Atmos. Sci.,* **43,** 802–822.

Blanchard, D. O., 1998: Assessing the vertical distribution of convective available potential energy. *Wea. Forecasting,* **13,** 870–877.

Bluestein, H. B., E. W. McCaul Jr., G. P. Byrd, and G. R. Woodall, 1988: Mobile sounding observations of a tornadic storm near the dryline: The Canadian TX storm of 7 May 1986. *Mon. Wea. Rev.,* **116,** 1790–1804.

Blyth, A. M., 1993: Entrainment in cumulus clouds. *J. Appl. Meteor.,* **32,** 626–641.

Bosart, L. F., and J. W. Nielsen, 1993: Radiosonde penetration of an undilute cumulonimbus anvil. *Mon. Wea. Rev.,* **121,** 1688–1702.

Braham, R. R., 1952: The water and energy budgets of the thunderstorm and their relation to thunderstorm development. *J. Meteor.,* **9,** 227–242.

Cotton, W. R., and R. A. Anthes, 1989: *Storm and Cloud Dynamics.* Academic Press, 883 pp.

Davies-Jones, R. P., 1974: Discussion of measurements inside high-speed thunderstorm updrafts. *J. Appl. Meteor.,* **13,** 710–717.

Ebert, E. E., and G. J. Holland, 1992: Observations of record cold cloud top temperatures in tropical cyclone Hilda (1990). *Mon. Wea. Rev.,* **120,** 2240–2251.

Emanuel, K. A., 1994: *Atmospheric Convection.* Oxford University Press, 580 pp.

Ferrier, B. S., and R. A. Houze Jr., 1989: One-dimensional time-dependent modeling of GATE cumulonimbus convection. *J. Atmos. Sci.,* **46,** 330–352.

Goodman, S. J., and H. Christian, 1993: Global observations of lightning. *Atlas of Satellite Observations Related to Global Change.* R. J. Gurney, J. L. Foster, and C. L. Parkinson, Eds., Cambridge University Press, 191–219.

——, D. E. Buechler, K. Knupp, D. Driscoll, and E. W. McCaul, 2000: The 1997–98 El Niño event and related wintertime lightning variations in the southeastern United States. *Geophys. Res. Lett.,* **27** (4), 541–544.

Heymsfield, G. M., and Coauthors, 1996: The EDOP radar system on the high-altitude NASA ER-2 aircraft. *J. Atmos. Oceanic Technol.,* **13,** 795–809.

Houze, R. A., Jr., 1993: *Cloud Dynamics.* Academic Press, 573 pp.

Igau, R. C., M. A. LeMone, and D. Wei, 1999: Updraft and downdraft cores in TOGA-COARE: Why so many buoyant downdraft cores? *J. Atmos. Sci.,* **56,** 2233–2245.

Johnson, R. H., and D. C. Kriete, 1982: Thermodynamic and circulation characteristics of winter monsoon tropical mesoscale convection. *Mon. Wea. Rev.,* **110,** 1898–1911.

——, T. M. Rickenbach, S. A. Rutledge, P. E. Ciesielski, and W. H. Schubert, 1999: Trimodal characteristics of tropical convection. *J. Climate,* **12,** 2397–2418.

Jorgensen, D. P., and M. A. LeMone, 1989: Vertical velocity characteristics of oceanic convection. *J. Atmos. Sci.,* **46,** 621–640.

——, E. J. Zipser, and M. A. LeMone, 1985: Vertical motion in intense hurricanes. *J. Atmos. Sci.,* **42,** 839–856.

Keenan, T. D., and R. E. Carbone, 1992: A preliminary morphology of precipitation systems in tropical northern Australia. *Quart. J. Roy. Meteor. Soc.,* **118** (504), 283–326.

Kummerow, C., W. Barnes, T. Kozu, J. Shiue, and J. Simpson, 1998: The tropical rainfall measuring mission (TRMM) sensor package. *J. Atmos. Oceanic Technol.,* **15,** 809–817.

Laing, A. G., and J. M. Fritsch, 1997: The global population of mesoscale convective complexes. *Quart. J. Roy. Meteor. Soc.,* **123,** 389–405.

LeMone, M. A., and E. J. Zipser, 1980: Cumulonimbus vertical velocity events in GATE. Part I: Diameter, intensity, and mass flux. *J. Atmos. Sci.,* **37,** 2444–2457.

——, ——, and S. B. Trier, 1998: The role of environmental shear and CAPE in determining the structure and evolution of mesoscale convective systems during TOGA COARE. *J. Atmos. Sci.,* **55,** 3493–3518.

Lucas, C., M. A. LeMone, and E. J. Zipser, 1994: Vertical velocity in oceanic convection off tropical Australia. *J. Atmos. Sci.,* **51** (21), 3183–3193.

——, E. J. Zipser, and M. A. LeMone, 1996: Reply. *J. Atmos, Sci.,* **53,** 1212–1216.

McCollum, J. R., A. Gruber, and M. B. Ba, 2000: Discrepancy between gauges and satellite estimates of rainfall in equatorial Africa. *J. Appl. Meteor.,* **39,** 666–679.

Miller, L. J., and J. C. Fankhauser, 1983: Radar echo structure, air motion and hail formation in a large stationary multi-cellular thunderstorm. *J. Atmos. Sci.,* **40,** 2339–2418.

——, J. D. Tuttle, and C. A. Knight, 1988: Airflow and hail growth in a severe northern high plains supercell. *J. Atmos. Sci.,* **45,** 736–762.

Mohr, K. I., and E. J. Zipser, 1996: Mesoscale convective systems defined by their 85 GHz ice scattering signature: Size and intensity comparison over tropical oceans and continents. *Mon. Wea. Rev.,* **124,** 2417–2437.

——, J. S. Famiglietti, and E. J. Zipser, 1999: The contribution to tropical rainfall with respect to convective system type, size, and intensity estimated from the ice scattering signature. *J. Appl. Meteor.,* **38,** 596–606.

Musil, D. J., A. J. Heymsfield, and P. L. Smith, 1986: Microphysical characteristics of a well-developed weak echo region in a High Plains supercell thunderstorm. *J. Climate Appl. Meteor.,* **25,** 1037–1051.

Nelson, S. P., 1983: The influence of storm flow structure on hail growth. *J. Atmos. Sci.,* **40,** 1965–1983.

Nesbitt, S. W., E. J. Zipser, and D. J. Cecil, 2000: A census of precipitation features in the tropics using TRMM: Radar, ice scattering, and lightning observations. *J. Climate,* **13,** 4087–4106.

Ooyama, K. V., 1990: A thermodynamic foundation for modeling the moist atmosphere. *J. Atmos. Sci.,* **47,** 2580–2593.

Orville, R. E., and R. W. Henderson, 1986: Global distribution of midnight lightning: September 1977 to August 1978. *Mon. Wea. Rev.,* **114,** 2640–2653.

Petersen, W. A., and S. A. Rutledge, 1998: On the relationship between cloud-to-ground lightning and convective rainfall. *J. Geophys. Res.,* **103,** 14 025–14 040.

Riehl, H. 1979: *Climate and Weather in the Tropics.* Academic Press, 611 pp.

——, and J. S. Malkus, 1958: On the heat balance in the equatorial trough zone. *Geophysica,* **6,** 503–538.

——, and J. Simpson, 1979: The heat balance of the equatorial trough zone, revisited. *Beitr. Phys. Atmos.,* **52,** 287–305.

Rutledge, S. A., E. R. Williams, and T. D. Keenan, 1992: The Down Under Doppler and Electricity Experiment (DUNDEE): Overview and preliminary results. *Bull. Amer. Meteor. Soc.,* **73,** 3–16.

Sax, R. I., 1969: The importance of natural glaciation on the modification of tropical maritime cumuli by silver iodide seeding. *J. Appl. Meteor.,* **8,** 92–104.

Simpson, J., and V. Wiggert, 1969: Models of precipitating cumulus towers. *Mon. Wea. Rev.,* **97,** 471–489.

——, R. F. Adler, and G. R. North, 1988: Proposed tropical rainfall measuring mission (TRMM) satellite. *Bull. Amer. Meteor. Soc.,* **69,** 278–295.

——, T. D. Keenan, B. Ferrier, R. H. Simpson, and G. J. Holland, 1993: Cumulus mergers in the Maritime Continent region. *Meteor. Atmos. Phys.,* **51,** 73–99.

——, C. Kummerow, W.-K. Tao, and R. F. Adler, 1996: On the Tropical Rainfall Measuring Mission (TRMM). *Meteor. Atmos. Phys.,* **60,** 19–36.

——, E. Ritchie, G. J. Holland, J. Halverson, and S. Stewart, 1997: Mesoscale interactions in tropical cyclone genesis. *Mon. Wea. Rev.,* **125,** 2643–2661.

Spencer, R. W., H. M. Goodman, and R. E. Hood, 1989: Precipitation retrieval over land and ocean with the SSM/I: Identification and characteristics of the scattering signal. *J. Atmos. Oceanic Technol.,* **6,** 254–273.

Vivekanandan, T., J. Turk, and V. N. Bringi, 1991: Ice water path estimation and characterization using passive microwave radiometry. *J. Appl. Meteor.,* **30,** 1407–1421.

Wei, D., A. M. Blyth, and D. J. Raymond, 1998: Buoyancy of convective clouds in TOGA COARE. *J. Atmos. Sci.,* **55,** 3381–3391.

Williams, E. R., S. A. Rutledge, S. G. Geotis, N. Renno, E. Rasmussen, and T. Rickenbach, 1992: A radar and electrical study of tropical "hot towers." *J. Atmos. Sci.,* **49,** 1386–1395.

Willis, P. T., and J. Hallett, 1991: Microphysical measurements from an aircraft ascending with a growing isolated maritime cumulus tower. *J. Atmos. Sci.,* **48,** 283–300.

Xu, K.-M., and K. A. Emanuel, 1989: Is the tropical atmosphere conditionally unstable? *Mon. Wea. Rev.,* **117,** 1471–1479.

Zipser, E. J., 1994: Deep cumulonimbus cloud systems in the Tropics with and without lightning. *Mon. Wea. Rev.,* **122,** 1837–1851.

——, and M. A. LeMone, 1980: Cumulonimbus vertical velocity events in GATE. Part II: Synthesis and model core structure. *J. Atmos. Sci.,* **37,** 2458–2469.

——, and K. Lutz, 1994: The vertical profile of radar reflectivity of convective cells: A strong indicator of storm intensity and lightning probability? *Mon. Wea. Rev.,* **122,** 1751–1759.

Chapter 6

Spaceborne Inferences of Cloud Microstructure and Precipitation Processes: Synthesis, Insights, and Implications

DANIEL ROSENFELD

Hebrew University of Jerusalem, Jerusalem, Israel

WILLIAM L. WOODLEY

Woodley Weather Consultants, Littleton, Colorado

ABSTRACT

Spaceborne inferences of cloud microstructure and precipitation-forming processes with height have been used to investigate the effect of ingested aerosols on clouds and to integrate the findings with past cloud physics research. The inferences were made with a method that analyzes data from National Oceanic and Atmospheric Administration Advanced Very High Resolution Radiometer (NOAA AVHRR) and Tropical Rainfall Measuring Mission Visible and Infrared Scanner (TRMM VIRS) sensors to determine the effective radius of cloud particles with height. In addition, the TRMM Precipitation Radar (PR) made it possible to measure the rainfall simultaneously with the microphysical retrievals, which were validated by aircraft cloud physics measurements under a wide range of conditions. For example, the satellite inferences suggest that vigorous convective clouds over many portions of the globe remain supercooled to near $-38°C$, the point of homogeneous nucleation. These inferences were then validated in Texas and Argentina by in situ measurements using a cloud physics jet aircraft.

This unique satellite vantage point has documented enormous variability of cloud conditions in space and time and the strong susceptibility of cloud microstructure and precipitation to the ingested aerosols. This is in agreement with past cloud physics research. In particular, it has been documented that smoke and air pollution can suppress both water and ice precipitation-forming processes over large areas. Measurements in Thailand of convective clouds suggest that the suppression of coalescence can decrease areal rainfall by as much as a factor of 2. It would appear, therefore, that pollution has the potential to alter the global climate by suppressing rainfall and decreasing the net latent heating to the atmosphere and/or forcing its redistribution. In addition, it appears that intense lightning activity, as documented by the TRMM Lightning Imaging Sensor (LIS), is usually associated with microphysically highly "continental" clouds having large concentrations of ingested aerosols, great cloud-base concentrations of tiny droplets, and high cloud water contents. Conversely, strongly "maritime" clouds, having intense coalescence, early fallout of the hydrometeors, and glaciation at warm temperatures, show little lightning activity. By extension these results suggest that pollution can enhance lightning activity.

The satellite inferences suggest that the effect of pollution on clouds is greater and on a much larger scale than any that have been documented for deliberate cloud seeding. They also provide insights for cloud seeding programs. Having documented the great variability in space and time of cloud structure, it is likely that the results of many cloud seeding efforts have been mixed and inconclusive, because both suitable and unsuitable clouds have been seeded and grouped together for evaluation. This can be addressed in the future by partitioning the cases based on the microphysical structure of the cloud field at seeding and then looking for seeding effects within each partition.

This study is built on the scientific foundation laid by many past investigators and its results can be viewed as a synthesis of the new satellite methodology with their findings. Especially noteworthy in this regard is Dr. Joanne Simpson, who has spent much of her career studying and modeling cumulus clouds and specifying their crucial role in driving the hurricane and the global atmospheric circulation. She also was a pioneer in early cloud seeding research in which she emphasized cloud dynamics rather than just microphysics in her seeding hypotheses and in her development and use of numerical models. It is appropriate, therefore, that this paper is offered to acknowledge Dr. Joanne Simpson and her many colleagues who paved the way for this research effort.

---

1. Introduction

This study builds on past research and incorporates new technology, findings, and insights to develop a cloud physics synthesis concerning the effect of ingested aerosols on clouds. The results are presented in the con-text of the 50-yr career of Dr. Joanne Simpson, during which our meteorological discipline has come full circle. It started with imperfect cloud models and inadequate cloud physics observations, much of it in the context of cloud seeding research efforts. It then moved to space, where space technology provided results and insights that had never been possible before. Now we are back where we started but with greatly improved technology, observations, and insights that benefit cloud seeding programs and make it possible to address the effect of pollution and natural aerosols on rainfall and on the global climate system.

The current interest has been on the role that high concentrations of tiny cloud condensation nuclei (CCN) play in generating highly reflective clouds by altering the cloud drop size distribution. Such clouds return a greater percentage of incoming radiation to space than they would have had they not ingested the high concentrations of small CCN. This CCN "pollution" is postulated as a possible cooling mechanism that might act to offset the global warming due to increased concentrations of carbon dioxide (CO_2). This is now called the Twomey effect (Twomey et al. 1987).

As shown herein, anthropogenic pollution also enters into the global equation. Smoke from biomass burning and from industry and urban pollution acts to suppress coalescence and ice processes in cloud. This has the potential to decrease and/or redistribute the precipitation and the global heating due to the precipitation. Thus the effect of anthropogenic aerosols does not stop at radiative processes; they can also have a profound detrimental impact on precipitation. These insights would not have been possible without the new spaceborne measurements that will be discussed here.

The opportunity to present new research results in a paper honoring the long and illustrious career of Dr. Joanne Simpson is a once-in-a-lifetime opportunity. We hope that we will be judged to have made the most of it. Joanne's career has been a long and fascinating journey for all who have had the pleasure and honor to travel it with her. This is recounted in this paper in the context of the research performed by its two authors, who owe much to Joanne for her help and inspiration. It begins with cloud seeding concepts and studies; this is followed by new research results, which document the apparent role of aerosols and coalescence in the production of rain from clouds.

2. Cloud seeding concepts and studies

Convective clouds figured prominently in the development and testing of a conceptual model for the reduction of hurricane wind speeds through glaciogenic seeding of deep convection in bands exterior to the hurricane eyewall. As Director of Project Stormfury, Dr. Joanne Simpson worked on this intriguing problem with her husband Bob and a cadre of other scientists. The key initial point in the conceptual model was the assumption that deep convective clouds in the hurricane remain supercooled through a considerable depth of the troposphere, making them suitable for glaciogenic seeding. Microphysical measurements later revealed, however, that these conditions are not met in the hurricane.

On the contrary, deep convection in the hurricane glaciates typically at temperatures $> -10°C$. Had this been known in advance, the Stormfury program might never have begun. This finding emphasizes the importance of cloud microphysical measurements.

Scientific spin-offs from Project Stormfury under Joanne Simpson soon proved to be more important than Stormfury itself. Joanne pursued development of her one-dimensional cloud model (Witt and Malkus 1959) in conjunction with randomized experimentation on supercooled convective clouds over the Caribbean and then in Florida. In a classic paper (Simpson et al. 1967) Joanne used her simple model to predict the growth response (called "seedability") of treated clouds and compared her predictions to what was observed following seeding (seeding effect). There was excellent agreement between the predicted and observed growths of the unseeded and seeded clouds, respectively. This paper validated Joanne's view that one of the best ways to understand clouds is to perturb their natural state through cloud seeding. It also gave scientific credibility to cloud seeding.

This separation of seeded and nonseeded clouds into two distinct populations based on model predictions was a very persuasive finding for the postulate that seeding with a glaciogenic agent could affect the dynamics of clouds through the heat released during glaciation. This model became known as the "dynamic seeding" conceptual model. Dr. Woodley worked initially with Joanne on its development and testing until she left for other scientific challenges at the University of Virginia and then at the National Aeronautics and Space Administration (NASA). He continued his study of the effect of cloud seeding on cloud structure and precipitation and was joined by Dr. Rosenfeld in 1980. They continue their collaboration to this day.

The development of cloud seeding concepts has been hindered by dynamical and microphysical uncertainties. The initial dynamic seeding conceptual model conceded that the seeding might make the seeded volume less precipitation efficient because of the expected high concentration of ice particles following seeding. It was assumed, however, that the presumed microphysical inefficiencies of converting cloud water into precipitation would be overwhelmed by increases in cloud size, duration, and rainfall. Although such changes have been observed following seeding (Simpson et al. 1967; Simpson and Woodley 1971), they provide no information regarding the efficiency of the cloud microphysical processes.

A second uncertainty with respect to cloud seeding experiments is how an effect of seeding in one cloud might be communicated to other nearby clouds. Joanne (Simpson 1980) argued persuasively that enhancement of downdrafts by seeding is the mechanism whereby an effect of seeding is communicated to other clouds. The downdrafts are postulated to result in regions of convergence between the downdraft outflows and the ambient flows or with outflows from other clouds. New clouds then grow in the convergent regions, resulting in the broadening and deepening of the precipitation area.

These ideas were incorporated into the dynamic seeding conceptual model (Woodley et al. 1982), which was revised further as of 1993 (Rosenfeld and Woodley 1993). By this time, the conceptual model had come full circle from its original version to postulate that seeding makes clouds more microphysically efficient by producing graupel, which grows preferentially at the expense of the cloud water. It is obvious that any model that purports to explain natural cloud processes and their alteration by seeding must address these interactive dynamical and microphysical processes.

3. The role of aerosols and coalescence in rain production

The cloud seeding conceptual model, as developed by the authors (Rosenfeld and Woodley 1993), predicts that the effect of seeding will be a function in part of the aerosols ingested by clouds and by their ability to develop raindrops through coalescence. This explains the keen interest of the authors in coalescence processes. They are of great importance to the Tropical Rainfall Measuring Mission (TRMM) as well, because coalescence leads to rainfall, leaving the heat of condensation irreversibly to the atmosphere. Such irreversible heat releases are characteristically larger in microphysically maritime clouds than in clouds of continental character, because they precipitate a larger fraction of their cloud water.

The fundamental difference between microphysically continental and maritime clouds is the drop size distribution near their bases. Clouds are considered quite maritime when the cloud-base droplet concentration is less than 100 droplets per cubic centimeter, whereas continental clouds have cloud-base droplet concentrations greater by about an order of magnitude. The terms "continental" and "maritime" originate from the observation that the drop size distribution near cloud base is governed by the distribution of the cloud CCN, continents being the source of high CCN concentrations and marine environments being the source of low CCN concentrations. It was recognized more than 40 years ago that continental clouds have larger "colloidal stability" than maritime clouds (Squires 1958). Laboratory experiments of relevance to coalescence in clouds in CCN-rich and -lean environments (Gunn and Phillips 1957) led to the suggestion that highly continental clouds, which form in a CCN-rich atmosphere, would produce less precipitation. It has been widely accepted since then that, in the absence of ice processes, microphysically continental clouds are less precipitation efficient and produce less rainfall than "clean" maritime clouds of the same size, because coalescence processes are weaker in continental clouds, resulting in a slower conversion of their water into precipitation.

These differences between continental and maritime clouds were the basis for the early hygroscopic seeding conceptual models aimed at augmenting precipitation by enhancing coalescence processes (Bowen 1952; Biswas and Dennis 1971; Cotton 1982). Braham (1964) has postulated that enhancement of coalescence in supercooled clouds can enhance the ice precipitation processes. The larger drops freeze faster and at warmer temperatures (Bigg 1953) and grow faster by riming than drops of comparable mass would have grown by coalescence (Johnson 1987; Pinsky et al. 1998). This likely explains the presence of so much ice in deep convection in the hurricane. The hygroscopic seeding conceptual model of recent experiments has been amended to include the positive impact of enhanced coalescence on mixed-phase precipitation processes (Mather et al. 1997; Cooper et al. 1997).

Modeling studies (Reisin et al. 1996) of the rainfall amounts from clouds with top temperatures of about $-20°C$ provided some quantitative assessment of the importance of coalescence to the production of rainfall from clouds. A simulated change of the CCN concentrations at 1% supersaturation from maritime ($100 \, cm^{-3}$) to continental ($1100 \, cm^{-3}$) resulted in an 86% decrease in the simulated rainfall.

The importance of coalescence in the production of rainfall from Thai convective cells has been investigated by Rosenfeld and Woodley using volume-scan radar observations from the Applied Atmospheric Research Resources Project (AARRP) 10-cm Doppler radar in northwest Thailand. The radar estimates of cell properties were partitioned using in situ observations of the presence or absence of detectable raindrops on the windshield of the project's Aero Commander seeder aircraft as it penetrated the updrafts of growing convective towers 200–600 m below their tops at about the $-8°C$ level [about 6.5 km mean sea level (MSL)]. Cells observed to contain detectable raindrops during these aircraft penetrations were found to have smaller first-echo depths than cells without observed raindrops when growing through the aircraft penetration level, where first-echo depth is the difference between first-echo height and the height of cloud base. Although cloud dynamics likely played a role, this faster formation of raindrops is thought to be due primarily to a rapid onset of coalescence in the convective cells.

Rain volumes produced by the aircraft-measured rain clouds were estimated, using 5-min sequences of S-band radar-volume scans. Each rain cell was represented by a reflectivity maximum, bounded at the 12-dBZ level or the line of minimum between it and a neighboring cell. The rain volumes were determined by time-area integration of the rain intensities of the tracked rain cells, using the methodology of long-tracked cells as described by Rosenfeld and Woodley (1993). Convective cells exhibiting a rapid onset of coalescence (category-

TABLE 6.1. Radar-estimated rain volumes ($m^3 \times 10^3$) for convective cells and convective cloud systems as a function of coalescence category (the statistical P values for the ratios were obtained from rerandomization procedures).

Convective type	Category 2	Category 1	Category 2/category 1	Rerandomization P value
Convective cells	267.9	130.6	2.05	0.016
Convective cloud systems	6204.4	2631.8	2.36	0.000

2 clouds with detectable raindrops when growing through the aircraft penetration level) produced over a factor of 2 more rainfall than comparably sized cells in which the onset of coalescence was slower (category-1 clouds with no detectable raindrops when growing through the aircraft penetration level). This was true also for convective cloud systems over an area covering 1964 km^2 (Table 6.1). Both ratios have impressive rerandomization (permutation) statistical P values, defining the probability that the result was obtained by chance. In making the radar-rainfall estimates, the rain intensities (R) were calculated from the reflectivity (Z) by using the Z–R relationship $Z = 300R^{1.5}$ on the 2-km altitude (cloud base) reflectivities. A systematic bias in the Z–R relation for clouds with active warm rainfall and other clouds would also induce a systematic difference in the radar-inferred rain volume. However, warm rain produces typically smaller reflectivities for the same rain intensity than clouds with mixed-phase precipitation processes. This observation underlies the selection of Z–R relationships for the Weather Surveillance Radar-1988 Doppler (WSR-88D) network, using for "summer deep convection" the relation $Z = 300R^{1.4}$ (Woodley and Herndon, 1970), and using for "tropical convective systems" the relation $Z = 250R^{1.2}$ (Rosenfeld et al. 1993). According to this, if there is any bias due to differences in the Z–R relations, it would work to underestimate the differences under the assumption of fixed Z–R relations.

It was also found that the rain production of convective cells increases with maximum echo-top height in both coalescence categories, but the increase is greater for clouds exhibiting a rapid onset of coalescence as shown in Fig. 6.1. The data are plotted at the center point of 2-km intervals of maximum echo-top height.

The mean rain-volume differences (dashed line) and ratios of mean rain volumes (solid curve) versus maximum precipitation echo-top height for cells growing on days with weak coalescence activity (category 1) and on days with strong coalescence activity (category 2) are shown in Fig. 6.2. Although the ratios of the rain volume between convective cells exhibiting a rapid onset of coalescence and those with slower coalescence activity are impressively high for the shallow convective cells, in which condensation–coalescence is the dominant precipitation mechanism, the volumetric rain differences are rather small.

On the other hand, for the deep convective cells, in which mixed-phase precipitation development processes are dominant, the rain-volume ratios are lower but still physically significant; and they are associated with volumetric rain differences that are important to the overall rainfall of the area. For example, deep tropical cumulonimbus cells (all cells with $H_{max} > 10$ km) that exhibit a rapid onset of coalescence produce about a factor of 2 more rain volume than cells with slower coalescence activity. This finding has high statistical significance. The volumetric rainfall difference is 253×10^3 m^3 per cell with a P value smaller than 0.03 from a "t" test, assuming unequal variances. Most important, life cycle analysis of convective cloud systems in which the convective cells reside showed that the results for the cell scale are preserved on the scale of cloud systems (Table 6.1).

These findings highlight the important role that co-

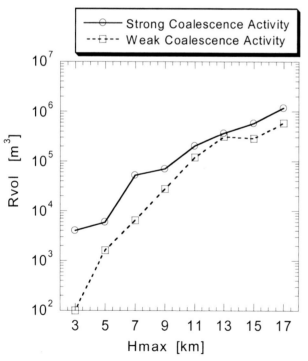

FIG. 6.1. The mean rain volumes (Rvol) of convective cells as a function of maximum precipitation echo-top height (Hmax) for cells growing on days with weak coalescence activity (category 1) and on days with strong coalescence activity (category 2). The data are plotted at the center point of 2-km intervals of maximum echo-top height. For the purpose of showing the trend on a logarithmic scale, a value of 10^2 m^3 was used instead of the zero value of Rvol for the 3-km maximum echo-top height interval for cells growing on days with weak coalescence activity (category 1).

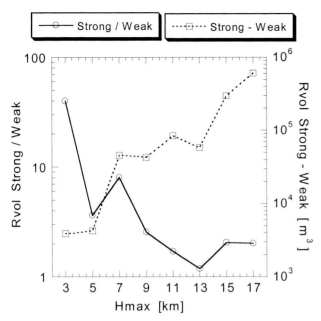

FIG. 6.2. Mean rain − volume differences (dashed line) and ratios of mean rain volumes (solid curve) vs max precipitation echo-top height for cells growing on days with weak coalescence activity (category 1) and on days with strong coalescence activity (category 2). The data are plotted at the center point of 2-km intervals of maximum echo-top height. For the purpose of showing the trend on a logarithmic scale, a value of 10^2 m³ was used instead of the zero value of Rvol for the 3-km maximum echo-top height interval for cells growing on days with weak coalescence activity (category 1).

alescence plays in the production of rain from both shallow and deep convective cells and cloud systems, and they provide valuable insights into natural cloud processes. Other meteorological factors (e.g., cloud dynamics) being equal, they also help in understanding regional differences in rainfall and their probable causes through the linkages between CCN, cloud droplet size distribution, coalescence, and rainfall amounts. These results also suggest that pollution, containing high concentrations of CCN that suppress coalescence, is likely to act detrimentally on rain-forming processes in clouds. If the pollution is extensive, it has the potential to affect the global climate. The effects of seeding should also be different in polluted clouds and nonpolluted clouds.

4. The role of satellites in cloud and precipitation physics

a. Overview

The microstructure of clouds and precipitation is typically determined by flying an instrumented cloud physics aircraft into the clouds for in situ measurements, although the National Center for Atmospheric Research (NCAR) Electra Doppler Radar (ELDORA) and NASA's ER2 Doppler radar (EDOP) on aircraft have been used for remote sensing of cloud properties. This permits only limited measurements in space and time.

Increasingly, however, the internal properties of clouds have been inferred remotely using either radar or satellite. This has allowed much greater time and space sampling. The combined use of all systems is the best of all worlds. The instrumented aircraft still provides the measurement standard while radar and satellites allow for a much expanded view.

Recent breakthroughs in the use of satellites to infer cloud microphysical properties have been especially noteworthy. Traditionally, the single visible and single IR channels on most satellites provided only very limited information about the cloud properties, such as visible reflectance, cloud-top temperature, and a crude estimate of vertically integrated cloud water. Additional channels on National Oceanic and Atmospheric Administration Advanced Very High Resolution Radiometer (NOAA AVHRR) and TRMM Visible and Infrared Scanner (VIRS) in the solar IR (1.6 µm), mid-IR (3.8 µm), and far IR (10.8 and 12.0 µm), and many more on the recently launched Moderate Resolution Imaging Spectroradiometer (MODIS) onboard of *Terra*, now make it possible to obtain additional information about a wide range of cloud properties. Among them are cloud thickness and cloud composition, that is, effective radius of cloud-top particles, and whether they are composed of water or ice.

The added multispectral information makes it possible to retrieve various properties of various types of clouds and surfaces. This can be used for classification of clouds into physically meaningful types. It can be also used to identify the clouds even over a background of ice and snow, or lower cloud layers. The clouds can be classified according to their composition: water, supercooled water, mixed phase, or ice clouds of various types. The evolution of the cloud composition and particle size in the vertical can be used most of the time to infer the precipitation properties: nonprecipitating, drizzle, warm rain, mixed phase, or ice precipitation processes.

The possibility of obtaining such information from multispectral analyses of satellite data opens a whole new range of potential applications, some of which were not contemplated only a few years ago. Although the potential for detection of supercooled water and icing conditions was already recognized (Schickel et al. 1994), other applications related to precipitation are just starting to emerge.

A particularly intriguing prospect is the possibility of detecting how cloud composition and precipitation are affected by natural (e.g., desert dust) and anthropogenic aerosols (e.g., smoke from burning vegetation, urban and industrial air pollution). Many of these potential products and new directions of research are currently in their early developmental stages. These new possibilities represent yet untapped potential from geostationary and orbiting satellites with multispectral capabilities that have been launched during recent years.

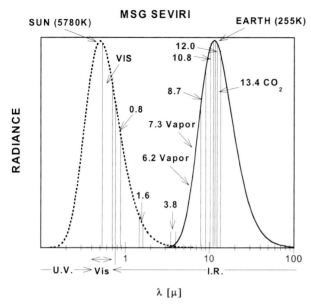

FIG. 6.3. The relation of the spectral bands of the MSG satellite with respect to the top-of-the-atmosphere solar and surface terrestrial radiation fluxes. Note that the 3.8-μm channel receives both solar and thermal radiation, and it is on the common minimum. Therefore, thermal emission must be deducted in obtaining solar reflectance at 3.8 μm.

b. Physical basis of the satellite methodology

The family of sensors on board the recent satellites has a family of spectral bands in the solar and terrestrial portion of the radiation spectrum. This is illustrated in Fig. 6.3 for the Meteosat Second Generation (MSG) satellite, which has 12 spectral bands. The situation for the Geostationary Operational Environmental Satellite (GOES), NOAA AVHRR, and TRMM VIRS is similar, but with some of the channels missing. These channels make it possible to measure additional parameters in addition to visible reflectance and the thermal emission temperature.

The fundamental reason for the added capabilities is the differences in the interaction of the radiation with cloud particles at different wavelengths. A very important contribution of the added multispectral capability comes from the dependence of the absorption of water and ice on the wavelength, as illustrated in Fig. 6.4. The absorption of the solar radiation by the cloud particles occurs within their volume (αr^3), whereas the scattering of the radiation occurs at their surface (αr^2). The absorption detracts from the overall backscattered radiation. This results in net reflected radiation to the satellite αr^{-1}. Therefore, the reflected solar radiation in absorbing wavelengths is inversely proportional to the particle size. Taking into account the relative angles of the sun, object, and sensor, it is possible to retrieve the effective radius of cloud particles, r_e. The effective radius is proportional to the sum of the volumes divided by the sum of the surface areas of the particles in the measurement volume of the satellite pixel. The large

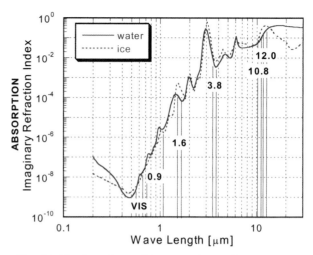

FIG. 6.4. The dependence of the absorption coefficients of water and ice on wavelength. The MSG channels in the atmospheric windows are denoted. Note that water and ice are practically nonabsorbing in the visible and strongly absorbing at 3.8 μm. The ice absorbs much more strongly than water at 1.6 μm. This makes 1.6 μm useful for separating ice from water clouds and surfaces.

absorption in the 1.6- and 3.8-μm bands makes it possible to use these channels for retrieving r_e, in contrast to the nonabsorbing visible channel (Arking and Childs 1985; Nakajima and King 1990).

In reality, the calculations of the effective radius need to take into account effects such as multiple scattering within the clouds, correction for absorption by gases, upwelling radiation from the surface through the clouds, possible multiple-layer clouds, and shadowing effects. Most of these errors increase with increasing solar zenith angles, limiting reasonably accurate retrievals only to times where the sun is more than about 20° above the horizon, and satellite zenith angle $<\sim$50°.

Retrieval of r_e has been applied extensively to marine stratocumulus, in order to measure the impact of aerosols on cloud microstructure and, through that, on the earth's radiation budget. Attention was directed to this subject shortly after launch of the first NOAA orbiting satellites with the AVHRR, revealing ship tracks consisting of polluted clouds with enhanced albedo and reduced droplet sizes (Coakley et al. 1987). An example of ship tracks in the north Pacific in shown in Fig. 6.5.

The pollution particles emitted from the ship stacks act as CCN, redistributing the fixed amount of cloud water that must condense into a larger number of smaller droplets (Radke et al. 1989). It was recognized then that the reduced droplet size suppressed the coalescence and drizzle, thus preventing loss of cloud water. The effects of larger amounts of cloud water in smaller droplets are to increase the cloud reflectance of solar radiation. Similar effects were observed over land, where the impact of smoke on shallow tropical convective clouds was shown to decrease the droplets effective radius from about 15 to 9 μm (Kaufman and Fraser 1997). The increased reflectance, mainly in the solar infrared, has

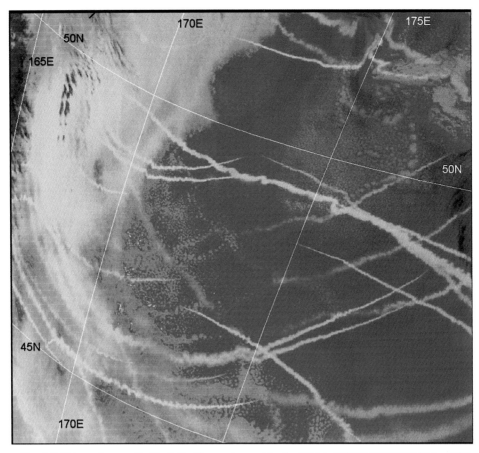

FIG. 6.5. Ship tracks over the North Pacific, as observed by the NOAA AVHRR 0325 GTC 22 Jun 2000. The color scheme is red for the visible, green for 3.7-μm reflectance component, and blue for temperature. The ship tracks with reduced particle size appear as cloud bands of enhanced green, or enhanced 3.7-μm reflectance. For full description of the color table see Table 6.2.

been considered as a cooling mechanism partially counteracting global warming due to the increased greenhouse gases (Houghton et al. 1994). However, the potential applications of 1.6- and 3.8-μm reflectance go much beyond the important subject of cloud-aerosol impacts on the earth's radiative budget. The reported (Kaufman and Fraser 1997) reduction of cloud droplets below the threshold of 14 μm, which is the minimum required size for effective coalescence (Rosenfeld and Gutman 1994), means that in addition to the radiative effects, the smoke also had a suppressive effect on precipitation. In the next sections we will see how the multispectral observations can provide us with qualitative (section 6.4.4) and quantitative (section 6.4.5) information about cloud microstructure and precipitation.

c. Multispectrally based cloud classification

The various channels provide different kinds of information. Whereas the interpretation of the visible and thermal infrared window channels is well known, the information content of the near-infrared (NIR, 1.6 and 3.8 μm) reflectance deserves more attention. For clouds composed of very small water droplets, the scattering dominates the absorption, and therefore the ratio of NIR/visible reflectance will be closest to unity. For increasingly large droplets the NIR reflectance decreases and the ratio decreases. Because ice absorbs more strongly than water at 3.8 μm, and even more so at 1.6 μm (See Fig. 6.4), the reflectance in these channels is even smaller for ice particles. Therefore, ratios of NIR/visible reflectance approaching 0 indicate ice surfaces or clouds. Therefore, the combination of the visible and NIR reflectance provide valuable information about the cloud optical thickness and composition. With the added temperature information from the thermal infrared window, a highly detailed cloud classification and characterization can be made. Rendering the three parameters in a red–green–blue color composite is natural for human perception, creating colors that correspond to the different cloud types, as described in Table 6.2 (Rosenfeld and Lensky 1998).

This qualitative classification can be readily used for a large number of applications, including the following:

TABLE 6.2. The red–green–blue color compositions and their interpretation.

Color components
- **VIS reflectance**.   redder: larger visible reflectance
- **3.7-mm reflectance**   greener: smaller cloud top particles
- **Temperature**   bluer: warmer tops

Color compositions
- White: **Low thick water or dust clouds, no precipitation**.
  - High red: Bright and thick clouds
  - High green: Small particles
  - High blue: Warm tops
- Red: **Deep precipitating clouds** (precipitation not necessarily reaching the ground)
  - High red: Bright and thick clouds
  - Low green: Large particles
  - Low blue: Cold tops
- Yellow: **Supercooled thick water clouds**
  - High red: Bright and thick clouds
  - High green: Small particles
  - Low blue: Cold tops
- Green: **Supercooled thin water cloud.** or small ice particles thin clouds
                   typically altocumulus, altostratus and rarely cirrus
  - High red: Bright and thick clouds
  - High green: Small particles
  - Low blue: Cold tops
- Black: **Large ice particles, thin clouds, often cirrus**
  - Low red: Dark and thin clouds
  - Low green: Large particles
  - Low blue: Cold tops
- Violet :   **Warm rain clouds**
  - High red: Bright and thick clouds
  - Low green: Large particles
  - High blue: Warm tops
- Pink :   **"Warm ice"** precipitating clouds, or snow on the ground
  - High red: Bright and thick clouds
  - Low green: Large particles
  - Medium blue: Mid tops (0 to -100 C)
- Blue: **Ground**
  - Low red: Dark and thin clouds
  - Low green: Large particles
  - High blue: Warm tops

Detection of aircraft icing conditions. Icing occurs when aircraft fly in supercooled water clouds. Such clouds should be detectable under most conditions according to Table 6.2.

Detection of fog or low clouds on the background of a snow-covered surface. Water clouds and snow or ice surfaces are very different according to Table 6.2. The frozen surface appears purple, and the water clouds appear white or yellow on the purple background.

Distinction between precipitating and nonprecipitating clouds. Precipitation-forming processes require that a cloud must be deep (highly reflective in the visible) and composed of large droplets or ice par-

ticles (small reflectance at 3.8 μm). This can be identified by the appropriate entries in Table 6.2.
Identification of air pollution. Low-level clouds composed of small droplets (large reflectance at 3.8 μm) are indicative of an air mass containing a large concentration of aerosols. The polluted clouds appear as relatively bright patches or stripes on the pink background of clouds with large droplets in clean air. The classic example for that is ship tracks, as they appear in Fig. 6.5. Such pollution tracks have been identified also in clouds over land areas (Rosenfeld 2000).

d. Quantitative multispectral methods

While human perception more readily absorbs large quantities of qualitative information, especially when presented with the appropriate method of visualization, greater benefits and deeper insights can come from quantitative analyses using computerized algorithms. Some of the methods and the added insights gained by them are presented in this section.

Some of the more important parameters that can be derived for each satellite pixel from the multispectral satellite data in daylight conditions are listed below.

- *Cloud reflectance in the visible (R), normalized for the solar zenith angle.* For Lambertian cloud surfaces (rarely actually realized) the reflectance equals the albedo. Under some assumptions, optical depth and liquid water path can be retrieved from the visible reflectance.
- *Cloud-top temperature (T).* This can be obtained from the thermal channel at 10.8 μm and can be made more accurate with a correction for water vapor, when using the 12.0 μm and water vapor channels.
- *Cloud particle effective radius r_e.* Using the 1.6- and 3.8-μm reflectance, it is possible to calculate the r_e of a cloud top. The calculation is based on the assumption that the cloud is composed of water drops, the cloud completely fills the pixel, and the cloud is thick; that is, a negligible amount of radiation from the surface at 1.6 or 3.8 μm is upwelling through the cloud top. If this is not the case, the surface reflectance must be known precisely (Arking and Childs 1985).
- *Cloud microphysical phase—water, mixed phase, or ice.* Clouds with $T > 0°C$ must be water clouds. However, only clouds at $T \leq -39°C$ can be determined unambiguously as ice on the single pixel level, because supercooled water clouds can exist down to $-38°C$ (Rosenfeld and Woodley, 2000).

Much more information about the cloud microstructure and precipitation-forming processes in convective clouds can be obtained from analyses of complete cloud clusters, residing in areas containing thousands of satellite pixels. The underlying assumption is that the microphysical evolution of a convective cloud can be rep-

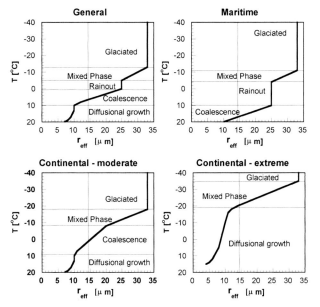

FIG. 6.6. The classification scheme of convective clouds into microphysical zones, according to the shape of the $T–r_e$ relations. Note that in extremely continental clouds r_e at cloud base is very small, the coalescence zone vanishes, mixed-phase zone starts at $T < -15°C$, and the glaciation can occur in the most extreme situation at the height of homogeneous freezing temperature of $-39°C$. In contrast, maritime clouds start with large r_e at their base, crossing the precipitation threshold of 14 μm a short distance above the base. The deep rainout zone is indicative of fully developed warm rain processes in the maritime clouds. The large droplets freeze at relatively high temperatures, resulting in a shallow mixed-phase zone and a glaciation temperature reached near $-10°C$.

resented by composition of the instantaneous values of the tops of convective clouds at different heights. This is based on the knowledge that cloud droplets form mainly at the base of convective clouds and grow with increasing height or decreasing T. The form of dependence of r_e on T contains vital information about the cloud and precipitation processes, as described below. The $T–r_e$ relations are obtained from an ensemble of clouds having tops covering a large range of T. Usually many pairs of $T–r_e$ for each 1°C interval are observed in a region containing a convective cloud cluster. The points with smaller r_e for a given T are typically associated with the younger cloud elements, whereas the larger r_e for the same T are associated with the more mature cloud elements, in which the droplet's growth had more time to progress by coalescence, and ice particles had more time to develop. Therefore, it is useful to plot not only the median value of $T–r_e$ relation, but also, say, the 15th and 85th percentiles (see Fig. 6.6), for representing the younger and more mature cloud elements within the measurement region.

Based on the shapes of the $T–r_e$ relations, Rosenfeld and Lensky (1998) defined the following five microphysical zones in convective clouds.

1) *Diffusional droplet growth zone.* Very slow growth

of cloud droplets with depth above cloud base, indicated by small $-dr_e/dT$.

2) *Droplet coalescence growth zone.* Large increase of the droplet growth with height, as depicted by large $-dr_e/dT$ at T warmer than freezing temperatures, indicating rapid cloud droplet growth with depth above cloud base. Such rapid growth can occur there only by drop coalescence.

3) *Rainout zone.* A zone where r_e remains stable between 20 and 25 μm, probably determined by the maximum drop size that can be sustained by rising air near cloud top, where the larger drops are precipitated to lower elevations and may eventually fall as rain from the cloud base. This zone is so named because droplet growth by coalescence is balanced by precipitation of the largest drops from cloud top. Therefore, the clouds seem to be raining out much of their water while growing. The radius of the drops that actually rain out from cloud tops is much larger than the indicated r_e of 20–25 μm, being at the upper end of the drop size distribution there.

4) *Mixed-phase zone.* A zone of large indicated droplet growth with height, occuring at $T < 0°C$, due to coalescence as well as to mixed-phase precipitation formation processes. Therefore, the mixed-phase and the coalescence zones are ambiguous at $0 < T < -39°C$. The conditions for determining the mixed-phase zone within this range are specified in Rosenfeld and Lensky (1998).

5) *Glaciated zone.* A nearly stable zone of r_e having a value greater than that of the rainout zone or the mixed phase zone at $T < 0°C$.

These zones are idealizations. Not all clouds conform to this idealized picture. The transition between the coalescence and mixed-phase zones, which are not separated by a rainout zone, cannot be determined and are therefore set arbitrarily to $-6°C$ in accordance with aircraft observations. The height of the glaciation zone can be overestimated in the cases of highly maritime clouds that grow through a deep rainout zone, because the scarcity of water in the supercooled portions of the clouds causes small ice particles, which sometimes can be mistaken for a mixed-phase cloud. Addition of more spectral bands can help in separating the water from the ice, irrespective of the particle size. On the other hand, in vigorous clouds with active coalescence the height of the glaciation zone can be underestimated, because the high amounts of large ice hydrometeors dominate the radiative properties of the clouds, even when they coexist with cloud supercooled water.

All these microphysical zones are defined only for convective cloud elements. Multilayer clouds start with small r_e at the base of each cloud layer. This can be used for distinguishing stratified from convective clouds by their microstructure. Typically, a convective cloud has a larger r_e than a layer cloud at the same height,

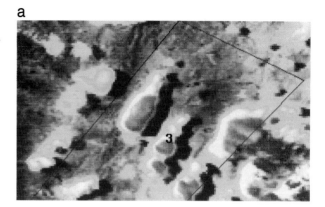

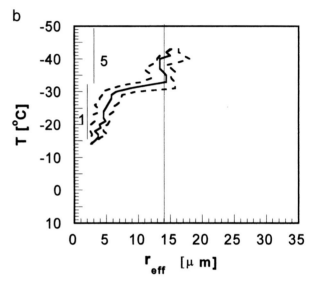

FIG. 6.7. (a) Wave clouds in area of about 300×200 km in the lee of the Andes, forming a supercooled water cloud at their leading edge (upper-left edge of the waves), as evident by the yellow color (see Table 6.2 for color interpretation). The image was taken by the NOAA AVHRR on 2008 UTC 7 Jan 1999. Abrupt freezing occurs at $-32°C$, as indicated by the red color and the jump to a larger indicated effective radius at $-32°C$. (b) The $T–r_e$ for (a). The microphysical zones are marked by the vertical bars where 1 = diffusional growth, and 5 = glaciated. Note that glaciation is indicated at $-32°C$.

because the convective cloud is deeper and contains more water in the form of larger drops.

In addition to the microphysical zones, it also can be determined that convective clouds start precipitating at $r_e > 14$ μm (Rosenfeld and Gutman 1994; Gerber 1996). This can be used quantitatively for improving the accuracy of rainfall measurements from space, as demonstrated by Lensky and Rosenfeld (1997) for the NOAA AVHRR.

Examples of the various microphysical zones have been obtained in clouds around the world. A simple case, showing the onset of glaciation, is the wave clouds shown in Fig. 6.7. The small scatter of the effective radius for any given T and the repetitive behavior of each of the clouds, showing the jump of r_e at $T = 32°C$,

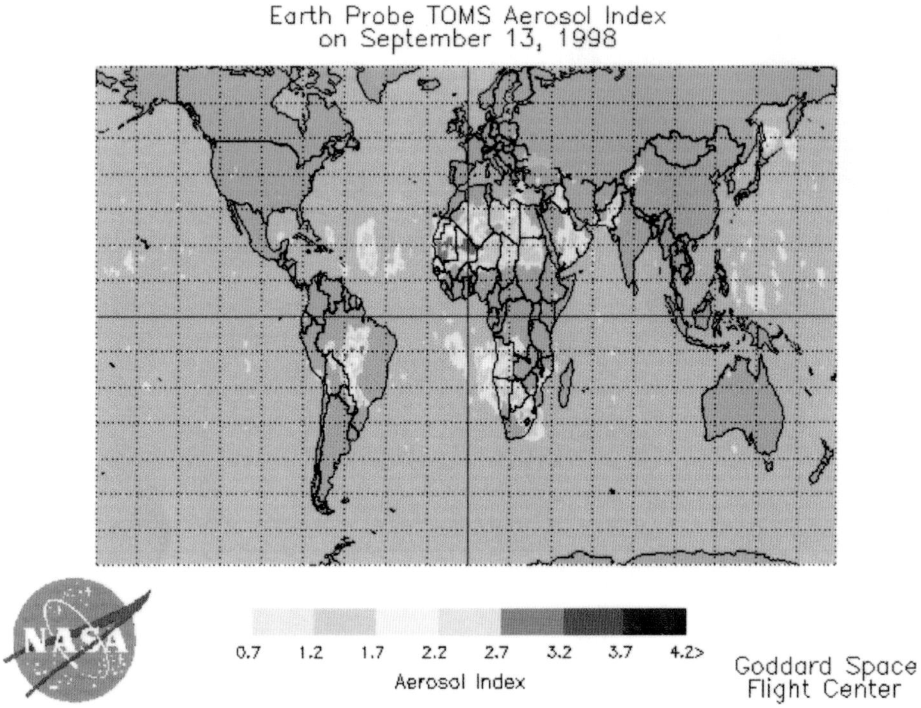

FIG. 6.8. Earth Probe TOMS Aerosol Index on 13 Sep 1998, showing the smoke from forest fires in the Amazon and Congo basins.

make a compelling case for inference of the glaciation temperature at this T.

Shallow convective clouds provide another simple example. Of interest here are clouds growing in a smoky atmosphere over the Amazon on 13 September 1998. The smoke aerosols are documented by the global plots of the Earth Probe Total Ozone Mapping Spectrophotometer (TOMS) Aerosol Index on this day (Fig. 6.8), showing pollution over the Amazon in the region of the growing clouds pictured in the processed TRMM image at 1915 GMT (Fig. 6.9a). The 230-km swath of the Precipitation Radar (PR) is bounded by the white lines. The white stippling shows the radar echoes superimposed on the clouds that have been colorized according to the scheme in Table 6.2. The yellow crosses represent lightning flashes as detected by the TRMM LIS, seen at the bottom of Fig. 6.9a just outside the TRMM PR swath.

The T–r_e plot for area 1 in Fig. 6.9a is shown in Fig. 6.9b. The 15th, 50th, and 85th percentiles for r_e versus T for the area shown are plotted. These plots can be thought of as typifying the clouds in the area. Note that the clouds have a deep zone of diffusional droplet growth overlain by a shallower zone (0°C to −10°C) of droplet growth dominated by coalescence. The mixed-phase zone begins near −10°C and continues to cloud top near −13°C. Note the 14-μm precipitation threshold is not reached until −10°C.

The vertical cross section for this case along the line in Fig. 6.9a is provided in Fig. 6.9c. Note that clouds with tops <6.5 km do not produce precipitation and in the two that exceed that height precipitation is quite weak. Considering the location of the clouds over the tropical Amazon, this means strong suppression of coalescence and ice precipitation–forming processes.

The next look is at a TRMM image of deep convective clouds growing in smoke over the Amazon two days later (Fig. 6.10a). The presentation is similar to that of Fig. 6.9a. These clouds have an even deeper zone of diffusional droplet growth overlain by a deep zone of mixed-phase processes (Fig. 6.10b). Glaciation is not indicated until −27°C. The cross section along the line in Fig. 6.10a is quite different from that for the shallower clouds. The reflectivity profiles are strong from virtually cloud base to top, indicating that ice processes had produced large, highly reflective ice particles in these clouds. Note that lightning activity (yellow crosses) is greater in this image than in Fig. 6.9a.

In contrast to the clouds studied on 13 and 15 September 1998 are those growing in clean air over the Amazon on 13 April 1998 (Fig. 6.11a). These clouds obviously are quite maritime in character with deep zones of coalescence and rainout overlain by a shallow mixed-phase zone with a deep glaciated zone above that (Fig. 6.11b). Complete glaciation is indicated at −7°C.

The vertical radar cross section (Fig. 6.11c) shows maximum reflectivities below 5 km and a greater distance between the visible cloud tops and their radar-measured tops than was the case for the deep continental clouds. Even very shallow clouds precipitate in agree-

(a)

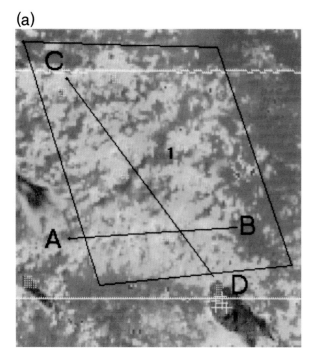

(b)   LBXD9809131915_tre_a

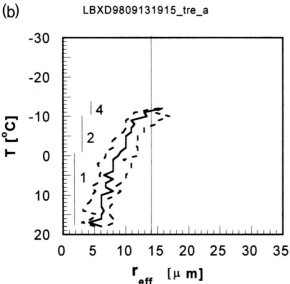

(c)

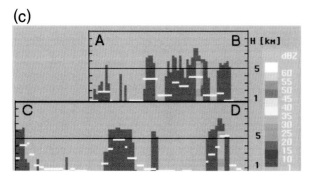

ment with the $T-r_e$ plot (Fig. 6.11b), which shows virtually the entire cloud depth having droplets that exceed the 14-μm precipitation threshold. Such clouds convert their water into precipitation very fast and efficiently, in contrast to the clouds in the polluted air (Figs. 6.9, 6.10).

In order to make the point that these clouds are highly maritime, even though they are more than 1000 km away from the closest ocean, a comparison with truly equatorial oceanic clouds is made, in the form of a TRMM image of deep cumulus and cumulonimbus clouds over the equatorial Pacific near Kwajalein at 0251 UTC on 5 November 1998 (Fig. 6.12a). The $T-r_e$ plot (Fig. 6.12b) is utterly different from that for the deep continental clouds growing in smoke over the Amazon (Fig. 6.10b) but quite similar to the plot for the deep maritime clouds over the Amazon (Fig. 6.11b). Already at cloud base, r_e exceeds the precipitation threshold of 14 μm. There is no zone of diffusional growth and the zone of coalescence extends from cloud base to 11°C. The rainout zone extends from 11° to −5°C and the mixed-phase zone from −5° to −12°C. Glaciation is complete at $T < -12$°C, which is even a little colder than the Amazon maritime case.

The vertical cross section of the clouds along the line in area 2 of Fig. 6.12a (Fig. 6.12c) is different from the vertical cross section taken through the continental clouds, which showed strong reflectivities extending to great heights. As with the Amazon maritime case, the extremely maritime clouds show the strongest reflectivities (>40 dBZ) below 5 km and the precipitation echoes average about 3 km below the visible cloud tops (Fig. 6.12c). This is due to the early formation and fallout of precipitation-sized particles, which deplete the upper portions of the clouds of water and thus of large and highly reflective particles. Differences in vertical velocities, which are typically stronger in continental clouds, can explain some of the observed differences. However, no dynamical difference can explain the total lack of precipitation from polluted tropical convective clouds that do not exceed the freezing level, and certainly they would not undergo rainout if they grew to cumulonimbus stature. These differences in cloud and echo structure, which may be due in part to differences in vertical velocity, can explain the lack of lightning discharges in maritime clouds (Zipser and Lutz 1994).

←

between the white lines. The white hatching is areas with radar-detected precipitation. The yellow crosses are lightning flashes, as detected by the TRMM LIS. The color scheme is given in Table 6.2. (b) The $T-r_e$ for area 1 of (a). The microphysical zones are marked by the vertical bars where 1 = diffusional growth, 2 = coalescence, and 4 = mixed-phase zones. The lines for the 15th, 50th, and 85th percentiles of pixels having r_e for a given T are plotted. Note that the 14-μm precipitation threshold is reached only at $T < -10$°C. (c) The vertical cross section of the clouds along the lines in (a). The two cross sections are delimited by the same characters here and in (a). Note that clouds tops reaching <7 km do not precipitate.

FIG. 6.9. (a) A TRMM image showing microphysically continental cumulus clouds growing in smoke over the Amazon, centered at 15°S, 60°W, at 1915 UTC 13 Sep 1998. The 230-km radar swath is delimited

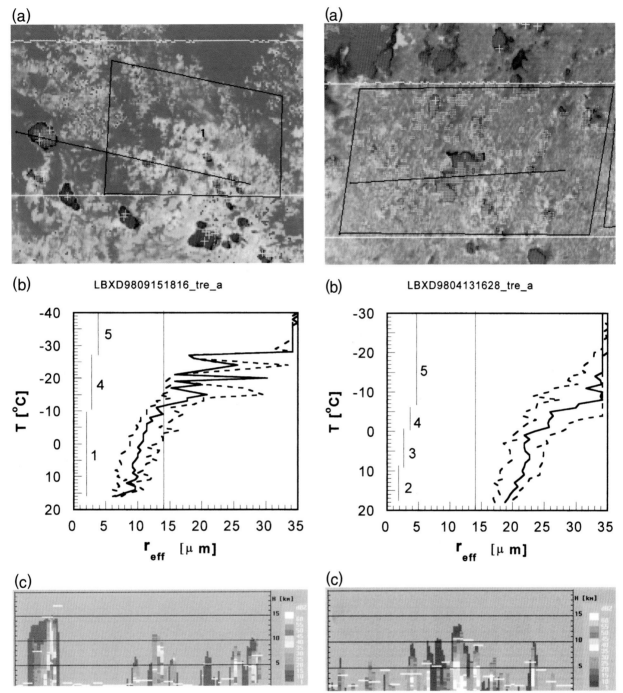

FIG. 6.10. (a) A TRMM image showing microphysically continental cumulus and cumulonimbus clouds growing in smoke over the Amazon, centered at 13°S, 61°W, at 1816 UTC 15 Sep 1998. The color scheme is given in Table 6.2. (b) The T–r_e for (a). The lines of the 15th, 50th, and 85th percentiles of pixels with r_e for a given T are plotted. The microphysical zones are marked by the vertical bars where 1 = diffusional growth, 4 = mixed phase, and 5 = glaciated. Note that glaciation is indicated at −27°C. (c) The vertical cross section of the clouds along the line in (a). Note that that the $Z > 40$ dBZ extends to $H = 10$ km, i.e., well above the 0°C isotherm.

FIG. 6.11. (a) A TRMM image showing microphysically maritime cumulus and cumulonimbus clouds growing in clean air over the Amazon, centered at 5°S, 59°W, at 1628 UTC on 13 Apr 1998. The color scheme is given in Table 6.2. (b) The T–r_e plot for (a). The microphysical zones are marked by the vertical bars 2 = coalescence, 3 = rainout, 4 = mixed phase, and 5 = glaciated zones. Note the deep rainout zone and the glaciation indicated at the relatively high $T = −7$°C. (c) The vertical cross section of the clouds along the line in (a). Note that the shallowest clouds already precipitate. Also note that the $Z > 40$ dBZ is confined to $H < 5$ km, i.e., below the 0°C isotherm.

(a)

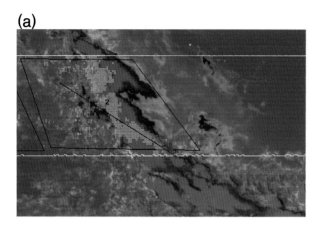

(b)

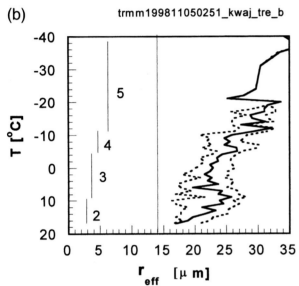

(c)

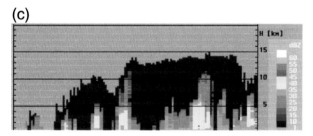

FIG. 6.12. (a) A TRMM image showing microphysically maritime cumulus and cumulonimbus clouds over the equatorial Pacific near Kwajalein, centered at 6°N, 167°W, at 0251 UTC 5 Nov 1998. The color scheme is given in Table 6.2. (b) The $T–r_e$ for (a). Note the great similarity of the microphysical zones to the clouds over the Amazon in the clean air, as shown in Fig. 6.11b. (c) The vertical cross section of the clouds along the line in (a). Note that the shallowest clouds already precipitate. Also note that the $Z > 40$ dBZ is confined to $H < 5$ km, i.e., below the 0°C isotherm.

These physical insights to the cloud and precipitation processes can be the basis for a large number of applications beyond those already mentioned in the qualitative part (section 6.4.3 of this paper), including the following.

• Quantitative estimation of precipitation from clouds not detectable by any other satellite method, including passive microwave, which is not sensitive to warm rain and weak to moderate precipitation of any kind over land (see Lensky and Rosenfield 1997).
• Detection of severe storms, based on the depth and microstructure of the convective cloud elements.
• Detection of the impact of air pollution on the precipitation forming processes, as done already by Rosenfeld (1999, 2000).
• Application to cloud seeding for precipitation enhancement. The identification of regions containing clouds with ample supercooled water and inefficient precipitation processes can be used for evaluating the potential for added precipitation by cloud seeding, and for directing the cloud seeding operations. Furthermore, the seeding signature can be identified as microphysical changes in the seeded clouds, as shown by Woodley et al. (2000). The observed cloud microstructure is a manifestation of the aerosols and instability of the air mass, which are both slowly changing properties. Therefore, characterization of the cloud microstructure, as obtained from an orbital satellite, is still quite useful in spite of the scarcity of the satellite overpasses, because it is valid for at least several hours from the time of the observation.
• Application to cloud seeding for hail suppression. The identification of clouds with dynamical and microphysical structure conducive for hail can be used for directing the hail suppression operations of cloud seeding. The experience shows that a window of 25 × 25 1-km pixels is sufficient for the analyses of small convective clusters. In cases of seeded clouds it is quite practical to seed most of the towers in such an area and create a seeding signature (Rosenfeld et al. 2001). The microphysical changes in the clouds due to seeding can be monitored and detected as well, as shown in Fig. 6.13.

5. Scientific potential

a. Documentation of the effect of smoke on cloud processes

Besides addressing the method and its use and validation, the paper by Rosenfeld and Lensky (1998) deals with a matter of global import in suggesting that the smoke from the burning of biomass on a large scale can suppress the regional rainfall from clouds ingesting the smoke.

A follow-up paper (Rosenfeld 1999) made use of TRMM PR data to show unequivocally that the smoke from the fires in Indonesia and Malaysia in the deep Tropics did indeed suppress the regional rainfall. The TRMM satellite was crucial to these findings. The VIRS sensors were used to determine that both coalescence and ice processes were suppressed in the clouds in-

(a)

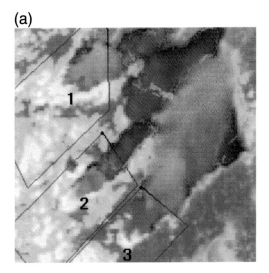

(b1)

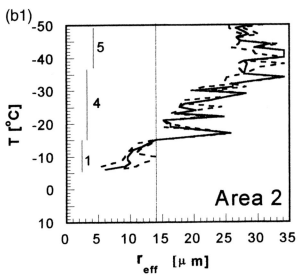

(b2)

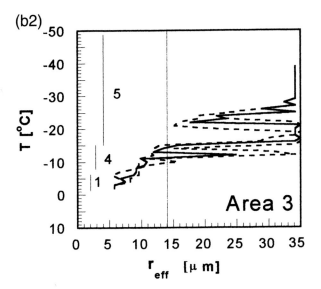

gesting the smoke and TRMM radar was used to show that rainfall was suppressed in the region of smoke.

b. The effect of urban and industrial pollution on clouds

Not much is known about the impact of aerosols from urban and industrial air pollution on precipitation. As was the case with the smoke from burning vegetation, it was assumed initially that industrial and urban pollution inhibited precipitation (Gunn and Phillips 1957). However, later reports of enhanced rainfall downwind of paper mills (Eagen et al. 1974) and over major urban areas (Braham 1981) suggested that giant CCN caused enhancement of the precipitation (Johnson 1982), but attempts to correlate the urban-enhanced rainfall to the air pollution sources failed to show any relationship (Gatz 1979). The most plausible explanation for the urban rain enhancement invokes the heat-island effect and increased friction, both of which would tend to increase the surface convergence, resulting in more cloud growth and rainfall over and downwind of the urban areas. On the other hand, the recent suggestion published in "Nature" (Cerveny and Balling 1998) that air pollution might enhance precipitation on the large scale in the northeastern United States and Canada and the speculative explanations for this effect would appear to confuse the issue. The truth is that rather little is known definitively about this subject.

Spaceborne measurements of ship tracks in marine stratocumulus provided the first evidence that effluents from ship stacks change cloud microstructure such that their water is redistributed into a larger number of smaller droplets (Coakley et al. 1987). Extrapolation of these observations to clouds that are sufficiently thick for precipitation (i.e., at least 2 km from base to top) would mean that the effluents have the potential to suppress precipitation.

Application of the imaging scheme of Rosenfeld and Lensky (1998) to the AVHRR onboard the NOAA orbiting weather satellites has now revealed numerous "ship track"–like features in clouds over land, emanating from major urban and industrial pollution sourc-

←

FIG. 6.13. (a) Hailstorm cumulonimbus clouds in an area of about 150 × 150 km over Alberta, Canada, as observed by the NOAA AVHRR overpass at 2215 UTC 10 Jul 1998. According to the color classification scheme presented in Table 6.2, the yellow cumuliform cloud elements indicate they are composed of supercooled droplets. The red areas are the anvils of the cumulonimbus tops. (b) The T–r_e for area 2 and 3 of (a). The T–r_e graphs for area 2 shows very deep supercooled water and mixed phase clouds with glaciation occurring only at −37°C. The same applies to area 1 (not shown). The clouds in area 3 were seeded with A_gI ice nuclei, and glaciated between −15° and −25°C in different parts of the cloud, as indicated by the T–r_e graph for that area. Hail was observed in the area on that day. However, clouds 1 and 2 occurred over remote areas where there were no ground-based observations.

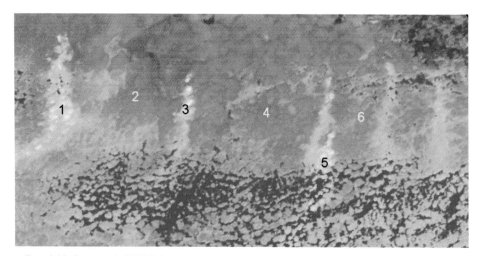

FIG. 6.14. Processed AVHRR image at 2012 UTC 20 Jul 2000 showing five pollution tracks in a supercooled cloud deck in north-central Manitoba Province, Canada. The dimensions of the image is 550 × 250 km.

es. These tracks are evident only in the near-infrared images with no hint of them in the visible images. Because the tracks evidently originate from pollution sources, they are named hereafter "pollution tracks." Such tracks in the clouds have been illustrated downwind of Istanbul, Turkey, in Canada, and downwind of major metropolitan and industrial areas in Australia (see Rosenfeld 2000). In addition, pollution tracks have been noted (not shown) by Rosenfeld and Woodley downwind of Houston, Texas, over Manila Bay downwind of Manila, Philippines, near Guadalajara and Monterey, Mexico, downwind of Seattle, Washington, and Winnipeg, Canada. The longer one looks, the more such

tracks are found. Pollution tracks seem to be a global phenomenon.

Most of the pollution tracks were detected readily because there were pristine unpolluted areas of cloud nearby. This is illustrated in the image at 2012 UTC on 20 July 1998 (Fig. 6.14) showing five pollution tracks in north-central Manitoba, Canada. These pollution plumes can be seen streaming to the south, staining the pristine cloud deck. The $T–r_e$ plots for three of the stronger pollution tracks and for three unpolluted areas between the tracks are given in Fig. 6.15. The three plots for the unpolluted areas have effective radii that exceed the 14-μm precipitation threshold, whereas the plots for the polluted areas have effective radii averaging 10 μm or less.

Farther south of the Canada pollution sources the unpolluted regions become harder to find. This is illustrated in the AVHRR image for the Great Lakes industrial area at 2007 UTC on 5 June 1998 (Fig. 6.16). The yellow and orange coloration suggests that the effective radius is small in this region and this is verified by the $T–r_e$ plots (Fig. 6.17) for the four areas shown in Fig. 6.16. Note that r_e averages 6–10 μm between +2° and −10°C. These values are similar to those found in the discrete pollution plumes farther north in Canada, only now the pollution is nearly continuous.

All of this is anecdotal, however, and does not make a quantitative case for the suppression of the rainfall by anthropogenic pollution. Rosenfeld (2000) made such a case for pollution tracks in Australia. Satellite retrievals of cloud microstructure clearly revealed plumes, embedded in extensive cloudy areas, in which the clouds contained small particle sizes. These plumes originated from major urban areas and industrial facilities such as power plants. The satellite retrievals in the polluted and unpolluted regions showed no coalescence in the polluted region and strong coalescence in the pristine

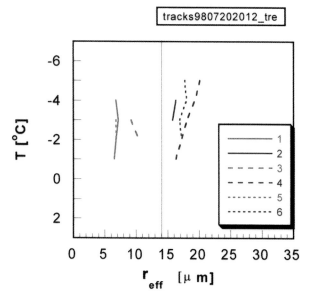

tracks9807202012_tre

FIG. 6.15. The effective radius in and outside of the pollution tracks shown in Fig. 6.14. The numbers in the inner box correspond to the numbered areas in Fig. 6.14.

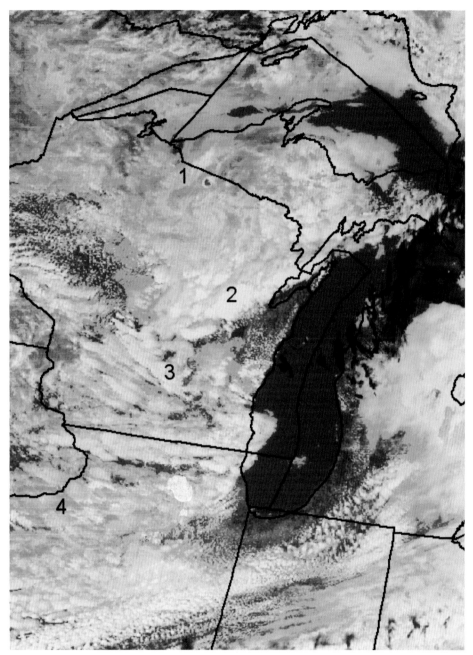

FIG. 6.16. The AVHRR image for the Great Lakes industrial area at 2007 UTC 5 Jun 1998.

clouds. In addition, the TRMM PR revealed that the plumes of polluted clouds were devoid of precipitation, whereas the ambient clouds had precipitation intensities exceeding 10 mm h^{-1}. Although producing no precipitation, the clouds in the plumes were as thick as the adjacent precipitating clouds and had no shortage of water. In addition, the PR detected a "brightband" signature, which is indicative of melting snow, in the adjacent precipitating unpolluted clouds, suggesting further that the pollution also suppresses the processes leading to the growth of ice particles to precipitation size.

It is suggested that the pollution suppresses precipitation in the same way as does smoke from burning vegetation. The ingested pollution aerosols nucleate many small cloud droplets, which remain small, thereby suppressing both warm rain processes and the formation and proliferation of ice in the clouds.

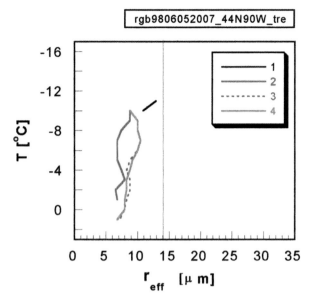

FIG. 6.17. The T–r_e plots for the four areas shown in Fig. 6.16. Note that r_e averages 6–10 μm between +2°C and −10°C.

c. Measurement of supercooled water to near −38°C

Adding to the credibility of the present satellite methodology was the discovery of large quantities (up to 1.9 gm m⁻³) of supercooled cloud liquid water to near −38°C (Fig. 6.18a). The slow growth of cloud droplets with height (Fig. 6.18b) and their large concentrations (Fig. 6.18c) show that under these conditions both warm and ice precipitation processes are suppressed. The mea-

surements were made in Texas on 13 August 1999 by the authors with an instrumented cloud physics jet aircraft in an attempt to validate the inferences by the present satellite methodology using TRMM data of supercooled liquid water to near the point of homogeneous nucleation.

The T–r_e plots for the TRMM satellite overpass at 2245 UTC (1745 CDT) on 13 August 1999 are provided in Fig. 6.19. This is within 10 min of the time the liquid water measurements were made from the jet aircraft. Note that the r_e increased steadily with decreasing T until −38°C, indicating that only at that temperature were the clouds mostly glaciated. Thus the satellite inferences and the in situ validation measurements were in general agreement on this day. Similar coincident measurements took place two days before, on 11 August 1999 (not shown), where the in situ and TRMM-measured glaciation temperatures were −36° and −38°C, respectively. Without the validating cloud physics measurements, however, it is doubtful that the satellite inferences would have been believed.

The authors made similar satellite and aircraft measurements near Mendoza, Argentina in January and February 2000 and observed supercooled cloud water contents up to 4 g m⁻³ near −38°C in Argentine clouds. Again, these aircraft measurements agreed with the indications from the satellite imagery.

6. Discussions

The Australian study is the first documented case linking urban and industrial air pollution to the reduction

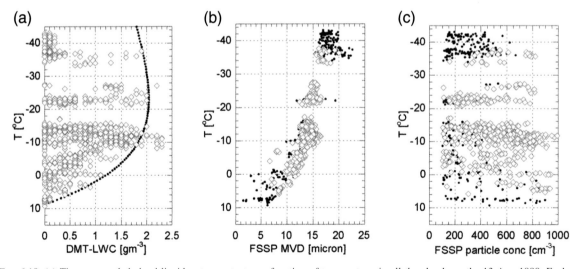

FIG. 6.18. (a) The supercooled cloud liquid water content as a function of temperature, in all the clouds on the 13 Aug 1999. Each point represents 1 s of measurements, or 150–200 m of cloud path. Note the abrupt decrease of the water at −38°C, indicating the point of homogeneous freezing. The dotted curve marks the 50% adiabatic water content, as a reference for the rate of consumption of cloud water to freezing and other processes. The black dots are those in which the hot-wire sensor detected less than 0.3 g m⁻³ liquid water contents. (b) The median volume diameter of the cloud particles measured by a Forward Scattering Spectrometer Probe (FSSP) as a function of temperature, in all the clouds on the 13 Aug 1999. Each point represents 1 s of data, or 150–200 m of cloud path, containing more than 100 cm⁻³ FSSP-measured cloud particles. The red symbols denote cloud segments with hot-wire measured liquid water contents ≥0.3 gm⁻³. (c) The same as Fig. 6.2 but for the FSSP-100 measured concentration of cloud particles.

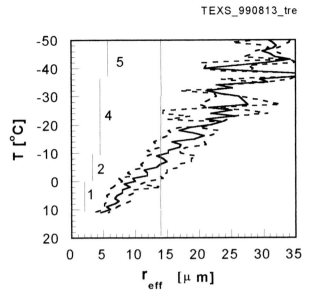

TEXS_990813_tre

FIG. 6.19. (a) The T–r_e plots for the TRMM satellite overpass at 2245 UTC (1745 CDT) 13 Aug 1999. The clouds profiled were those in which the aircraft measurements were made. Note that the r_e increased steadily with decreasing T until about $-38°C$, indicating that only at that temperature were the clouds mostly glaciated.

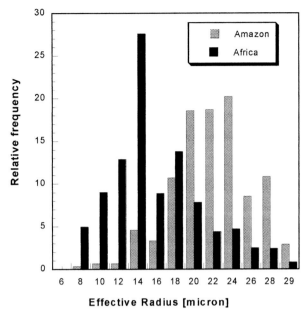

FIG. 6.20. Plot of the relative frequency of effective cloud radius for clouds over the African Congo (black bars) and the Amazon (crosshatched bars). (After McCollum et al. 1999.)

of the precipitation, pinpointing both the sources and the affected clouds. It was made possible by the newly acquired capabilities to observe both cloud microstructure and precipitation over large areas from the TRMM satellite. It might seem strange that some of the most obvious pollution signatures occur in Australia, which is probably the least polluted inhabited continent. The pollution signature on rainfall is most evident in Australia, perhaps because it is seen against a background of pristine clouds, whereas in most other places, such as the areas surrounding the Great Lakes in the United States and Canada (see previous section), the clouds are already polluted on a very large scale.

The results of the Australian and Indonesian cases are consistent with the findings of Gribbin (1995), indicating a decrease of rainfall in the Tropics. A look at multispectral satellite data over the past couple of years suggests, however, that cloud microstructure and rainfall are being affected comparably by pollution over many other regions over the globe, including the Tropics.

An intriguing case in point is the disparity in rainfall between the African Congo and the Amazon basin of South America even though satellite rainfall estimates for the two regions, based on their infrared (IR) presentations and on passive microwave rainfall algorithms, are comparable (McCollum et al. 2000). Rainfall is nearly 50% less in the Congo than in the Amazon basin. The rainfall from Congo clouds might be less than the rainfall from Amazon clouds because they have (a) less precipitable water with which to work, and/or (b) lesser precipitation efficiency, and/or (c) weaker forcing.

McCollum et al. (2000) used the present satellite technique to produce frequency plots for effective cloud radii (r_e) in the African Congo and the Amazon basin (Fig. 6.20). Note that the Amazon plot is shifted substantially toward larger effective radii relative to the plot for the Congo. As mentioned earlier, an effective radius threshold of 14 μm (Rosenfeld and Gutman 1994) is used typically to identify which clouds have developed coalescence and are precipitating and which are not. Examining the areas under the curves in Fig. 6.20, it appears that Amazon clouds in the sample have at least some coalescence over 90% of the time, while the coalescence frequency for Congo clouds is about 50%. This is a large difference.

McCollum et al. (2000) have already suggested the possibility that the Congo versus Amazon rainfall differences are, at least in part, due to less efficient clouds over the Congo, which are not able to form raindrops by coalescence as readily as the clouds over the Amazon. The microphysical observations of McCollum et al. (2000) are far from being climatologically representative for the whole area and all seasons. Therefore, they cannot be used as evidence that the indicated differences in the rainfall between central Africa and the Amazon are mainly due to aerosol-induced microphysical differences. However, the new findings relating coalescence to rain production reported in section 6.3 are consistent with this suggestion, as already made by McCollum et al. (2000) themselves. Although this could result from a number of causes, the nature of the CCN ingested by the clouds in the two regions is a plausible explanation. The air mass supplying Amazon clouds comes typically from the Caribbean and Atlantic and is likely more maritime in terms of the sizes and concentrations of the input CCN than the source of air for Congo clouds, which includes the Sahara, east Africa, and the Kalahari deserts in addition to the Atlantic Ocean. The ingestion of anthropogenic pollution by the clouds in the African Congo also may play a role in the lesser precipitation efficiency relative to those in the Amazon.

This view is supported by the global plots of the Earth Probe TOMS Aerosol Index (Fig. 6.7). Virtually any other selection would have shown essentially the same pattern. Africa is an anomaly when it comes to pollutants and aerosols, especially on the northern and southern margins of equatorial Africa. There are almost always fires and dust storms on the fringes of the intertropical convergence zone and they move north and south with the seasons. One can readily imagine these aerosols being drawn toward the equator into the migratory weather disturbances that move from east to west across the African continent much of the year. These aerosols certainly have the potential to influence the rainfall. The Amazon basin on the other hand has less aerosol pollution relative to Africa, although the example presented in Fig. 6.7 is one of the exceptions. It seems possible, therefore, that the natural and anthropogenic aerosols are acting to decrease the rainfall over tropical Africa.

Collectively, the Amazon basin and the African Congo constitute a significant fraction of the world's deep Tropics. If natural and/or anthropogenic pollution is shown to account for the differences in rainfall and lightning between the two regions, its effect on climate might be quantified through computer simulations.

Until this is done, we will have to live with the clear implications that smoke from biomass burning and from urban and industrial sources are highly detrimental to the efficiency of both water and ice precipitation–forming processes in convective clouds. The resulting excess water vapor remaining in the atmosphere may eventually precipitate elsewhere, causing redistribution of the large-scale precipitation and latent heating, which means alterations of the driving force behind atmospheric circulations. Hence, the question arises whether microphysical processes can influence the large-scale or even global climate.

Aircraft validation of satellite inferences of deep supercooling in vigorous convective clouds has implications for the modeling of convective clouds, for deliberate modification of cloud processes, and for studies of global climate. The validation has been made in Texas and Argentina, but the satellite imagery indicates that such supercooling takes place over many other regions in the world. This would seem to fly in the face of conventional wisdom and at variance with most modeling simulations that show glaciation in continental convective clouds at warmer temperatures. For example, full glaciation was parameterized to occur naturally at $-20°C$ in clouds in the High Plains Experiment (HI-

PLEX) (Orville 1996), which took place in the same area where the clouds were shown to fully glaciate only at $-38°C$ (Rosenfeld and Woodley 2000). This supercooling indicates that such clouds have low precipitation efficiency and may be good candidates for seeding intervention.

Deep supercooling is also probably a manifestation of feedback from natural and anthropogenic pollution. Polluted clouds have suppressed coalescence and ice processes and likely are characterized by supercooling to great depths. If this is so, the convective clouds in tropical Africa must be much more supercooled than those in the South American Amazon. Such regional differences may be changing with time with implications for the global climate.

7. Conclusions

We are back where Joanne Simpson began, but with new tools and new insights. We know much more about microphysical processes in clouds. Maritime clouds are much different from continental clouds. They ingest fewer aerosols, produce raindrops through coalescence, and average up to 100% more rain volume per convective cell than continental clouds. Highly maritime clouds are poor candidates for both glaciogenic and hygroscopic seeding, because they produce raindrops at temperatures $>0°C$, which readily freeze at temperatures near $-5°C$, producing ice splinters and rapid glaciation of the cloud. Continental clouds, on the other hand, have been shown to carry their cloud water in many cases to near $-38°C$ before freezing. Such clouds are highly inefficient and should be good candidates for glaciogenic and/or hygroscopic seeding intervention.

Pollution has a major impact on cloud processes. New satellite imagery and new methods to assimilate and analyze it have documented the microphysical structure of clouds over large space scales and timescales. This new look has demonstrated the profound impact that natural and anthropogenic aerosols have on the rainforming processes in clouds. The implications are potentially enormous. The increasing pollution must be affecting global rainfall detrimentally. At the very least, pollution must be redistributing the rainfall.

Anthropogenic pollution from industry and urbanization also acts negatively on the rain-forming processes in clouds. In the cities this negative effect may be overcompensated by increased convergence and rainfall due to dynamic factors, including the "heat island" and increased surface roughness from the clustering of large structures (Braham 1981). Away from these dynamic enhancements, however, the effect of pollution on rainfall should be negative. It is quite possible, therefore, that the combination of smoke from burning vegetation, and urban and industrial pollution is having an effect on the global climate by decreasing and/or redistributing the rainfall, especially where precipitation originates from convective clouds. This has serious potential implications for the availability of water resources, which might be compromised, especially in the most densely populated areas of the Tropics and subtropics, where people are dependent on these waters for their livelihood.

The changes in precipitation distribution must be associated with respective changes in latent heat release, which drives the global circulation. Therefore, a change in the global circulation is a likely outcome. If so, these changes are already with us today, and the climate system would have been different without today's anthropogenic emissions of aerosols.

In getting back to where we started, we have learned much along the way. Tropical convective clouds are just as important to the world climate as they were when Joanne Simpson began her illustrious meteorological career, but other factors have intervened. Our results, and those of the Metropolitan Meteorological Experiment (METROMEX) years ago, indicate that inadvertent weather modification is a reality (Braham 1981). The more we look, the more prevalent it appears to be. The challenge now is to understand and quantify its extent and its effect on clouds, precipitation, and on local and global climates. Deliberate weather modification through cloud seeding pales in comparison to the modifications being accomplished inadvertently, and it may some day have to be employed to offset the unintended modifications (e.g., hygroscopic seeding of clouds to offset the detrimental effects of pollution on rain-forming processes). Many of the past advances in cloud physics technology and insights have been made in the context of cloud seeding experiments. Future progress is likely to be accelerated now that pollution and its probable effect on the global climate has been brought to the attention of the scientific community.

Acknowledgments. We offer a special thanks to Dr. Joanne Simpson for her help and inspiration through the years. We have seen time and again in our trek through life how much difference just one person can make in the lives of others. Joanne was that person in our professional lives. Because words often fail us, as in this acknowledgment, we have produced this paper in her honor in the hope that it will provide her some excitement and new ways to look at the meteorological problems that have intrigued her during her distinguished career.

This study was partially funded by the Israeli Space Agency.

APPENDIX

Glossary

AVHRR: Advanced Very High Resolution Radiometer onboard the NOAA satellites
CCN: cloud condensation nuclei
EDOP: NASA's ER2 Doppler radar
ELDORA: NCAR Electra Doppler Radar
HIPLEX: High Plains Cooperative Experiment
LIS: TRMM Lightning Imaging Sensor
Meteosat: European geostationary meteorological satellite
METROMEX: Metropolitan Meteorological Experiment
MODIS: Moderate Resolution Imaging Spectroradiometer
MSG: METEOSAT Second Generation
NASA: National Aeronautics and Space Administration
NCAR: National Center for Atmospheric Research
NOAA: National Oceanic and Atmospheric Administration
PR: TRMM Precipitation Radar
TOMS: Total Ozone Mapping Spectrometer
VIRS: TRMM Visible and Infrared Sensor

REFERENCES

Arking, A., and J. D. Childs, 1985: Retrieval of cloud cover parameters from multispectral satellite images. *J. Climate Appl. Meteor.,* **24** (4), 322–333.

Bigg, E. K., 1953: The formation of atmospheric ice crystals by the freezing of droplets. *Quart. J. Roy. Meteor. Soc.,* **79**, 510–519.

Biswas, K. R., and A. S. Dennis, 1971: Formation of rain shower by salt seeding. *J. Appl. Meteor.,* **10**, 780–784.

Bowen, E. G., 1952: A new method of stimulating convective clouds to produce rain and hail. *Quart. J. Roy. Meteor. Soc.,* **78**, 37–45.

Braham, R. R., Jr., 1964: What is the role of ice in summer rain showers? *J. Atmos. Sci.,* **21**, 640–645.

——, 1981: Summary of urban effects on clouds and rain. *METROMEX: A Review and Summary, Meteor. Monogr.,* No. 40, Amer. Meteor. Soc., 141–152.

Cerveny, R. S., and R. C. Balling Jr., 1998: Weekly cycles of air pollutants, precipitation and tropical storm intensity in the coastal NW Atlantic region (Letter to Nature). *Nature,* **394**, 561–563.

Coakley, J. A., R. L. Bernstein, and P. R. Durkee, 1987: Effects of ship-stack effluents on cloud reflectivity. *Science,* **237**, 1020–1022.

Cooper, W. A., R. T. Bruintjes, and G. K. Mather, 1997: Calculations pertaining to hygroscopic seeding with flares. *J. Appl. Meteor.,* **36**, 1449–1469.

Cotton, W. R., 1982: Modification of precipitation from warm clouds—A review. *Bull. Amer. Meteor. Soc.,* **63**, 146–160.

Eagen, R. C., P. V. Hobbs, and L. F. Radke, 1974: Particle emissions from a large Kraft paper mill and their effects on the microstructure of warm clouds. *J. Appl. Meteor.,* **13**, 535–552.

Gatz, D. F., 1979: Investigation of pollutant source strength rainfall relationships at St. Louis. *J. Appl. Meteor.,* **18**, 1245–1251.

Gerber, H., 1996: Microphysics of marine stratocumulus clouds with two drizzle modes. *J. Atmos. Sci.,* **53**, 1649–1662.

Gribbin, J., 1995: Rain moves north in the global greenhouse. *New Sci.,* 18.

Gunn, R., and B. B. Phillips, 1957: An experimental investigation of the effect of air pollution on the initiation of rain. *J. Meteor.,* **14**, 272–280.

Houghton, J. T., L. G. Meira Filho, J. Bruce, Hoesung Lee, B. A. Callander, E. Haites, N. Harris, and K. Maskell, 1994: *Climate Change 1994—Radiative forcing of climate change and an Evaluation of the IPCC IS92 Emission Scenarios. Reports of Working Groups I and II of the Intergovernmental Panel on Climate Change.* Cambridge University Press, 339 pp.

Johnson, D. B., 1982: Role of giant and ultragiant aerosol particles in warm rain initiation. *J. Atmos. Sci.,* **39,** 448–460.

——, 1987: On the relative efficiency of coalescence and riming. *J. Atmos. Sci.,* **44,** 1671–1680.

Kaufman, Y. J., and R. S. Fraser, 1997: The effect of smoke particles on clouds and climate forcing. *Science,* **277,** 1636–1638.

Lensky, I. M., and D. Rosenfeld, 1997: Estimation of precipitation area and rain intensity based on the microphisical properties retrieved from NOAA AVHRR data. *J. Appl. Meteor.,* **36,** 234–242.

Mather, G. K., D. E. Terblanche, F. E. Steffens, and L. Fletcher, 1997: Results of the South African cloud seeding experiments using hygroscopic flares. *J. Appl. Meteor.,* **36,** 1433–1447.

McCollum, J. A., A. Gruber, and M. B. Ba, 2000: Discrepancy between gauges and satellite estimates of rainfall in equatorial Africa. *J. Appl. Meteor.,* **39,** 666–679.

Nakajima, T., and M. D. King, 1990: Determination of the optical thickness and effective particle radius of clouds from reflected solar radiation measurements. Part I: Theory. *J. Atmos. Sci.,* **47,** 1878–1893.

Orville, H. D., 1996: A review of cloud modeling in weather modification. *Bull. Amer. Meteor. Soc.,* **77,** 1535–1555.

Pinsky, M. B., A. P. Khain, D. Rosenfeld, and A. Pokrovsky, 1998: Comparison of collision velocity differences of drops and graupel particles in a very turbulent cloud. *Atmos. Res.,* **49,** 99–113.

Radke, L. F., J. A. Coakley, and M. D. King, 1989: Direct and remote sensing observations of the effects of ships on clouds. *Science,* **246,** 1146–1149.

Reisin, T., S. Tzivion, and Z. Levin, 1996: Seeding convective clouds with ice nuclei or hygroscopic particles: A numerical study using a model with detailed microphysics. *J. Appl. Meteor.,* **35,** 1416–1434.

Rosenfeld, D., 1999: TRMM observed first direct evidence of smoke from forest fires inhibiting rainfall. *Geophys. Res. Lett.,* **26,** 3105.

——, 2000: Suppression of rain and snow by urban and industrial air pollution. *Science,* **287,** 1793–1796.

——, and W. L. Woodley, 1993: Effects of cloud seeding in west Texas: Additional results and new insights. *J. Appl. Meteor.,* **32,** 1848–1866.

——, and G. Gutman, 1994: Retrieving microphysical properties near the tops of potential rain clouds by multispectral analysis of AVHRR data. *J. Atmos. Res.,* **34,** 259–283.

——, and M. I. Lensky, 1998: Space-borne based insights into precipitation formation processes in continental and maritime convective clouds. *Bull. Amer. Meteor. Soc.,* **79,** 2457–2476.

——, and W. L. Woodley, 2000: Convective clouds with sustained highly supercooled liquid water down to −37.5°C. *Nature,* **405,** 440–442.

——, D. B. Wolff, and D. Atlas, 1993: General probability-matched relations between radar reflectivity and rain rate. *J. Appl. Meteor.,* **32,** 50–72.

——, W. L. Woodley, and T. Krauss, 2001: Satellite Observations of the microstructure of natural and seeded severe hailstorms in Argentina and Alberta. Preprints, *15th Conf. on Planned and Inadvertent Weather Modification,* Albuquerque, NM, Amer. Meteor. Soc., 68–74.

Schickel, K. P., H. E. Hoffmann, and K. T. Kriebel, 1994: Identification of icing water clouds by NOAA AVHRR satellite data. *Atmos. Res.,* **34** (1–4), 177–183.

Simpson, J., 1980: Downdraft as linkages in dynamic cumulus seeding effects. *J. Appl. Meteor.,* **19,** 477–487.

——, and W. L. Woodley, 1971: Seeding cumulus in Florida: New 1970 results. *Science,* **173,** 117–126.

——, G. W. Brier, and R. H. Simpson, 1967: Stormfury cumulus seeding experiment 1965: Statistical analysis and main results. *J. Atmos. Sci.,* **24,** 508–521.

Squires, P., 1958: The microstructure and colloidal stability of warm clouds. *Tellus,* **10,** 256–271.

Twomey, S., R. Gall, and M. Leuthold, 1987: Pollution and cloud reflectance. *Bound.-Layer Meteor.,* **41,** 335–348.

Witt, G., and J. Malkus, 1959: The evolution of a convective element: A numerical calculation. *The Atmosphere and the Sea in Motion,* R. Bolin, Ed., Oxford University Press, 425–439.

Woodley, W., and A. Herndon, 1970: A raingauge evaluation of the Miami reflectivity–rainfall rate relation. *J. Appl. Meteor.,* **9,** 258–264.

——, J. Jordan, J. Simpson, R. Biondini, J. A. Flueck, and A. Barnston, 1982: Rainfall results of the Florida Area Cumulus Experiment. *J. Appl. Meteor.,* **21,** 139–164.

——, A. G. Barnston, J. A. Flueck, and R. Biondini, 1983: The Florida Area Cumulus Experiment's second phase (FACE-2). Part II: Replicated and confirmatory analyses. *J. Appl. Meteor.,* **22,** 1529–1540.

——, D. Rosenfeld, and A. Strautins, 2000: Identification of a seeding signature in Texas using multi-spectral satellite imagery. *J. Wea. Mod.,* **32,** 37–51.

Zipser, J. E., and K. R. Lutz, 1994: The vertical profile of radar reflectivity of convective cells: A strong indicator of storm intensity and lightning probability? *Mon. Wea. Rev.,* **122,** 1751–1759.

Chapter 7

Isotopic Variations and Internal Storm Dynamics in the Amazon Basin

ISABELLA ANGELINI, MICHAEL GARSTANG, STEPHEN MACKO, AND ROBERT SWAP

Department of Environmental Sciences, University of Virginia, Charlottesville, Virginia

DEREK STEWART

Department of Physics, University of Virginia, Charlottesville, Virginia

HILLÂNDIA B. CUNHA

Instituto Nacional Pesquisas da Amazônia, Manaus, Brazil

ABSTRACT

Rainwater samples taken every 10 min, protected from fractionation by a hydrocarbon layer and collected every 12 h, are subjected to isotopic analyses to obtain a time series of oxygen and deuterium values through successive rain events in the eastern and central Amazon basin. Satellite imagery is used to characterize the rain events, and rain rates from recording rain gauges are used to delineate changes in internal rain production within each storm.

Three clear isotopic signals are seen in the storm systems examined. These three responses consist of depletion of heavy isotopes by as much as −6.7‰ in a single storm, depletion followed by enrichment, and little change in the isotopic signal. Each of these changes in isotopic content of the rainwater can be related to the internal rain-rate production, evaporation/condensation processes together with the implied convective/stratiform circulations of the storm. The storm-related isotopic results suggest, in addition to illuminating the internal dynamics of these storm systems, that sampling of rain from any given rain-producing system can yield significantly different isotopic values. Conclusions about the large-scale hydrologic cycle and the sources and pathways followed by water contained within rain must take these internal storm variations in isotopic values into account.

1. Introduction

Stable isotope geochemistry has proven a valuable tool in determining pathways in hydrological cycles for a wide variety of environments (Ingraham et al. 1991; Friedman et al. 1992). Water isotopes have been used previously to estimate the amount of recycling in river basins (Begemann and Libby 1957; Salati et al. 1979; Victoria et al. 1991). These geochemical techniques have allowed for the establishment of key relationships between groundwater and precipitation.

Three stable isotopes of oxygen and two stable isotopes of hydrogen and water molecules yield nine different mass combinations. Slight differences in mass lead to fractionation during mass-dependent physical processes (Faure 1986). During evaporation, small differences in molecular mass enrich the remaining water in oxygen (^{18}O) and deuterium (D) relative to the original liquid. Drops falling as rain after being subjected to evaporation are enriched in ^{18}O and D. On a global scale, these small isotopic variations leave unique fingerprints on water from different regions.

Meteoric water samples are described in terms of the water reference Vienna Standard Mean Ocean Water (VSMOW). Generally, continental precipitation and freshwater are relatively depleted in heavier isotopes in comparison to oceanic water (Hoefs 1987). The relative degree of depletion is expressed in terms of the δ value for a water sample:

$$\delta X = \left(\frac{R_{sample} - R_{VSMOW}}{R_{VSMOW}} \right) \times 10^3,$$

where X is the heavy isotope of the element under study, R is the isotope ratio, and VSMOW is the standard isotopic ratio for oceanic water (Faure 1986).

Traditionally, differences in the isotopic composition of meteoric samples from various geographic locations have been explained in terms of the Rayleigh distillation model (Hoefs 1987; Faure 1986; Dansgaard 1964). While this model presents a simplistic representation of storm dynamics, it has been able to explain a wide variety of global meteoric isotope patterns (Dansgaard 1964). The Rayleigh distillation model considers a storm as consisting of an isolated reservoir of water vapor (Faure 1986). There is no inflow from the surrounding atmosphere or ground-based water vapor. Although the difference in mass between the isotopes is slight, precipitation from this reservoir will preferentially select the heavier isotopes. To a ground-based observer, the initial precipitation will be enriched in the heavy isotopes (D and ^{18}O). As the storm progresses, the concentration of heavy isotopes remaining in the storm water vapor will decrease. This results in the isotopic concentration of the resulting precipitation becoming more depleted as time progresses.

This model has found success in explaining average meteoric isotope values on a global scale. Meteoric water becomes more depleted in heavy isotopes as the sampling location moves further from the coast in a *continental* effect (Hoefs 1987). The Rayleigh distillation model explains this as a preferential rainout of heavy isotopes as storms move inland. This selection of heavy isotopes for precipitation also leads to more isotopically depleted meteoric samples at the poles than the equator.

While the Rayleigh model can explain global trends in the isotopic distribution of precipitation, the suitability for analysis of single storm events must be addressed. It has been noted that the Rayleigh model neglects fractionation processes that occur as rain falls (Gedzelman and Arnold 1994). Evaporation during descent and isotopic exchange with surrounding water vapor can affect isotopic signatures. This leads to the Rayleigh distillation model overestimating the isotopic depletion of meteoric water. In geographic locations with high humidity such as the Amazon basin, isotopic exchange with atmospheric water vapor can have significant effects on the isotopic composition of sampled meteoric water. Water vapor is most enriched near the earth's surface and becomes exponentially more depleted with altitude (Araguás-Araguás et al. 2000). The presence of this effective reservoir of enriched water vapor under the cloud base can lead to initial enrichment of precipitation through isotopic exchange. In general, the removal of heavy isotopes from the cloud and atmospheric vapor results in the isotopic composition of meteoric water becoming more depleted with increasing rainfall totals in a phenomenon termed the *amount effect* (Gedzelman and Arnold 1994; Dansgaard 1964).

The primary application of this isotopic technique has been to analyze bulk precipitation consisting of composite samples of meteoric water from different storms for the span of a week or more (Victoria et al. 1991; Salati et al. 1979). Although these samples provide an average isotopic composition for a particular region, they are unable to delineate contributions from single storm events. These samples are also subject to physical processes such as fractionation and contamination during the long periods between sampling. Post-collection fractionation can lead to inaccurate isotopic compositions and can significantly influence the interpretation of the hydrological cycle in a region. In addition, from a meteorological standpoint, these isotopic samples provide us with little information on the internal dynamics of individual storm events. The bulk sampling technique commonly used is simply too coarse to provide higher-resolution information.

Relatively high-resolution sampling of rainwater is used in this study to provide a window into the internal processes that drive storm events. Temporal changes in isotopic composition of rainwater can then be related to both differences in storm types and internal differences within a given storm. In this study, we report on a sampling campaign conducted in 1999 in the Amazon River basin as part of the Tropical Rainfall Measuring Mission Large-Scale Biosphere Atmosphere (TRMM LBA) field program. Sequential samplers in combination with tipping buckets were used at three sites along the Amazon River to provide high time resolution data on rain rates and isotopic composition of individual storms. These ground-based observations were complemented by satellite imagery of the area, which was used to classify and track these individual storms in the basin. This project is the first of its kind to sample and categorize storm systems based on intrastorm isotopic analysis over an entire ecological region.

Previous research

Several researchers have examined the isotopic variation of a single storm event by collecting high time resolution samples. However, there has been no comprehensive effort to classify storms in terms of isotopic variations. Previous studies found that the isotopic composition in an individual storm event can vary significantly. Many storms show a characteristic "v-shaped" curve with δ values dropping after the beginning of the storm to a minimum in the middle of the rain event and subsequently recovering to the initial values (Rindsberger et al. 1990; Schirmer 1995). Short showers are characterized by relatively constant isotopic composition (Dansgaard 1953; Matsuo and Friedman 1967). Periods of intense rainfall have also shown some correlation to very depleted rain isotopic values (Ehhalt et al. 1963). The highest sampling rate in these studies was one sample every 10 min for a single storm in Heidelberg, Germany (Ehhalt et al. 1963).

Analysis of sequential samples from individual storm events has also been conducted on one occasion in Manaus in the central Amazon River basin (Matsui et al. 1983). This study provided data on the fluctuation of deuterium concentrations within four isolated rain events. However, this work provided no discussion or model for the underlying physical processes responsible for the observed variations.

Gedzelman and Arnold (1994) developed a two-dimensional, kinematic, bulk cloud microphysical model that incorporated stable isotope physics. This model predicts significant depletion as a function of increasing rainfall. The model also examined the depleted isotopes found in hurricanes and thunderstorms and related them to internal recirculation processes.

During the Atmospheric Boundary Layer Experiment (ABLE) 2A and 2B expeditions to the Amazon River basin, storm events were placed into three main categories: coastal occurring systems (COS), basin occurring systems (BOS), and local occurring systems (LOS) (Greco et al. 1990; Swap 1990). The COS structures, as seen in Fig. 7.1a, originate in the sea-breeze convergence zone on the northeast coast of Brazil and propagate into the region as squall lines (Cohen et al. 1995). The BOS structures are large slow-moving mesosynoptic-scale systems with multiple regions of convective activity (Fig. 7.1b). The LOS systems consist of small local convective systems (Fig. 7.1c). These storms have significantly different characteristics both in terms of size and speed of propagation. On the basis of these previously demonstrated differences in their physical characteristics and areas of origin, we expected to see clear differences in isotopic signatures and variations of the three storm types.

2. Methodology

The study was designed to have a sampling system that could span the wide range in length scales necessary to characterize storms in the Amazon River basin. A nested three-tiered system consisting of satellite imagery, tipping bucket data loggers, and sequential isotopic sampling of rainwater was used. This methodology allowed for a synoptic overview of storm activity along with detailed synchronized characterization of storm nature and intrastorm processes.

a. Site description

Data and samples were collected during the wet season (18 Jan–18 Feb 1999) at three sites in Brazil along 1400 km of the Amazon River. Belém, situated at the mouth of the Amazon, provided water samples representative of incoming oceanic water vapor flux. Santarém, located 800 km from the coast, served as the middle sampling site. The final site was located in Manaus, 1400 km from the coast, near the confluence of the Rio Negro and the Rio Solimões. The total change in elevation from the coast to the central Amazon basin station is less than 100 m. The sampling site at each location was chosen to provide consistency and sampling representative of the region. Sites in clearings well removed from trees and buildings were chosen to avoid any local contamination. However, equipment and sampling difficulties effectively eliminated Santarém from this analysis. The results presented here are, therefore, based entirely upon measurements made at Belém and Manaus. The sampling problem at Santarém stems, in large part, from the fact that the westward-propagating coastal-originating squall lines decay during the night reaching a nocturnal minimum in the vicinity of San-

tarém. Sampling depends more critically at Santarém than at the other two locations on the precise location and timing of the precipitation features relative to the location of the sampling site.

b. Satellite imagery

Infrared data from the imager on the *Geostationary Operational Environmental Satellite 8* (*GOES-8*) was used for storm classification and system point of origin. Images were collected courtesy of the TRMM LBA project and were provided by the Goddard Mesoscale Processes TRMM Division (T. Richenbach 2001, personal communication). This database of satellite images provided the backdrop for analysis of the sequential samples in terms of storm type and origin.

Satellite imagery of rain events at the two sites were analyzed and sorted in terms of the ABLE 2A and 2B characterization method (Greco et al. 1990). This approach categorizes storms based on size, shape, and point of origin (Figs. 7.1a–c). Analysis of satellite imagery and coincident ground-based pluviometer data were used in charaterizing rain-bearing storms from our sampling period. As a check, times of occurrence for the different storm types in this study were compared with the results from previous work (Greco et al. 1990) and found to be in general agreement.

c. Tipping buckets

Tipping buckets with Starlogger dataloggers were used at each of the three sites. These gauges provided detailed temporal information on precipitation in the region, with an accuracy of 0.25 mm min^{-1} and time resolution of 1 s. These gauges provided vital rain-rate information that was used in connection with isotopic variations to examine internal storm dynamics.

d. Sequential sampler

The sequential sampler formed the core of the suite of sampling techniques and provided the water samples necessary for isotopic analysis (Fig. 7.2). The apparatus was a modification of an automatic sequential rainfall sampler created by Gray et al. (1974). In order to prevent the evaporation of collected rainwater, which would compromise the original isotopic signature, each collection tube contained a layer of mineral oil roughly 10 mm in height. The use of a hydrocarbon layer to prevent the isotopic fractionation of hydrologic samples is a standard technique that has been used by a number of researchers (Gray et al. 1974; Adar and Long 1987; Cortes and Farvolden 1989; Friedman et al. 1992; Scholl et al. 1996). Precipitation samples were taken at 10-min increments for a 24-h period to cover the sampling of every rain event. Sample collections occurred every 12 h, where each sample was placed directly into a 20-ml borosilicate glass scintillation vial with a poly cone-

(a)

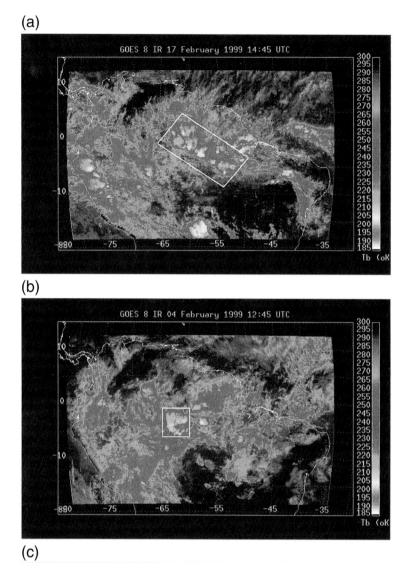

(b)

(c)

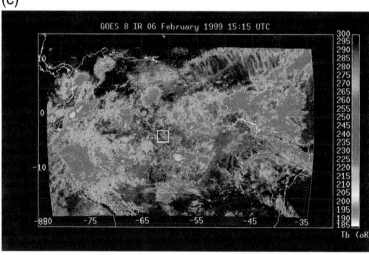

FIG. 7.1. (a) Example of COS highlighted in white box, 1445 UTC 17 Feb 1999;
(b) example of BOS highlighted in white box, 1245 UTC 4 Feb 1999; (c) example of
LOS highlighted in white box, 1515 UTC 6 Feb 1999.

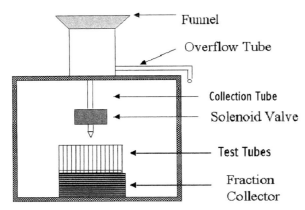

Fwnel

Overflow Tube

Collection Tube

Solenoid Valve

Test Tubes

Fraction Collector

FIG. 7.2. Diagram of sequential sampler and components.

shaped liner cap for transport to the United States. As an extra preventive measure, each bottle cap was wrapped in electrical tape to reduce the potential for evaporation and air exchange.

e. The ^{18}O analysis

Isotopic ratios of ^{18}O were examined using a Micromass VG Isogas Prism triple-collector mass spectrometer. Oxygen isotope concentrations were determined using the Epstein and Mayeda (1953) equilibrium/extraction method. The procedure reproducibility for the oxygen analysis was typically better than $\pm 0.2\%$; sample data are reported relative to the standard VSMOW.

3. Results

Three general isotopic patterns were identified during the study. Large organized storm events generally possessed much more variation in isotopic composition than small events. The isotopic pattern in these cases could be classified either as a v-shaped curve (Figs. 7.3a,b) or a gradual descent to more depleted isotopic δ values (Fig. 7.4a). These patterns were found in both COS and BOS structures. The isotopic content of small and medium LOS systems stayed constant (Figs. 7.5a,b).

a. Intrastorm isotopic variations for large-scale storms

Several large and medium storms showed a gradual drop in isotopic values. Many BOS systems followed this pattern toward more depleted isotopic values as time progressed. A 4 February 1999 storm in Manaus is representative (Fig. 7.4a) and shows a dramatic drop toward more depleted isotopic concentrations throughout the duration of the storm.

Many large organized storms are characterized by a v-shaped curve in isotopic composition. This trend can be seen in two COS systems over Belém and Manaus (Figs. 7.3a,b). These systems consist of a concentrated convective region at the leading edge of the squall line

followed by a trailing stratiform region that is usually multilayered. Examining the rain rate during the storm event supports this characterization. Precipitation in the Amazon River basin can be classified in terms of rain rate (Garstang et al. 1988; Garstang and Fitzjarrald 1999).

Table 7.1 summarizes the rain rates, total amount of rain, and the duration of the rain for each of the six storms shown in Figs. 7.3–7.5. A threshold of 0.50 mm min^{-1} is used to separate convective from stratiform rainfall. The structure and associated rain rates in the large Amazon basin rain systems would appear to differ significantly from those described by Houze and Zipser over both the tropical oceans and temperate land areas (Houze 1977, 1993; Leary and Houze 1979; Gamache and Houze 1982, 1983, 1985; Wei and Houze 1987; Zipser 1969). Three distinct regions are present within the Amazon basin systems: a convective region with rain rates that can exceed 0.75 mm min^{-1}, a multilayered stratiform region with rain rates that can reach 0.50 mm min^{-1}, and a single-layered stratiform region with rain rates generally less than 0.25 mm min^{-1} (Garstang et al. 1988; Greco et al. 1990, 1994). The percentage of the total system rainfall ascribed to the convective class ranges between 40% and 70%, multilayered region between 10% and 20%, and in the single-layered stratiform region between 20% and 50%.

Tokay and Short (1996) and Tokay et al. (1999), using distrometers and profiler data from rain systems in the far western Pacific (warm pool), found essentially the same partitioning between convective and stratiform rain. They describe a four-category system consisting of shallow convection, deep convection, mixed convective–stratiform, and stratiform. Their mixed convective–stratiform corresponds to our multilayered stratiform region. For the maritime systems examined by Tokay et al. (1999), all rain rates greater than 10 mm h^{-1} (0.17 mm min^{-1}) were considered to be convective.

High variability in both time and space of the rainfall is observed so that any selected rain rate can only be regarded as a guide to a general upper or lower limit. The rain rates shown in Figs. 7.3 and 7.4 suggest an upper limit of 0.25 mm min^{-1} as a division between the convective and stratiform regions of the storm. Even so, this limit is exceeded for short times in both the 23 January and 4 February (Figs. 7.3b, 7.4a) storms. Rain rates in much of the stratiform regions of all the storms are, however, much less than 0.25 mm min^{-1}.

The isotopic concentration for the 2 February COS system also undergoes a significant transition as the event progresses (Fig. 7.3a). The initial precipitation samples contained the highest δ values observed for the system. Twenty minutes into the storm event, the isotopic composition drops dramatically to more depleted values. Over the next 200 min, the isotopic composition gradually rises to values slightly lower than those found at the initiation of the event. Fluctuations in the isotopic composition can also be seen during

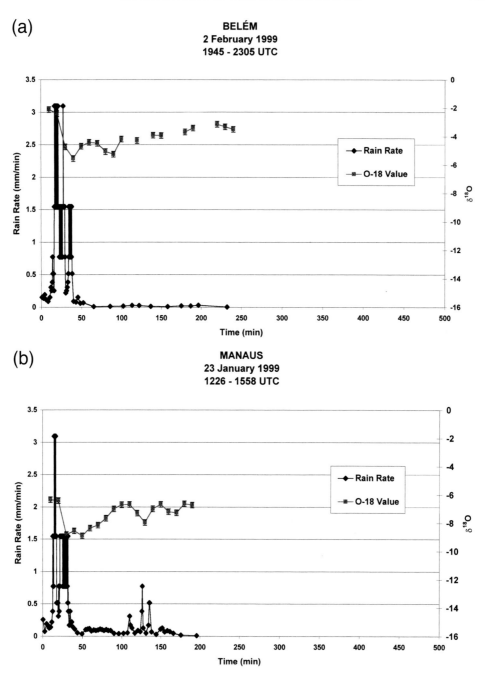

FIG. 7.3. (a) Intrastorm [18]O and rain-rate variation of a large organized COS at Belém, 2 Feb 1999; (b) intrastorm [18]O and rain-rate variation of a large organized COS at Manaus, 23 Jan 1999.

this rise and appear to coincide with the small increases in rain rate later in the storm (Fig. 7.3b). These smaller fluctuations in isotopic composition vary from storm to storm; the common characteristic of several storms is the initial dip in isotopic composition and subsequent rise to more enriched isotopic values. While this v-shaped pattern was observed primarily in COS systems, it was also observed in one of the BOS systems measured.

b. Intrastorm isotopic variations for local unorganized storms

The small and medium unorganized LOS storms measured during the study show a much different isotopic temporal evolution. These storms (Figs. 7.5a,b) are characterized by relatively constant isotopic composition. Analysis of the temporal frequency of storms each day indicates that a significant portion of the small storms

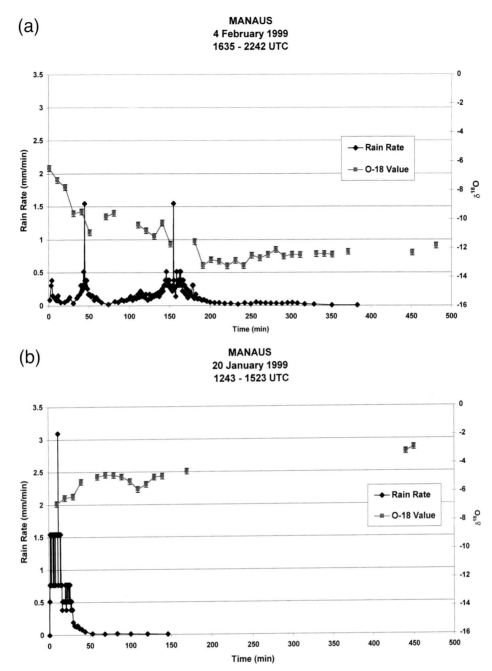

FIG. 7.4. (a) Intrastorm [18]O and rain-rate variation of a large organized BOS at Manaus, 4 Feb 1999; (b) intrastorm [18]O and rain-rate variation of a large organized BOS at Manaus, 20 Jan 1999.

form in the late afternoon and early evening hours. This suggests that these small storms are a product of local heating and evaporation as opposed to larger-scale (mesoscale to synoptic scale) organization. The rain rates in these storms are generally much lower than those found in large systems from the same region. In our study, the maximum rain rate for a small storm in Belém was 0.52 mm min^{-1}, while a large storm from the same site had rain rates that reached 3.09 mm min^{-1}.

4. Discussion

The isotopic patterns observed show that there are clear trends in associations between storm type and isotopic variations. While general trends occur, there are several factors that can lead to storm variability. These systems possess life cycles that can affect internal processes. A mature squall line will have dramatically different rain rates and isotopic variations than a decaying

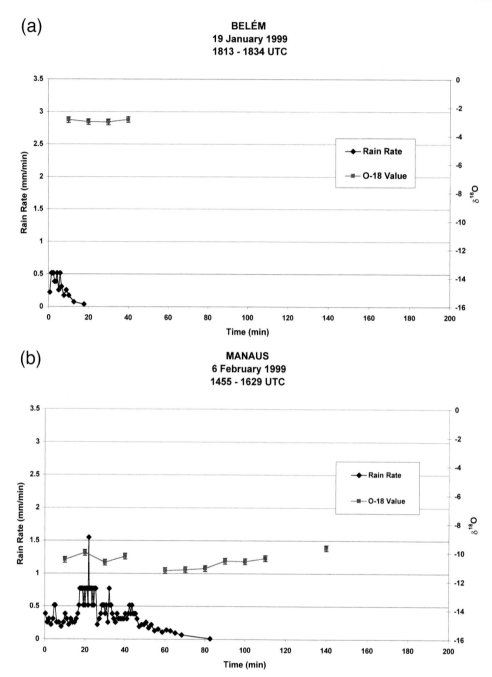

FIG. 7.5. (a) Intrastorm [18]O and rain-rate variation of a small unorganized LOS at Belém, 19 Jan 1999; (b) intrastorm [18]O and rain-rate variation of a small unorganized LOS at Manaus, 6 Feb 1999.

storm. Since BOS structures consist of multiple convective cells embedded in a stratiform region, rain rates and isotopic variations are also dependent on the direction of the storm and the relative position of the sampling site. An interpretation and the possible use of these observations to measure storm efficiency and intrastorm transport follows below.

a. Coastal squall lines

The classical Rayleigh distillation model (Fig. 7.6) predicts that condensation during the storm initiation will preferentially select heavier isotopes for water droplet formation (Faure 1986). This leads to more enriched δ values at storm commencement. In the absence of

TABLE 7.1. Rain rate, amounts, and duration for the six storms shown in Figs. 7.3–7.5.

	Convective			Stratiform			Storm		
	Average rate (mm min⁻¹)	Total amt. (mm)	Total time (min)	Average rate (mm min⁻¹)	Total amt. (mm)	Total time (min)	Average rate (mm min⁻¹)	Total amt. (mm)	Total time (min)
Belém 2 Feb 1999	1.582	25.8	18.5	0.176	8.77	213.7	1.19	34.57	232.2
Manaus 23 Jan 1999	1.466	20.89	16.8	0.179	16.25	178.5	0.883	37.15	195.3
Manaus 4 Feb 1999	1.554	0.52	1	0.189	34.57	381.7	0.21	35.09	382.7
Manaus 20 Jan 1999	1.345	21.93	18.6	0.315	8.51	126.9	1.05	30.44	145.5
Belém 19 Jan 1999	0	0	0	0.324	3.87	17.7	0.324	3.87	17.7
Manaus 6 Feb 1999	0.815	4.90	7	0.327	17.54	75.8	0.433	22.44	82.8

mass convergence of water vapor containing heavy isotopes, the ¹⁸O composition within the region of cloud decreases and the precipitation samples gradually become more depleted with time. This model explains the initial drop to more depleted values in coastal squall lines but provides no mechanisms to explain the later rise to more enriched values. The Rayleigh distillation model alone is unable to explain the isotopic evolution of these large storms. Additional mechanisms capable of producing the subsequent trend toward enriched precipitation samples must be present.

One possible explanation for the v-shaped curves in isotopic composition could lie in the structure of the overall storm system. Ground-based and aerial observations of these large synoptic systems show common characteristics (Houze 1977, 1989; Garstang et al. 1988, 1994). Fuel-rich air in terms of high moisture and sensible heat (high equivalent potential temperature θ_e) in the boundary layer and lower troposphere (<4 km) enters the storm over the leading edge outflow boundary. Storm-relative inflow of drier, colder air (low θ_e) descends in the storm forming the cold outflow of the gust front and the low θ_e air now occupying the boundary

layer in the wake of the storm (Garstang and Fitzjarrald 1999, p. 299). The relative balance of high θ_e inflow and low θ_e outflow maintains or signals the demise of the storm. In the convective region, cloud-base altitudes may be well below 1 km. In the stratiform region, cloud-base altitudes range from <2 km in the multilayered region to >10 km in the single-layered region. The stratiform region at high altitudes contains much more of its precipitable water in the form of ice crystals than in the convective region (Tao and Simpson 1989).

The time required for the growth of precipitation-size particles (ice as well as water) in convective clouds is much less than in stratiform clouds (Tokay and Short 1996). Vertical velocities in stratiform regions are less than the terminal velocities of ice crystals and snow (1–3 m s⁻¹). Ice crystal growth is by aggregation and riming. In the convective region, drop growth is primarily by accretion (differential fall velocities, collisions, and coalescence) and by fracturing. Significant differences in isotopic fractionation can occur in these two different regions of precipitation formation (Stewart 1975; Miyake et al. 1968; Ehhalt et al. 1963; Kinzer and Gunn 1951). As the ice crystal or drop descends, it is subject to evaporation as it passes through layers of drier cloud-free air (<100% RH) (Stewart 1975; Araguás-Araguás et al. 2000). Evaporation preferentially selects lighter isotopes to form water vapor, leaving the droplet more enriched in ¹⁸O. The degree of enrichment depends on the raindrop descent time and the number of cloud layers and the heights of these clouds within the stratiform region (Miyake et al. 1968).

The evolution of drop radius during a typical large synoptic event can also have important effects on the isotopic composition of the meteoric water sampled. Since evaporation of these drops can be treated as a diffusion problem, the ratio of drop surface area to volume is directly related to the rate of evaporation for a drop (Kinzer and Gunn 1951). Miyake et al. (1968) found that as raindrop size increased, both the fraction of the total water vapor evaporated and the amount of isotopic en-

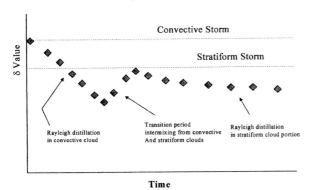

FIG. 7.6. Illustration of two-component Rayleigh model.

richment in the raindrop decreased. Gedzelman and Arnold (1994) found that the rate of change of the isotope ratio of rain, R_r, due to diffusion was inversely proportional to the square of the drop radius:

$$\frac{dR_r}{dt} = \frac{\gamma}{a^2}\left[S\left(R_v - \frac{R_r}{\alpha_w}\right) + \frac{1-S}{1+b}R_r\left(D_{rat} - \frac{1}{\alpha_w}\right)\right],$$

where R_v is the isotope ratio of the surrounding water vapor, α_w is the fractionation factor for water, S is the saturation ratio, b denotes a thermal inertia to evaporation, and

$$\gamma = \frac{3f\Psi'\rho_s}{\rho_{H_2O}},$$

where f is the ventilation factor, ρ_s is the saturated vapor density, and Ψ' is the diffusion coefficient. The prime denotes the quantity for the heavy isotope. The final variable, D_{rat}, is a ratio of product of ventilation factors and diffusion coefficients for the light and heavy isotopes. Since raindrop diameter increases with rain rate in storms (Miyake et al. 1968), raindrops in periods of high rain rate experience less evaporation than those found in periods of low rain rate. Woodcock and Friedman (1963) found such an inverse relationship between drop size and deuterium content in samples from Hawaiian storms. Tokay and Short (1996), however, found for tropical maritime systems in the far western Pacific that the average drop size distributions of stratiform and convective systems for the same rain rate shows relatively more larger drops in stratiform rain. Their results (Tokay and Short 1996, Table 3, p. 309) show that this tendency is reversed at the highest rain rates. The structure of large synoptic systems in the Amazon basin is, however, characterized by a marked transition from high rain rates in convective cells to lower rain rates in the stratiform region (Garstang et al. 1988; Garstang and Fitzjarrald 1999). Very high rain rates and, hence, a large fraction of larger drops and shorter descent times act to reduce the evaporation effect in the convective region of the storm. Isotopic enrichment of precipitation due to differences in drop size distribution in the stratiform versus the convective region may therefore play an additional role in enriching the rainfall from the stratiform region. The greater effect leading to enrichment in the stratiform region is derived from prolonged and/or successive evaporation during descent.

A complementary process that could elevate the observed isotopic values in the stratiform rain collected at the surface would be restoration within the stratiform region of heavy isotopes associated with air from the source region. If the bulk of the inflow of moisture-rich air supplying the water vapor of these large Amazon basin systems is derived from the boundary layer or lower troposphere (≤ 4 km or approximately 700 hPa), ingestion of this air into the stratiform region of the storm without depletion of the heavy isotopes is not easily envisioned (Swap et al. 1992). It is equally dif-

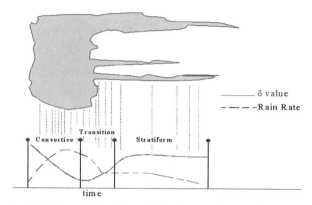

FIG. 7.7. Proposed isotopic model for large-scale synoptic systems.

ficult to call upon any other sources of water vapor that are supplying moisture to the stratiform region that are enhanced rather than depleted of heavy isotopes.

Through the combination of these processes, a model evolves that describes the observed isotopic variations in rainwater (Fig. 7.7). During the first portion of the storm, the convective core sweeps air nearly vertically aloft, resulting in rapid condensation of water vapor. The resulting large rain rates produce rainwater enriched in heavy isotopes. A combination of Rayleigh distillation and high rain rates depletes the isotopic content at the end of the convective phase of the storm. As the stratiform region is entered, the depletion processes are arrested and enrichment begins. Raindrops falling through the stratified layers are enriched from evaporation and equilibrate through isotopic exchange. The resulting meteoric water reflects the enriched isotopic composition of the stratiform rainfall.

This isotopic model corresponds well with the experimental observations of coastal squall lines. During the ABLE 2A and 2B expeditions, radar images of squall lines provided another window into the squall line structure (Garstang et al. 1994). The convective and stratiform regions observed had strong reflectivity signals, but with a region of lower reflectivity between the convective and stratiform structures. This region most likely corresponds to the transition period observed in the present study where the isotopic values climb to more enriched values. Changes in the mixing ratio during the passage of a squall line also support the present isotopic model. During the passage of a coastal squall line over Embrapa on 26 April 1987, there is a significant drop in the mixing ratio after the convective region moves over the sampling site (Garstang et al. 1994). During the period of stratiform rain, the mixing ratio remains low, indicating surface outflow of low and midtroposphere air, a local reduction in humidity, and the possibility of enhanced raindrop evaporation during descent.

b. Basin occurring systems

Due to their large size and slow speeds, BOSs can often spend considerable time situated over a single region. The distribution of convective cells in a stratiform layer results in the rainfall distribution in the region being peaked spatially at locations under the convective cells. Depending on the position of the sampling site in relation to the storm, two isotopic patterns are possible. In the first, the sampling location is situated under a convective cell for the entire span of the rain event. Since the rainfall is highly localized for a long period of time and converts a significant portion of the initial water vapor, the classical Rayleigh distillation model is well suited to model the isotopic patterns. The Rayleigh distillation model prediction of a gradual drop to more depleted isotopic values is observed. The 4 February 1999 storm, which lasted from 1635 to 2242 UTC, is an example (Fig. 7.4a). This storm formed in the basin and produced precipitation in Manaus for 6 h and 21 min. The convective rain lasted for only 1 min while the stratiform rain lasted for 6 h and 20 min. At a mean rainfall rate of 1.56 mm min^{-1} the convective portion of the storm contributed 0.52 mm, and at a mean rainfall rate of 0.189 mm min^{-1} the stratiform portion of the storm contributed 34.57 mm, for a total storm rainfall of 35.09 mm. The isotopic value started at -6.5% and dropped to -11.8%. This was the largest isotopic storm depletion event observed in the study period. The long duration and high amount of isotopic depletion support the hypothesis that although only a small fraction of the convective rain region was sampled, the sampling site was never far removed from either the convective rain region or the multilayered stratiform region.

There were also situations in which the convective cell moved outside of the range of the sampling site to be replaced by the surrounding stratiform region. This transition is reflected in the isotopic record of the rain event. As was discussed in the examination of the COS storm dynamics, stratiform regions should possess precipitation much more enriched than the neighboring convective regions. This trend is observed in the 20 January 1999 Manaus BOS system, where there is a correspondence between rain rates and isotopic composition (Fig. 7.4b). During the initial sampling period, the sampler is under a convective cell with high rain rates and depleted isotopic values. As the convective cell moves away, a stratiform region replaces it. A corresponding climb to more enriched isotopic values and a drop in rain rate serve to mark this transition. Since both the COS and BOS structures are defined in terms of regions of convective and stratiform precipitation, it is not surprising that isotopic variations from these two storm types can, at times, be similar.

c. Local occurring systems

The lack of isotopic variation in small storms (LOS) indicates that there is no significant change in storm structure or composition during such rain events. Manaus, on 6 February 1999, shows only a slight trend toward more depleted values as the storm progresses

(Fig. 7.5b). The Rayleigh distillation model for the isotopic composition of water vapor and the resulting precipitating liquid water is often expressed in terms of the fraction of source water vapor remaining. However, this form of the expression does not show how the isotopic composition will vary explicitly with time. The fraction of cloud water vapor remaining should be related to the inflow of water vapor into the storm and to the rain rate for the storm system. Low rain rates indicate that a significant portion of the initial water vapor still remains irrespective of any change in inflow rates, while high rain rates suggest that much of the initial water vapor has been converted into precipitation and must be replaced by inflow.

This conclusion is strengthened by consideration of storm efficiency. Storm efficiency is defined as the ratio between the amount of water precipitated and the total amount of water available to the storm. Large storms are relatively efficient ($\sim70\%$), while small storms are usually inefficient ($\sim30\%$; Ulanski and Garstang 1978a,b).

The lack of significant depletion due to Rayleigh distillation in the 19 January 1999 Belém storm could be the result of the low rain rates observed in the storm. This results in small changes in the fraction of remaining water vapor and small depletion of the isotopic composition of precipitation samples as well. This study's observations are consistent with the work of other researchers who found that the δ values remained constant throughout small storms (Miyake et al. 1968; Dansgaard 1953). It also indicates that analysis of isotopic variations can provide an important measure of storm efficiency in individual rain events.

5. Conclusions

This study shows that a relationship exists between temporal isotopic variations in rainwater and the composition and structure of individual storm events across the Amazon basin. These isotopic fluctuations provide crucial information on the internal storm dynamics. In the case of the large propagating coastal squall lines observed in the region, changes in isotopic composition can be related to the transition from convective to stratiform regions. For small storms (LOS) generated by daily local solar heating, the lack of isotopic fluctuations during the rain event are indicative of the system's low efficiency. Storm efficiency, speed, and composition are reflected in the resulting isotopic patterns. The low storm efficiency and speed of these local systems results in very little isotopic variations. Large synoptic-scale and mesoscale systems, such as propagating squall lines or large stationary or slow-moving systems, can be composed of distinct regions of convective and stratiform precipitation. The shape and speed of these systems will lead to different isotopic patterns. However, as noted in the case of the v-shaped curve, there are cases where similar isotopic patterns from storms of different origins can be observed. Nevertheless, the large storm systems of the Amazon basin may produce up to 80% of the annual total rainfall, meaning that the characteristics of these storms dominate the hydrologic cycle (Greco et al. 1990).

The observed time-dependent behavior of the isotopic signal in the large rain-producing systems of the Amazon basin suggests that these systems typically consist of discrete regions of intense convective activity followed or surrounded by large regions of multilayered stratiform clouds. Such stratiform regions must be maintained by mesoscale circulations sufficient to support precipitation processes but only at a moderate to light rain rate. Only small raindrops are maintained, which undergo successive evaporation and condensation, resulting in increasing the heavy isotopes in the raindrops and depleting the isotopic content of the remaining water vapor.

Shifts in isotopic composition of meteoric samples as a function of storm size, propagation, and internal dynamics complicate the use of isotopes as tracers for the sources and pathways of water vapor. This has a direct bearing on isotopic sampling in the field. A site that samples portions of both the convective and stratiform portions of a propagating squall line will have a markedly different average isotopic signal than a site that only samples the stratiform region. Assessment of the hydrologic cycle in a region such as the Amazon basin, where evapotranspiration from vegetation may play a major role, is a complex problem in sampling, processing, and analysis. The methods described in this contribution will be applied in a subsequent paper to address the relative role played by the vegetation in the central Amazon basin.

The combination of isotopic analysis of relatively frequent sequential samples with more traditional methods of rain rates and satellite imagery opens up a new avenue of meteorological investigation. Combination of isotopic analysis with trace chemicals in rainwater is a logical future step. Application of these methods to temperate and other tropical regions will advance the understanding of the hydrologic cycle and the part played by water in the global heat and energy budget.

Acknowledgments. This project was supported under the NASA TRMM LBA program. *GOES-8* imagery was provided from the NASA Goddard Space Flight Center archives. The authors wish to thank Dr. E. Cutrim at Western Michigan University and Dr. D. Nechet at the Universidade Federal do Para. The research presented here forms part of the work being conducted by Ms. Isabella Angelini toward her doctoral degree in Environmental Sciences at the University of Virginia.

REFERENCES

Adar, E., and A. Long, 1987: Oxygen-18 and deuterium distribution in rainfall, runoff and groundwater in a small semi-arid basin: The Aravaipa Valley in the Sonora Desert, Arizona. IAEA Rep. SM-299/135, 15 pp.

Araguás-Araguás, L., K. Froehlich, and K. Rozanski, 2000: Deuterium and oxygen-18 isotope composition of precipitation and atmospheric moisture. *Hydrol. Proc.,* **14,** 1341–1355.

Begemann, F., and W. F. Libby, 1957: Continental water balance, ground water inventory, and storage times, surface ocean mixing rates and world-wide circulation patterns from cosmic-ray and bomb tritium. *Geochim. Cosmochim. Acta,* **12,** 277–296.

Cohen, J., M. Dias, and C. Nobre, 1995: Environmental conditions associated with Amazonian squall lines: A case study. *Mon. Wea. Rev.,* **123,** 3163–3174.

Cortes, A., and R. N. Farvolden, 1989: Isotope studies of precipitation and groundwater in the Sierra de Las Cruces, Mexico. *J. Hydrol.,* **107,** 147–153.

Dansgaard, W., 1953: The abundance of ¹⁸O in atmospheric water and water vapor. *Tellus,* **5,** 461–469.

——, 1964: Stable isotopes in precipitation. *Tellus,* **4,** 436–468.

Ehhalt, D., K. Knott, J. F. Nagel, and J. C. Vogel, 1963: Deuterium and oxygen 18 in rain water. *J. Geophys. Res.,* **68,** 3775–3781.

Epstein, S., and T. Mayeda, 1953: Variation of O18 content of waters from natural sources. *Geochim. Cosmochim. Acta,* **4,** 213–224.

Faure, G., 1986: *Principles of Isotope Geology.* John Wiley and Sons, 589 pp.

Friedman, I., G. I. Smith, J. D. Gleason, A. Warde, and J. M. Harris, 1992: Stable isotope composition of waters in southeastern California: 1. Modern precipitation. *J. Geophys. Res.,* **97,** 5795–5812.

Gamache, J. F., and R. A. Houze, 1982: Mesoscale air motions associated with a tropical squall line. *Mon. Wea. Rev.,* **110,** 118–135.

——, and ——, 1983: Water budget of a mesoscale convective system in the tropics. *J. Atmos. Sci.,* **40,** 1835–1850.

——, and ——, 1985: Further analysis of the composite wind and thermodyanic structure of the 12 September GATE squall line. *Mon. Wea. Rev.,* **113,** 1241–1259.

Garstang, M., and D. R. Fitzjarrald, 1999: *Observations of Surface to Atmosphere Interactions in the Tropics.* Oxford University Press Inc., 405 pp.

——, H. Massie, J. Halverson, S. Greco, and J. Scala, 1994: Amazon coastal squall lines. Part I: Structure and kinematics. *Mon. Wea. Rev.,* **122,** 608–622.

——, and Coauthors, 1988: Trace gas exchanges and convective transports over the Amazonian rainforest. *J. Geophys. Res.,* **93,** 1528–1550.

Gedzelman, S. D., and R. Arnold, 1994: Modeling the isotopic composition of precipitation. *J. Geophys. Res.,* **99,** 10 455–10 471.

Gray, J., K. D. Hage, and H. W. Mary, 1974: An automatic sequential rainfall sampler. *Rev. Sci. Instr.,* **45,** 1517–1519.

Greco, S., R. Swap, M. Garstang, S. Ulanski, M. Shipham, R. Harriss, R. Talbot, M. Andreae, and P. Artaxo, 1990: Rainfall and surface kinematic conditions over central Amazonia during ABLE 2B. *J. Geophys. Res.,* **95,** 17 001–17 014.

——, J. Scala, J. Halverson, H. L. Massie, W.-K. Tao, and M. Garstang, 1994: Amazon coastal squall lines. Part II: Heat and moisture transports. *Mon. Wea. Rev.,* **122,** 623–635.

Hoefs, J., 1987: *Stable Isotope Geochemistry.* Springer-Verlag, 241 pp.

Houze, R. A., Jr., 1977: Structure and dynamics of a tropical squall line system. *Mon. Wea. Rev.,* **105,** 1540–1567.

——, 1989: Observed structure of mesoscale convective systems and implications for large-scale heating. *Quart. J. Roy. Meteor. Soc.,* **115,** 425–461.

——, 1993: *Cloud Dynamics.* Academic Press, 573 pp.

Ingraham, N. L., B. F. Lyles, R. L. Jacobson, and J. W. Hess, 1991: Stable isotopic study of precipitation and spring discharge in southern Nevada. *J. Hydrol.,* **125,** 243–258.

Kinzer, G. D., and R. Gunn, 1951: The evaporation, temperature and thermal relaxation time of freely falling waterdrops. *J. Meteor.,* **8,** 71–83.

Leary, C. A., and R. A. Houze Jr., 1979: Melting and evaporation of hydrometeors in precipitation from the anvil clouds of deep tropical convection. *J. Atmos. Sci.,* **36,** 669–679.

Matsui, E., E. Salati, M. N. Ribeiro, C. M. Reis, A. C. Tancredi, and J. R. Gat, 1983: Precipitation in the central Amazon Basin: The isotopic composition of rain and atmospheric moisture at Belém and Manaus. *Acta Amazonica,* **13,** 307–369.

Matsuo, S., and I. Friedman, 1967: Deuterium content in fractionally collected rainwater. *J. Geophys. Res.,* **72,** 6374–6376.

Miyake, Y., O. Matsubaya, and C. Nishihara, 1968: An isotopic study on meteoric precipitation. *Pap. Meteorol. Geophys.,* **19,** 243–266.

Rindsberger, M., Sh. Jaffe, Sh. Rahamim, and J. R. Gat, 1990: Patterns of the isotopic composition of precipitation in time and space: Data from the Israeli Storm Water Collection Program. *Tellus,* **42B,** 263–271.

Salati, E., A. Dall'Olio, E. Matsui, and J. R. Gat, 1979: Recycling of water in the Amazon Basin: An isotopic study. *Water Resour. Res.,* **15,** 1250–1258.

Schirmer, T., 1995: Die Zusammensetzung (D, ¹⁸O und ausgewählte Inhaltsstoffe) von Einzelniederschlagsereignissen Göttingens und Clausthal-Zellerfelds für den Zeitraum vom Mai 93 bis zum Marz 94. Diplomarbeit, University of Göttingen.

Scholl, M. A., S. E. Ingebritsen, C. J. Janik, and J. P. Kauahikaua, 1996: Use of precipitation and groundwater isotopes to interpret regional hydrology on a tropical volcanic island: Kilauea Volcano Area, Hawaii. *Water Resour. Res.,* **32,** 3525–3537.

Stewart, M. K., 1975: Stable isotope fractionation due to evaporation and isotopic exchange of falling waterdrops: Applications to atmospheric processes and evaporation of lakes. *J. Geophys. Res.,* **80,** 1133–1146.

Swap, R. J., August 1990: The nature and origin of central Amazonian wet season rainfall. M.S. thesis, Department of Environmental Sciences, University of Virginia, 116 pp.

——, M. Garstang, S. Greco, R. Talbot, and P. Kållberg, 1992: Saharan dust in the Amazon Basin. *Tellus,* **44B,** 133–149.

Tao, W.-K., and J. Simpson, 1989: Modeling study of a tropical squall-type convective line. *J. Atmos. Sci.,* **46,** 177–202.

Tokay, A., and D. A. Short, 1996: Evidence from tropical raindrop spectra of the origin of rain from stratiform versus convective clouds. *J. Appl. Meteor.,* **35,** 355–371.

——, ——, C. R. Williams, W. L. Ecklund, and K. S. Gage, 1999: Tropical rainfall associated with convective and stratiform clouds: Intercomparison of disdrometer and profiler measurements. *J. Appl. Meteor.,* **38,** 302–320.

Ulanski, S. L., and M. Garstang, 1978a: The role of surface divergence and vorticity in the life cycle of convective rainfall. Part I: Observations and analysis. *J. Atmos. Sci.,* **35,** 1047–1062.

——, and ——, 1978b: The role of surface divergence and vorticity in the life cycle of convective rainfall. Part II: Descriptive model. *J. Atmos. Sci.,* **35,** 1063–1069.

Victoria, R. L., L. A. Martinelli, J. Mortatti, and J. Richey, 1991: Mechanisms of water recycling in the Amazon Basin: Isotopic insights. *Ambio,* **20,** 384–387.

Wei, T., and R. A. Houze Jr., 1987: The GATE squall line of 9–10 August 1974. *Adv. Atmos. Sci.,* **4,** 85–92.

Woodcock, A. H., and I. Friedman, 1963: The deuterium content of raindrops. *J. Geophys. Res.,* **68,** 4477–4483.

Zipser, E. J., 1969: The role of organized unsaturated downdrafts in the structure and decay of an equatorial disturbance. *J. Appl. Meteor.,* **8,** 799–814.

Chapter 8

Cloud Models: Their Evolution and Future Challenges

WILLIAM R. COTTON

Department of Atmospheric Science, Colorado State University, Fort Collins, Colorado

ABSTRACT

A review of convective cloud modeling spanning the period from the days of the NOAA Experimental Meteorology Laboratory (EML) in the late 1960s to 2000 is presented. The intent is to illustrate the evolution of cloud models from the one-dimensional parcel-type models to the current generation of three-dimensional convective storm models and cloud ensemble models. Moreover, it is shown that Dr. Joanne Simpson played a pivotal role in the evolution of cloud models from the very first models to current generation cloud ensemble models. It is also shown that the first concept of the Regional Atmospheric Modeling System (RAMS) began while Drs. Cotton and Pielke worked under Dr. Simpson's supervision at EML. It is then illustrated how far cloud modeling has come with recent applications of RAMS to atmospheric research and numerical weather prediction. The chapter concludes with an outline of the major limitations of current generation convective cloud models.

1. Introduction

In this paper I review progress in mathematical/numerical modeling of convective clouds for the period beginning in the late 1960s to the present. We will see that throughout this period, Joanne Simpson was, in a supervisory capacity, either directly or indirectly involved in the development and applications of cloud models of a variety of formulations and dimensions.

During this same period I became trained in the then state-of-the-art cloud models at The Pennsylvania State University and began development and application of cloud models, while under Dr. Joanne Simpson's direction at the National Oceanic and Atmospheric Administration Experimental Meteorology Laboratory (NOAA EML) in Coral Gables, Florida. In 1975, equipped with intuition and experience gained from working with Joanne and participating in several field campaigns under her supervision, I joined the faculty of the Department of Atmospheric Science at Colorado State University, where I have directed a cloud modeling research program to the present. It is with this perspective that I have prepared this review.

2. Entraining bubble/plume cumulus models

The earliest form of cumulus model is the one-dimensional entraining bubble or plume model. Stommel (1947) postulated that cumulus clouds laterally entrain environmental air as the buoyant cloud rises through the cloud environment. Joanne was an early pioneer of the "bubble" concept (Ludlam and Scorer 1953; Malkus and Scorer 1955; Woodward 1959; Levine 1959; Malkus

1960), while a second school viewed a cumulus cloud as a steady-state jet of rising air (Schmidt 1947; Stommel 1947; Squires and Turner 1962). Figure 8.1 illustrates the features of these two models. Malkus (1960) formulated the entrainment rate as

$$\mu c = \frac{1}{M_c}\frac{dM_c}{dz} = \frac{b}{R},$$

where M is the cloud mass, z is the vertical distance, b is a dimensionless coefficient, and R is the radius of the thermal. As shown by Morton et al. (1956) and Malkus and Williams (1963), the same form of rise rate equation and entrainment rate formulation is applicable to both the bubble and plume models. The coefficient b, however, differs with b being 0.2 for a jet, where entrainment is restricted to being lateral, and 0.6 for a bubble, where entrainment is more three-dimensional about the bubble.

Computations with these models amounts to solving a simple marching problem solution as the parcel is followed in a Lagrangian sense, rising through the conditionally unstable environment.

The Simpson and Wiggert (1969) model is probably the most famous of the family of these cloud models. As a graduate student at The Pennsylvania State University, I became acquainted with the "other" model of the time, the Weinstein and Davis (1968) cloud model that was designed around the concept of a steady-state jet, whereas the Simpson and Wiggert model assumed that a bubble prevailed. The Weinstein and Davis model was heavily used by a number of research groups and students because the report on the model included the complete FORTRAN code. For my dissertation, I mod-

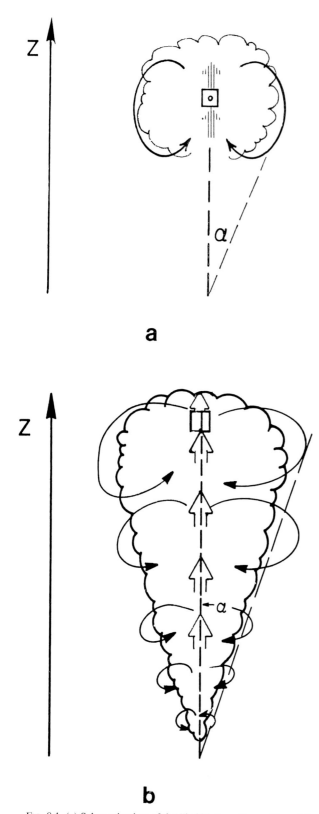

a

b

FIG. 8.1. (a) Schematic view of the "bubble" or "thermal" model of lateral entrainment in cumuli; (b) schematic view of the "steady-state jet" model of lateral entrainment in cumuli.

ified the basic dynamics of the Weinstein and Davis model slightly and then added a completely new microphysics module that contained some 21 classes of pristine ice crystals including their mass and major dimensions, rain, and graupel/hail (Cotton 1972a,b). Shortly after graduating I joined the staff at the EML and Joanne often referred to my model as being a "trolley car with a jet engine." Since then I have often used this phrase when reviewing articles in which very detailed bin-resolving microphysics is inserted in extremely crude dynamical frameworks.

The entraining bubble and plume models were used extensively for cloud seeding applications and still remain the foundation of cumulus parameterization schemes, such as the Arakawa and Schubert (1974) scheme.

The lateral entrainment models have come under considerable criticism over the years. Warner (1970) for example, claimed that models based on the lateral entrainment principle cannot simultaneously predict cloud-top height and liquid water content. This led to a rather heated debate in the literature (Simpson 1971, 1972; Cotton 1971). Joanne Simpson and Jack Warner then decided that it would be prudent to have one of the EML scientists (myself) visit Warner's research group and interact with them. Here I was less than two years out of graduate school, and Joanne gave me the opportunity of the lifetime to spend six months with my family in Australia. There I collected datasets from Warner's group's archives, participated in numerous discussions with scientists at Commonwealth Scientific and Industrial Research Organisation (CSIRO) and Monash University, presented lectures at Monash University, and participated in a field program flying in the CSIRO-instrumented DC-3 near Bundaberg, Queensland, Australia.

Another limitation of one-dimensional models, in general, is that they cannot simulate the interactions of cumulus clouds with vertical wind shear. Figure 8.2 illustrates that cumulus clouds rising through wind shear become quite asymmetrical exhibiting a nearly adiabatic, unmixed updraft core on their upshear flanks (Heymsfield et al. 1978) and strong mixing on their downshear flanks. The intensity of entrainment in shear is believed to be stronger than that in a weakly sheared environment.

In summary, the simple lateral entrainment models have the following limitations.

- They cannot simulate a complete precipitation cycle.
- The averaged properties across a bubble or plume are not representative of cumulus clouds, as the departures from the mean are very large.
- Lateral entrainment is a gross over-simplification of the three-dimensional (lateral and top) entrainment process and the range of eddy scales involved.
- They cannot represent properly the complexities of cloud interactions with vertical wind shear.

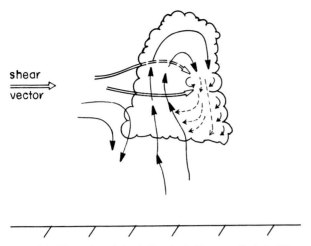

FIG. 8.2. Illustration of circulations in shallow cumuli; dashed lines depict parcel motions following mixing in the downshear flank of the cloud.

3. The 1.5D time-dependent cloud models

An extension of the simple marching-problem one-dimensional parcel model is the time-dependent one-dimensional cloud model first formulated by Asai and Kasahara (1967). Based on the axisymmetric cylindrical equations averaged across the cloud width, this class of model predicts the cloud-averaged updraft (downdraft) and diagnoses the compensating environmental subsidence (ascent). Hence this class of model is sometimes referred to as a 1.5D cloud model. Being time-dependent this model can simulate the entire precipitation life cycle, albeit constrained to a single chimney. Entrainment/detrainment is modeled by both turbulent mixing, using concepts similar to the classic entrainment models, and dynamic entrainment/detrainment diagnosed from a mass continuity equation for the rigid cylindrical geometry of the model. This diagnosed inflow/outflow through the cloud boundaries is exaggerated by the constraint of the cloud to the rigid cylindrical walls of the model. Other researchers (Ryan and Lalousis 1979; Pointin 1985; Wang 1983; Ferrier and Houze 1989) have permitted the radius of such a model to vary in time and height. Working with a case study obtained during my visit to Warner's group, I performed simulations with both the classic entrainment model and with a version of the 1.5D model (Cotton 1975). This work revealed that both the classic 1D entrainment model and the 1.5D cloud model share the same feature reported by Warner, that they cannot simultaneously predict observed cloud-top height and observed liquid water content. This is probably a result of the common feature in both models that cloud properties averaged across the cloud width are not representative of the scales of motion contributing to cloud momentum (cloud ascent) or scales of motion driving entrainment processes.

In summary, the 1.5D cloud model has the following limitations.

- Entrainment is modeled similar to bubble/jet models plus "dynamic entrainment" driven by mass continuity through a rigid cylindrical wall or a nonturbulence elastic boundary.
- Properties averaged across the cloud width are not representative of cloud dynamics governing cloud ascent and entrainment processes.
- Interactions with wind shear cannot be explicitly represented.
- Cloud interactions with gravity waves cannot be represented.

4. Axisymmetric and slab-symmetric two-dimensional models

Early numerical nonhydrostatic cloud models were two dimensional in either an axisymmetric or slab-symmetric geometry. Joanne (Malkus and Witt 1959) along with Ogura (1963) were the early pioneers in two-dimensional models. Later, Vic Wiggert, working under Joanne's supervision, experimented with the Murray (1970) cloud model and evaluated its applicability to the cloud seeding experiments in the Florida Area Cumulus Experiment (FACE). Entrainment and mixing processes in these models are simulated by subgrid-scale turbulence schemes as well as explicitly resolved motions. Murray (1970) referred to the explicitly resolved entrainment as dynamic entrainment. The scale of motion of the resolved entrainment, however, is dependent upon the grid spacing of the model. In contrast to the 1.5D models, the 2D models do not constrain the cloud boundaries to the wall of a fixed radius cylinder, but instead the model is free to determine its own cloud boundaries, subject to the constraints of the selected grid spacing and the all or nothing saturation scheme (i.e., subgrid partial cloudiness is not permitted).

There is a trade-off with the two geometries of the 2D models. The axisymmetric model is able to simulate full two-dimensional horizontal convergence of moist air in the subcloud layer and full two-dimensional entrainment/detrainment in the cloud layer. The slab-symmetric model, on the other hand, is restricted to only one-dimensional subcloud entrainment and one-dimensional entrainment/detrainment in the cloud layer. On the other hand, the axisymmetric model cannot simulate interactions of clouds with shear or cloud–cloud interactions in the horizontal. Of course, the slab-symmetric model is limited to unidirectional shear in its simulations. As shown by Ogura (1963), Murray (1970), and Soong and Ogura (1973), the updraft in an axisymmetric model grows more vigorously than in a slab-symmetric model for the same environmental conditions; the ratio of the maximum speed of compensating subsidence outside the cloud to the maximum updraft speed in a slab-symmetric model is significantly larger than in an axisymmetric model; and the compensating subsidence in a slab-symmetric model spreads in space more widely than in an axisymmetric model.

Perhaps owing to the more realistic visual appearance of the clouds and to the ability to interact with wind shear, 2D slab-symmetric models have been used extensively to simulate clouds and cloud systems (Orville and Sloan 1970; Liu and Orville 1969; Hill 1974; Tao and et al. 1987; Tao and Simpson 1989; Ogura 1975; Fovell and Ogura 1988; Fovell 1991; Fovell et al. 1992).

In summary, some of the limitations of two-dimensional models are as follows.

- Axisymmetric models cannot represent cloud interactions with wind shear.
- Slab-symmetric models cannot represent cloud interactions with directionally varying wind fields and entrainment/detrainment is only one-dimensional.
- As in all numerical cloud models, the scales of motion controlling resolved entrainment are limited to the grid spacings of the model, which for the most part is a few hundred meters.

5. Three-dimensional cloud models

Three-dimensional cloud modeling began in the early to mid-1970s (Steiner 1979; Miller and Pearce 1974; Pastushkov 1973; Cotton and Tripoli 1978; Klemp and Wilhelmson 1978a,b). These early 3D models were severely limited in resolution and domain size by the computers available to us at the time. To put it into perspective, my first 3D cloud simulation attempts were done under Dr. Simpson's direction on an old Univac 1108 computer that was moved to Miami after the Geophysical Fluid Dynamics Laboratory (GFDL) acquired a more modern computer. This computer had only 64 000 words of memory, or about 4 MB, and about a 1-Mips processor, and probably cost more than $1 million. My current wafer-thin notebook computer, by contrast, has 64 MB of memory and a 330-MHz processor, weighs 2.5 lbs, and cost about $2000. The Univac 1108 and its drum mass-storage device took up a large room. In order to run the model on this machine, I hired Otis Brown, who was a graduate student at the time at the University of Miami and is now the Dean of the Rosenstiel School of Marine and Atmospheric Sciences of the University of Miami, to design and implement an asynchronous input/output (I/O) scheme between the computer memory and the drums. All that was in memory in the computer was a five-column stencil and that stencil had to be scanned across the entire model domain while inputting/outputting data to and from the drum. Unfortunately, the volume of data transfers exceeded the error rates of the I/O system so that I could never get more than a few minutes of simulated time before the pointers to the stencil were lost and the model would crash.

Fortunately, since that time computers have advanced in speed and memory capacity so that 3D simulations are commonplace and even used in the graduate classroom environment. As will be discussed later, Grabowski et al. (1998) have performed 3D cloud ensemble model simulations over the tropical Atlantic with 2-km grid spacing over a 400- ×400-km domain. This is about a factor of 10 larger than 3D cloud simulations common only a decade ago. In addition in my group at Colorado State University (CSU), we have performed interactive nested-grid 3D simulations of mesoscale convective systems and tornadic storms in which the parent grid covered about the northern end of Australia or one-third of the United States and a cloud resolving grid had spacing of 1.6–2.0 km (Alexander and Cotton 1998; Bernardet and Cotton 1998; Nachamkin and Cotton 2000; Finley et al. 2001). In the latter simulation, additional nests were spawned to 0.1 km, which permitted explicit representation of tornado circulations. All those calculations were performed on high-performance desktop workstations rather than supercomputers.

Still there remain major limitations of the current generation of 3D convective cloud models.

- The horizontal and vertical grid spacing is still marginal to simulate the details of the major convective drafts and eddies responsible for entrainment/detrainment processes. Grid spacing of 1–2 km barely resolve the dominant drafts in supercell storms and severe convective storms, but for ordinary cumulonimbi and tropical cumulonimbi grid spacing of a couple of hundred meters is needed. Likewise for towering cumuli, grid spacing of a few tens of meters is probably needed to simulate entrainment processes properly; not much different than used in large eddy simulation (LES) models of boundary layer clouds.
- The cloud microphysics used in current generation convective storm models is not much more than slight generalizations of the Kessler (1969) bulk cloud parameterizations. Detailed bin-resolving models are just now making their way into application in LES models of the cloudy boundary layer (i.e., Stevens et al. 1998; Khairoutdinov and Kogan 1999) and simple 3D cumulus model configurations (Kogan 1991; Kogan and Shapiro 1996).

6. Cloud ensemble or cloud resolving models (CRMs)

Cloud ensemble models or CRMs are cloud models with sufficient resolution to resolve (at least crudely) the structures of individual clouds (e.g., cumulus clouds) and run over a spatial domain large enough to contain many clouds and for a time long enough to include many cloud lifetimes (Randall et al. 1996). Tao and Simpson (1989, 1993), Tao et al. (1987), and Simpson and Tao (1993) have been among the pioneers in the development of cloud ensemble models and their applications. Typically CRMs are two-dimensional with grid spacings between 1 and 2 km and domains of 400 km and more. Generally CRMs contain horizontally homogeneous ocean surfaces, but Tripoli and Cotton (1989a,b) applied a CRM to the Rocky Mountain west to simulate the development of mesoscale convective systems. In addition, Costa et al. (2001) coupled a CRM to a dynamic ocean model to simulate interactions between convective clouds and the upper ocean layer. Storm-produced mixing by wind gusts, and rain-induced freshwater lenses were simulated. Tao et al. (1987) and Grabowski et al. (1998) have shown that statistics derived from 2D CRMs are nearly the same as obtained from 3D simulations. This is probably a consequence of the contraints placed on CRMs by the prescribed large-scale forcing, which is consistent with the underlying assumptions in cumulus parameterization schemes such as the Arakawa and Schubert (1974) scheme.

Some of the limitations of CRMs are as follows.

- The same limitations of 2D slab-symmetric models apply to 2D CRMs.
- Generally, large-scale forcing does not include clouds, which can lead to substantial moisture biases.
- CRMs are strictly one-way in their interactions with the large-scale, and thus the ensemble statistics are always constrained by the prescribed large-scale forcing.
- Cloud microphysics in CRMs, like most 3D cloud models, is quite primitive, being generalizations of the Kessler approach.

7. The Regional Atmospheric Modeling System concept and its evolution

While working on the development of the 3D cloud model at the EML, Joanne also gave me the opportunity to participate in FACE as an aircraft observer in the NOAA DC-6. This involved several hundred hours of flying back and forth across the Florida peninsula sampling rain showers and measuring boundary layer fluxes and winds. This experience, plus looking at real-time radar plots, convinced me that the convection that was being seeded over Florida was strongly influenced by mesoscale circulations associated with the Florida sea breeze. I proposed to Joanne that we hire a young scientist to develop a 3D mesoscale model of the Florida sea breeze to complement my cloud modeling work. Joanne enthusiastically supported this proposal and gave me permission to hire someone. My first thought was to call my old office mate at Penn State, Roger Pielke. Roger was quite excited at the chance to work and be paid full time and do research that could eventually form the basis of his Ph.D. dissertation.

Roger and I then shared an office at EML where we often talked about merging our two models (my nonhydrostatic cloud/storm model and his hydrostatic mesoscale model) into one comprehensive modeling system that could explicitly simulate interactions between

the mesoscale and the cloud scale. But the time was not right, as neither the computer resources nor our model codes were sufficiently mature to implement such a combined model. Shortly, Roger and Joanne moved to the University of Virginia and I moved to CSU. Then in the winter of 1982, Roger joined the faculty at CSU and after a few years we began the construction of a new model, which contained features of my nonhydrostatic cloud model and Roger's hydrostatic mesoscale model. This model eventually became known as the Regional Atmospheric Modeling System (RAMS), making use of the mascot of the CSU athletic teams, the Rams.

Summaries of RAMS are given in Pielke et al. (1992), Pielke (1984), Pielke and Pearce (1994), Nicholls et al. (1995), and Cotton et al. (2002). It is now only a nonhydrostatic model as few users opted for the hydrostatic option. Model development on cloud microphysics, radiation, cumulus parameterization, and turbulence closures has been largely under my direction, while Roger's group has concentrated on surface land processes in RAMS. To illustrate what this model has become from its conceptualization at EML, a summary of the recent applications of RAMS are as follows:

- large eddy simulations (Hadfield et al. 1991; Hadfield et al. 1992; Walko et al. 1992; Stevens et al. 1996; Feingold et al. 1996; Duda et al. 1996; Eastman et al. 1998; Stevens et al. 1998; Feingold et al. 1999; Jiang et al. 2000, 2001; Harrington et al. 2000; Jiang and Cotton 2000);
- simulations of convective storms and tornadoes (Ryan et al. 1990; Ziegler et al. 1995; Grasso and Cotton 1995; Grasso 1996; Finley et al. 2001; Bernardet et al. 2000);
- simulations of mesoscale convective systems (Schmidt and Cotton 1990; Cram et al. 1992a,b; Olsson and Cotton 1997a,b; Blanchard et al. 1998; Bernardet and Cotton 1998; Alexander and Cotton 1998; Olsson et al. 1998; Nachamkin and Cotton 2000; Bernardet et al. 2000);
- regional climate studies and land-use effects on climate (Pielke et al. 1993; Mukabana and Pielke 1996; Copeland et al. 1996; Pielke et al. 1997a,b; Stohlgren et al. 1998; Greene et al. 1999; Pielke et al. 1999a–c; Liston and Pielke 2000; Eastman et al. 2001; Lu et al. 2000);
- air pollution (Lyons et al. 1995; Eastman et al. 1995; Moran and Pielke 1996a,b; Uliasz et al. 1996; Pielke et al. 1997a,b; Pielke and Uliaz 1998);
- mesoscale numerical weather prediction (Cotton et al. 1994, 1995; Gaudet and Cotton 1998);
- tropical cyclone modeling (Nicholls and Pielke 1995; Eastman 1995; Eastman et al. 1996; Nicholls et al. 1996);
- simulations of cirrus clouds (Heckman and Cotton 1993; Mitrescu 1998; Wu 1999; Wu et al. 2000; Cheng et al. 2001);

- extreme precipitation estimation (Ashby 2000; Abbs and Ryan 1997; Abbs 1999);
- cloud ensemble simulations (Tripoli and Cotton 1989a,b; Costa et al. 2001);
- development of parameterizations (Weissbluth and Cotton 1993; DeMott et al. 1994; Lee et al. 1995; Mocko and Cotton 1995; Zeng and Pielke 1995a,b; Meyers et al. 1997; Pielke et al. 1997a,b; Walko et al. 2000);
- terrain-forced mesoscale flows (Nicholls et al. 1991; Peterson et al. 1991; Meyers and Cotton 1992; Bossert and Cotton 1994a,b; Pielke et al. 1995; Snook and Pielke 1995; Taylor et al. 1998; Chase et al. 1999); and
- flow around buildings (Pielke et al. 1995; Nicholls et al. 1995).

It is interesting that our current generation prototype real-time forecast model runs on a cluster of 23 Pentium PC processors and its finest grid spacing is 3 km over a 120- $\times$ 120-km domain, which is marginally sufficient to explicitly represent large cumulonimbi. With this finest grid nested in a U.S.-scale parent grid, 48-h forecasts can be accomplished in about 5 h. This is rather remarkable considering the cost of the computer system was only $21 000 and it out performs $10 million supercomputers of a decade ago. Explicit cumulus modeling is now possible in real time!

8. Limitations of current cloud models

We have seen that there have been major advances in computer power from the days of EML in the early 1970s. This has provided us the opportunity to perform simulations and even forecasts of clouds and clouds systems that we could have hardly dreamed possible then. But still, computational power does not mean we can simulate all physical and dynamical processes in convective clouds properly. The following is a list of things that, in my opinion, we still cannot model well.

- *Initial broadening of cloud droplet spectra in warm clouds.* There remains much active debate about the importance of such processes as inhomogeneous mixing (Latham and Reed 1977; Baker and Latham 1979; Baker et al. 1980, 1984) and turbulent entity mixing (Telford and Chai 1980; Telford et al. 1984), small-scale vortex influences on cloud droplet concentrations and supersaturations (Shaw et al. 1998); turbulence influences on collision and coalescence (Khain and Pinsky 1995, 1997; Pinsky and Khain 1997a–c; Pinsky et al. 1998, 1999), ultragiant nuclei (Johnson 1982; Feingold et al. 1999), and radiation broadening (Roach 1976; Stephens 1983; Ackerman et al. 1995; Austin et al. 1995; Harrington et al. 2000). All these processes are subgrid scale in current generation models and will remain so for some time to come. Parameterizations of these processes must be developed and implemented in future cumulus models.

- *Prediction of ice particle concentrations.* We have learned a great deal about sources of discrepancy between ice-forming nuclei (IFN) concentrations and ice crystal concentrations, such as ice multiplication by the rime-splinter process (Hallet and Mossop 1974; Mossop 1978). We have also developed improved methods of IFN detection with continuous flow diffusion chambers (Rogers 1982; Al-Naimi and Saunders 1985), and empirical formulas derived from those measurements (Meyers et al. 1992). When we put all that information in our cloud models, predictions of ice crystal concentrations are often within a factor of 10 of observations for many cloud systems, but there still remain a number of situations in which high ice crystal concentrations form so rapidly that no existing theory can account for them (Hobbs and Rangno 1985; Rangno and Hobbs 1991, 1994).

- *Details of hydrometeor spectra evolution.* As we have seen, most cloud models today still employ simplified bulk microphysics models in which particle size spectra are specified by a prescribed basis function. A few models of rather simple cloud systems use bin-resolving microphysics in which the evolution of hydrometeor size spectra are predicted. For more complex clouds like cumulonimbi with mixed-phase hydrometeors, it is still too costly to use bin-resolving microphysics models. In the case of hail storms, some models use a hybrid approach in which hailstones are modeled with bins, while bulk microphysics is used on other hydrometeor species (Farley and Orville 1986; Johnson et al. 1994). Even in these models, the physics in the bin models is restricted to continuous accretion approximations. Uncertainties and inadequate representation of hydrometeor size spectra, habits, and densities can lead to errors in the simulation of storm precipitation rates, downdrafts, cold pools, and, as a consequence, cloud and storm propagation.

- *Interactions with dynamically active aerosol.* Current generation cloud models generally have universal formulas for the concentrations of cloud condensation nuclei (CCN) and IFN. In some cases, these formulas are modified to exhibit a lapse in concentrations with height. But in many situations, clouds may reside near the boundaries of air masses having very different aerosol concentrations, such as in coastal areas where the flow may be offshore with high CCN concentrations and then revert to onshore flow with very low concentrations. The behavior of the cloud systems is quite different depending on those aerosol properties. Likewise, the atmosphere is often vertically stratified in aerosol concentrations just as it is with temperature and moisture. Entrainment of aerosol can thus alter the microstructure and macrostructure of convective clouds substantially.

- *Quantitative simulations of entrainment rates.* We have seen that, through the years, both the qualitative and quantitative aspects of entrainment in cumuli have been a major source of debate. One would hope that as cloud models are refined to grid spacings of a few tens of meters, entrainment processes would be explicitly simulated properly. For large cumuli and cumulonimbi, we are still many years from achieving this end. Even so, the complexity of entrainment processes is such that we are by no means guaranteed that the models will properly represent entrainment.

These are just a few model deficiencies that come to mind. There are many others that relate to finite difference operators and physical processes, and so forth, but the ones listed above are the most important.

9. Conclusions

In summary, since the early 1970s we have made substantial progress in our ability to simulate clouds and cloud systems, and models have made substantial contributions to the fundamental understanding of clouds and cloud systems and their impacts on weather and climate. Still there are a number of deficiencies that must be resolved in the coming years. As a consequence, we still cannot simulate explicitly the dynamical and microphysical response of cumuli to cloud seeding with a high degree of confidence. Moreover, the list of model deficiencies given above severely limits our ability to simulate the chain of physical processes that is associated with the dynamic seeding hypothesis. Thus in spite of major advances in cloud models and computational resources, we are still unable to resolve many of the issues that Joanne Simpson sought and fought in the days of EML.

Acknowledgments. Brenda Thompson is gratefully acknowledged for her help with processing the manuscript and gathering the numerous reference citations. This research was partially supported by the National Science Foundation under Grants ATM-9904128 and ATM-9900929 and by the Colorado Agricultural Experiment Station under Project Number COL00692.

REFERENCES

Abbs, D., 1999: A numerical modeling study to investigate the assumptions used in the calculation of probable maximum precipitation. *Water Resour. Res.,* **35,** 785–796.
——, and B. F. Ryan, 1997: Numerical modelling of extreme precipitation events. Res. Rep. 131, Urban Water Research Association of Australia, CSIRO, 71 pp.
Ackerman, A. S., O. B. Toon, and P. V. Hobbs, 1995: A model for particle microphysics, turbulent mixing, and radiative transfer in the stratocumulus-topped marine boundary layer and comparisons with measurements. *J. Atmos. Sci.,* **52,** 1204–1236.
Alexander, G. D., and W. R. Cotton, 1998: The use of cloud-resolving simulations of mesoscale convective systems to build a convective parameterization scheme. *J. Atmos. Sci.,* **55,** 408–419.
Al-Naimi, R., and C. P. R. Saunders, 1985: Measurements of natural deposition and condensation-freezing ice nuclei with a continuous flow chamber. *Atmos. Environ.,* **19,** 1871–1882.
Arakawa, A., and W. H. Schubert, 1974: Interaction of a cumulus cloud ensemble with the large-scale environment. *J. Atmos. Sci.,* **31,** 674–701.

Asai, T., and A. Kasahara, 1967: A theoretical study of the compensating downward motions associated with cumulus clouds. *J. Atmos. Sci.,* **24,** 487–496.
Ashby, C. T., 2000: Sensitivity of simulated flash flood environment evolution to soil moisture initialization. M.S. thesis, Dept. of Atmospheric Science, Colorado State University, 117 pp.
Austin, P. H., S. Siems, and Y. Wang, 1995: Constraints on droplet growth in radiatively cooled stratocumulus. *J. Geophys. Res.,* **100,** 14 231–14 242.
Baker, M. B., and J. Latham, 1979: The evolution of droplet spectra and the rate of production of embryonic raindrops in small cumulus clouds. *J. Atmos. Sci.,* **36,** 1612–1615.
——, R. G. Corbin, and J. Latham, 1980: The influence of entrainment on the evolution of cloud droplet spectra: I. A model of inhomogeneous mixing. Quart. *J. Roy. Meteor. Soc.,* **106,** 581–598.
——, R. E. Breidenthal, T. W. Choularton, and J. Latham, 1984: The effects of turbulent mixing in clouds. *J. Atmos. Sci.,* **41,** 299–304.
Bernardet, L. R., and W. R. Cotton, 1998: Multiscale evolution of a derecho-producing MCS. *Mon. Wea. Rev.,* **126,** 2991–3015.
——, L. D. Grasso, J. E. Nachamkin, C. A. Finley, and W. R. Cotton, 2000: Simulating convective events using a high-resolution mesoscale model. *J. Geophys. Res.,* **105,** 14 963–14 982.
Blanchard, D. O., W. R. Cotton, and J. M. Brown, 1998: Mesoscale circulation growth under conditions of weak inertial instability. *Mon. Wea. Rev.,* **126,** 118–140.
Bossert, J. E., and W. R. Cotton, 1994a: Regional-scale flows in mountainous terrain. Part I: A numerical and observational comparison. *Mon. Wea. Rev.,* **122,** 1449–1471.
——, and ——, 1994b: Regional-scale flows in mountainous terrain. Part II: Simplified numerical experiments. *Mon. Wea. Rev.,* **122,** 1472–1489.
Chase, T. N., R. A. Pielke Sr., T. G. F. Kittel, J. S. Baron, and T. J. Stohlgren, 1999: Potential impacts on Colorado Rocky Mountain weather due to land use changes on the adjacent Great Plains. *J. Geophys. Res.,* **104,** 16 673–16 690.
Cheng, W. Y. Y., T. Wu, and W. R. Cotton, 2001: Large-eddy simulations of the 26 November 1991 FIRE II cirrus case. *J. Atmos. Sci.,* **58,** 1017–1034.
Copeland, J. H., R. A. Pielke, and T. G. F. Kittel, 1996: Potential climatic impacts of vegetation change: A regional modeling study. *J. Geophys. Res.,* **101,** 7409–7418.
Costa, A. A., R. L. Walko, W. R. Cotton, and R. A. Pielke Sr., 2001: SST sensitivities in multiday TOGA COARE cloud-resolving simulations. *J. Atmos. Sci.,* **58,** 253–268.
Cotton, W. R., 1971: Comments on "On steady-state one-dimensional models of cumulus convection." *J. Atmos. Sci.,* **28,** 647–648.
——, 1972a: Numerical simulation of precipitation development in supercooled cumuli, Part I. *Mon. Wea. Rev.,* **100,** 757–763.
——, 1972b: Numerical simulation of precipitation development in supercooled cumuli, Part II. *Mon. Wea. Rev.,* **100,** 764–784.
——, 1975: On parameterization of turbulent transport in cumulus clouds. *J. Atmos. Sci.,* **32,** 548–564.
——, and G. J. Tripoli, 1978: Cumulus convection in shear flow—Three-dimensional numerical experiments. *J. Atmos. Sci.,* **35,** 1503–1521.
——, and R. A. Anthes, 1989: *Storm and Cloud Dynamics.* International Geophysics Series; Vol. 44, Academic Press, 883 pp.
——, G. Thompson, and P. W. Mielke Jr., 1994: Real-time mesoscale prediction on workstations. *Bull. Amer. Meteor. Soc.,* **75,** 349–362.
——, J. F. Weaver, and B. A. Beitler, 1995: An unusual summertime downslope wind event in Fort Collins, Colorado, on 3 July 1993. *Wea. Forecasting,* **10,** 786–797.
——, and Coauthors, 2002: RAMS 2001. Current status and future directions. *Meteor. Atmos. Phys.,* in press.
Cram, J. M., R. A. Pielke, and W. R. Cotton, 1992a: Numerical simulation and analysis of a prefrontal squall line. Part I: Observations and basic simulation results. *J. Atmos. Sci.,* **49,** 189–208.

——, ——, and ——, 1992b: Numerical simulation and analysis of a prefrontal squall line. Part II: Propagation of the squall line as an internal gravity wave. *J. Atmos. Sci.*, **49**, 209–225.

DeMott, P. J., M. P. Meyers, and W. R. Cotton, 1994: Parameterization and impact of ice initiation processes relevant to numerical model simulations of cirrus clouds. *J. Atmos. Sci.*, **51**, 77–90.

Duda, D. P., G. L. Stephens, B. B. Stevens, and W. R. Cotton, 1996: Effects of aerosol and horizontal inhomogeneity on the broadband albedo of marine stratus: Numerical simulations. *J. Atmos. Sci.*, **53**, 3757–3769.

Eastman, J. L., 1995: Numerical simulation of Hurricane Andrew—Rapid intensification. Preprints, *21st Conf. on Hurricanes and Tropical Meteorology,* Miami, FL, Amer. Meteor. Soc., 111–113.

——, R. A. Pielke, and W. A. Lyons, 1995: Comparison of lake-breeze model simulations with tracer data. *J. Appl. Meteor.*, **34**, 1398–1418.

——, M. E. Nicholls, and R. A. Pielke, 1996: A numerical simulation of Hurricane Andrew. Presented at *Second Int. Symp. on Computational Wind Engineering,* Fort Collins, CO.

——, R. A. Pielke, and D. J. McDonald, 1998: Calibration of soil moisture for lager eddy simulations over the FIFE area. *J. Atmos. Sci.*, **55**, 1131–1140.

——, M. B. Coughenour, and R. A. Pielke, 2001: The effects of CO_2 and landscape change using a coupled plant and meteorological model. *Global Change Biol.*, **7**, 797–815.

Farley, R. D., and H. D. Orville, 1986: Numerical modeling of hailstorms and hailstone growth. Part I: Preliminary model verification and sensitivity tests. *J. Climate Appl. Meteor.*, **25**, 2014–2035.

Feingold, G., S. M. Kreidenweis, B. Stevens, and W. R. Cotton, 1996: Numerical simulations of stratocumulus processing of cloud condensation nuclei through collision-coalescence. *J. Geophys. Res.*, **101** (D16), 21 391–21 402.

——, W. R. Cotton, S. M. Kreidenweis, and J. T. Davis, 1999: The impact of giant cloud condensation nuclei on drizzle formation in stratocumulus: Implications for cloud radiative properties. *J. Atmos. Sci.*, **56**, 4100–4117.

Ferrier, B. S., and R. A. Houze Jr., 1989: One-dimensional time-dependent modeling of GATE cumulonimbus convection. *J. Atmos. Sci.*, **46**, 330–352.

Finley, C. A., W. R. Cotton, and R. A. Pielke, 2001: Numerical simulation of tornadogenesis in a high-precipitation supercell. Part I: Storm evolution and transition into a bow echo. *J. Atmos. Sci.*, **58**, 1597–1629.

Fovell, R., 1991: Influence of the Coriolis force in two-dimensional model storm. *Mon. Wea. Rev.*, **119**, 606–630.

——, and Y. Ogura, 1988: Numerical simulation of a midlatitude squall line in two dimensions. *J. Atmos. Sci.*, **45**, 3846–3879.

——, D. Durran, and J. R. Holton, 1992: Numerical simulations of convectively generated stratospheric gravity waves. *J. Atmos. Sci.*, **49**, 1427–1442.

Gaudet, B., and W. R. Cotton, 1998: Statistical characteristics of a real-time precipitation forecasting model. *Wea. Forecasting,* **13**, 966–982.

Grabowski, W. W., X. Wu, M. W. Moncrieff, and W. D. Hall, 1998: Cloud-resolving modeling of cloud systems during Phase III of GATE. Part II: Effects of resolution and the third spatial dimension. *J. Atmos. Sci.*, **55**, 3264–3282.

Grasso, L. G., 1996: Numerical simulation of the May 15 and April 26, 1991 tornadic thunderstorms. Ph.D. dissertation, Colorado State University, 151 pp.

——, and W. R. Cotton, 1995: Numerical simulation of a tornado vortex. *J. Atmos. Sci.*, **52**, 1092–1203.

Greene, E. M., G. E. Liston, and R. A. Pielke Sr., 1999: Relationships between landscape, snowcover depletion, and regional weather and climate. *Hydrol. Proc.*, **13**, 2453–2466.

Hadfield, M. G., W. R. Cotton, and R. A. Pielke, 1991: Large-eddy simulations of thermally forced circulations in the convective boundary layer. Part I: A small-scale, circulation with zero wind. *Bound.-Layer Meteor.*, **57**, 79–114.

——, ——, and ——, 1992: Large-eddy simulations of thermally forced circulations in the convective boundary layer. Part II: The effect of changes in wavelength and wind speed. *Bound.-Layer Meteor.*, **58**, 307–327.

Hallett, J., and S. C. Mossop, 1974: Production of secondary ice particles during the riming process. *Nature,* **249**, 26–28.

Harrington, J. Y., G. Feingold, W. R. Cotton, and S. M. Kreidenweis, 2000: Radiative impacts on the growth of a population of drops within simulated summertime Arctic stratus. *J. Atmos. Sci.*, **57**, 766–785.

Heckman, S. T., and W. R. Cotton, 1993: Mesoscale numerical simulation of cirrus clouds—FIRE case study and sensitivity analysis. *Mon. Wea. Rev.*, **121**, 2264–2284.

Heymsfield, A. J., D. N. Johnson, and J. E. Dye, 1978: Observations of moist adiabatic asent in northeast Colorado cumulus congestus clouds. *J. Atmos. Sci.*, **35**, 1689–1703.

Hill, G. E., 1974: Factors controlling the size of cumulus clouds as revealed by numerical experiments. *J. Atmos. Sci.*, **31**, 646–673.

Hobbs, P. V., and A. L. Rangno, 1985: Ice particle concentrations in clouds. *J. Atmos. Sci.*, **42**, 2523–2549.

Jiang, H., and W. R. Cotton, 2000: Large-eddy simulation of shallow cumulus convection during BOMEX: Sensitivity to microphysics and radiation. *J. Atmos. Sci.*, **57**, 582–594.

——, ——, J. O. Pinto, J. A. Curry, and M. J. Weissbluth, 2000: Cloud resolving simulations of mixed-phase Arctic stratus observed during BASE: Sensitivity to concentration of ice crystals and large-scale heat and moisture advection. *J. Atmos. Sci.*, **57**, 2105–2117.

——, G. Feingold, W. R. Cotton, and P. G. Duynkerke, 2001: Large-eddy simulations of entrainment of cloud condensation nuclei into the Arctic boundary layer: 18 May 1998 FIRE/SHEBA case study. *J. Geophys. Res.*, **106**, 15 113–15 122.

Johnson, D. B., 1982: The role of giant and ultragiant aerosol particles in warm rain initiation. *J. Atmos. Sci.*, **39**, 448–460.

——, P. K. Wang, and J. M. Straka, 1994: A study of microphysical processes in the 2 August 1981 CCOPE supercell storm. *Atmos. Res.*, **33**, 93–123.

Kessler, E., III, 1969: On the distribution and continuity of water substance in atmospheric circulation. *Meteor. Monogr.,* Amer. Meteor. Soc., No. 32, Amer. Meteor. Soc., 84 pp.

Khain, A. P., and M. B. Pinsky, 1995: Drop inertia and its contribution to turbulent coalescence in convective clouds. Part I: Drop fall in the flow with random horizontal velocity. *J. Atmos. Sci.*, **52**, 196–206.

——, and ——, 1997: Turbulence effects on the collision kernel. II: Increase of the swept volume of colliding drops. *Quart. J. Roy. Meteor. Soc.*, **123**, 1543–1560.

Khairoutdinov, M. F., and Y. L. Kogan, 1999: A large eddy simulation model with explicit microphysics: Validation against aircraft observations of a stratocumulus-topped boundary layer. *J. Atmos. Sci.*, **56**, 2115–2131.

Klemp, J. B., and R. B. Wilhelmson, 1978a: The simulation of three-dimensional convective storms dynamics. *J. Atmos. Sci.*, **35**, 1070–1096.

——, and ——, 1978b: Simulations of right- and left-moving storms produced through storm splitting. *J. Atmos. Sci.*, **35**, 1097–1110.

Kogan, Y. L., 1991: The simulation of a convective cloud in a 3-D model with explicit microphysics. Part I: Model description and sensitivity experiments. *J. Atmos. Sci.*, **48**, 1160–1189.

——, and A. Shapiro, 1996: The simulation of a convective cloud in a 3D model with explicit microphysics. Part II: Dynamical and microphysical aspects of cloud merger. *J. Atmos. Sci.*, **53**, 2525–2545.

Latham, J., and R. L. Reed, 1977: Laboratory studies of the effects of mixing on the evolution of cloud droplet spectra. *Quart. J. Roy. Meteor. Soc.*, **103**, 297–306.

Lee, T. J., and R. A. Pielke, and P. W. Mielke Jr., 1995: Modeling the clear-sky surface energy budget during FIFE87. *J. Geophys. Res.,* **100**, 25 585–25 593.

Levine, J., 1959: Spherical vortex theory of bubble-like motion in cumulus clouds. *J. Meteor.,* **16,** 653–662.

Liston, G. E., and R. A. Pielke, 2000: A climate version of the Regional Atmospheric Modeling System. *Theor. Appl. Climatol.,* **66,** 29–47.

Liu, J. Y., and H. D. Orville, 1969: Numerical modeling of precipitation and cloud shadow effects on mountain-induced cumuli. *J. Atmos. Sci.,* **26,** 1283–1298.

Lu, L., R. A. Pielke, G. E. Liston, W. J. Parton, D. Ojima, and M. Hartman, 2000: Implementation of a two-way interactive atmospheric and ecological model and its application to the central United States. *J. Climate,* **14,** 900–919.

Ludlam, F. H., and R. S. Scorer, 1953: Convection in the atmosphere. *Quart. J. Roy. Meteor. Soc.,* **79,** 94–103.

Lyons, W. A., C. J. Tremback, and R. A. Pielke, 1995: Applications of the Regional Atmospheric Modeling System (RAMS) to provide input to photochemical grid models for the Lake Michigan Ozone Study (LMOS). *J. Appl. Meteor.,* **34,** 1762–1786.

Malkus, J. S., 1960: Recent developments in studies of penetrative convection and an application to hurricane cumulonimbus towers. *Cumulus Dynamics,* C. E. Anderson, Ed., Pergamon Press, 65–84.

——, and R. S. Scorer, 1955: The erosion of cumulus towers. *J. Meteor.,* **12,** 43–57.

——, and R. T. Williams, 1963: On the interaction between severe storms and large cumulus clouds. *Meteor. Monogr.,* No. 5, Amer. Meteor. Soc., 59–64.

——, and G. Witt, 1959: The evolution of a moist convective element. A numerical calculation. *The Atmosphere and the Sea in Motion,* B. Bolvin, Ed., Rockefeller Institute Press, 425–439.

Meyers, M. P., and W. R. Cotton, 1992: Evaluation of the potential for wintertime quantitative precipitation forecasting over mountainous terrain with an explicit cloud model. Part I: Two-dimensional sensitivity experiments. *J. Appl. Meteor.,* **31,** 26–50.

——, P. J. DeMott, and W. R. Cotton, 1992: New primary ice nucleation parameterizations in an explicit cloud model. *J. Appl. Meteor.,* **31,** 708–721.

——, R. L. Walko, J. Y. Harrington, and W. R. Cotton, 1997: New RAMS cloud microphysics parameterization. Part II: The two-moment scheme. *Atmos. Res.,* **45,** 3–39.

Miller, M. J., and R. P. Pearce, 1974: A three-dimensional primitive equation model of cumulonimbus convection. *Quart. J. Roy. Meteor. Soc.,* **100,** 133–154.

Mitrescu, C., 1998: Cloud-resolving simulations of tropical cirrus clouds. M.S. thesis, Dept. of Atmospheric Science, Colorado State University, 85 pp.

Mocko, D. M., and W. R. Cotton, 1995: Evaluation of fractional cloudiness parameterizations for use in a mesoscale model. *J. Atmos. Sci.,* **52,** 2884–2901.

Moran, M. D., and R. A. Pielke, 1996a: Evaluation of a mesoscale atmospheric dispersion modeling system with observations from the 1980 Great Plains mesoscale tracer field experiment. Part I: Datasets and meteorological simulations. *J. Appl. Meteor.,* **35,** 281–307.

——, and ——, 1996b: Evaluation of a mesoscale atmospheric dispersion modeling system with observations from the 1980 Great Plains mesoscale tracer field experiment. Part II: Dispersion simulations. *J. Appl. Meteor.,* **35,** 308–329.

Morton, B. R., G. Taylor, and J. S. Turner, 1956: Turbulent gravitational convection from maintained and instantaneous sources. *Proc. Roy Soc. London,* **234,** 1–23.

Mossop, S. C., 1978: The influence of drop size distribution on the production of secondary ice particles during graupel growth. *Quart. J. Roy. Meteor. Soc.,* **104,** 323–330.

Mukabana, J. R., and R. A. Pielke, 1996: Investigating the influence of synoptic-scale monsoonal winds and mesoscale circulations on diurnal weather patterns over Kenya using a mesoscale numerical model. *Mon. Wea. Rev.,* **124,** 224–243.

Murray, F. W., 1970: Numerical models of a tropical cumulus clouds with bilateral and axial symmetry. *Mon. Wea. Rev.,* **98,** 14–28.

Nachamkin, J. E., and W. R. Cotton, 2000: Interactions between a developing mesoscale convective system and its environment. Part II: Numerical simulation. *Mon. Wea. Rev.,* **128,** 1225–1244.

Nicholls, M. E., and R. A. Pielke, 1995: A numerical investigation of the effect of vertical wind shear on tropical cyclone intensification. Preprints, *21st Conf. on Hurricanes and Tropical Meteorology,* Miami, FL, Amer. Meteor. Soc., 339–341.

——, ——, and W. R. Cotton, 1991: A two-dimensional numerical investigation of the interaction between sea breezes and deep convection over the Florida Peninsula. *Mon. Wea. Rev.,* **119,** 298–323.

——, ——, J. L. Eastman, C. A. Finley, W. A. Lyons, C. J. Tremback, R. L. Walko, and W. R. Cotton, 1995: Applications of the RAMS numerical model to disperson over urban areas. *Wind Climate in Cities,* J. E. Cermak et al., Eds., Kluwer Academic, 435–463.

——, J. L. Eastman, and R. A. Pielke, 1996: A numerical simulation of Hurricane Hugo. Presented at *Proc. Second Int. Symp. on Computational Wind Engineering,* Fort Collins, CO.

Ogura, Y., 1963: The evolution of a moist convective element in a shallow, conditionally unstable atmosphere: A numerical calculation. *J. Atmos. Sci.,* **20,** 407–424.

——, 1975: On the interaction between cumulus clouds and the larger-scale environment. *Pageoph.,* **113,** 869–889.

Olsson, P. Q., and W. R. Cotton, 1997a: Balanced and unbalanced circulations in a primitive equation simulation of a midlatitude MCC. Part I: The numerical simulation. *J. Atmos. Sci.,* **54,** 457–478.

——, and ——, 1997b: Balanced and unbalanced circulations in a primitive equation simulation of a midlatitude MCC. Part II: Analysis of balance. *J. Atmos. Sci.,* **54,** 481–497.

——, J. Y. Harrington, G. Feingold, W. R. Cotton, and S. Kreidenweis, 1998: Exploratory cloud-resolving simulations of boundary layer Arctic stratus clouds. Part I: Warm season clouds. *Atmos. Res.,* **47–48,** 573–597.

Orville, H. D., and L. J. Sloan, 1970: Effects of higher order advection techniques on a numerical cloud model. *Mon. Wea. Rev.,* **98,** 7–13.

Pastushkov, R. S., 1973: The effect of vertical wind shear on the development of convective clouds. *Izv. Acad. Sci. URSS Atmos. Oceanic Phys.,* **9,** 5–11.

Peterson, T. C., L. O. Grant, W. R. Cotton, and D. C. Rogers, 1991: The effect of decoupled low-level flow on winter orographic clouds and precipitation in the Yampa River Valley. *J. Appl. Meteor.,* **30,** 368–386.

Pielke, R. A., 1984: *Mesoscale Meteorological Modeling.* Academic Press, 612 pp.

——, and R. A. Pearce, Eds.,1994: *Mesoscale Modeling of the Atmosphere, Meteor. Monogr.* No. 25, 167 pp.

——, and M. Uliasz, 1998: Use of meteorological models as input to regional and mesoscale air quality models—Limitations and strengths. *Atmos. Environ.,* **32,** 1455–1466.

——, and Coauthors, 1992: A comprehensive meteorological modeling system—RAMS. *Meteor. Atmos. Phys.,* **49,** 69–91.

——, J. H. Rodriguez, J. L. Eastman, R. L. Walko, and R. A. Stocker, 1993: Influence of albedo variability in complex terrain on mesoscale systems. *J. Climate,* **6,** 1798–1806.

——, J. Eastman, L. D. Grasso, J. Knowles, M. Nicholls, R. L. Walko, and X. Zeng, 1995: Atmospheric vortices. *Fluid Vortices,* S. Green, Ed., Kluwer Academic, 617–650.

——, T. J. Lee, J. H. Copeland, J. L. Eastman, C. L. Ziegler, and C. A. Finley, 1997a: Use of USGS-provided data to improve weather and climate simulations. *Ecol. Appl.,* **7,** 3–21.

——, X. Zeng, T. J. Lee, and G. A. Dalu, 1997b: Mesoscale fluxes over heterogeneous flat landscapes for use in larger scale models. *J. Hydrol.,* **190,** 317–336.

——, G. E. Liston, L. Lu, and R. Avissar, 1999a: Land-surface influences on atmospheric dynamics and precipitation. *Integrating Hydrology, Ecosystem Dynamics, and Giogeochemistry in Complex Landscapes,* J. Tenhunen and P. Kabat, Eds., John Wiley and Sons, 105–116.

——, R. L. Walko, L. Steyaert, P. L. Vidale, G. E. Liston, and W. A. Lyons, 1999b: The influence of anthropogenic landscape changes on weather in south Florida. *Mon. Wea. Rev.*, **127**, 1663–1673.

——, G. E. Liston, J. L. Eastman, L. Lu, and M. Coughenour, 1999c: Seasonal weather prediction as an initial value problem. *J. Geophys. Res.*, **104**, 19 463–19 479.

Pinsky, M. B., and A. P. Khain, 1997a: Turbulence effects on the collision kernel. I: Formation of velocity deviations of drops falling within a turbulent three-dimensional flow. *Quart. J. Roy. Meteor. Soc.*, **123**, 1517–1542.

——, and ——, 1997b: Formation of inhomogeneity in drop concentration induced by the inertia of drops falling in a turbulent flow, and the influence of the inhomogeneity on the drop-spectrum broadening. *Quart. J. Roy. Meteor. Soc.*, **123**, 165–186.

——, and ——, 1997c: Turbulence effects on droplet growth and size distribution in clouds—A review. *J. Aerosol Sci.*, **28**, 1177–1214.

——, ——, D. Rosenfeld, and A. Pokrovsky, 1998: Comparison of collision velocity differences of drops and graupel particles in a very turbulent cloud. *Atmos. Res.*, **49**, 99–113.

——, ——, and M. Shapiro, 1999: Collisions of small drops in a turbulent flow. Part 1: Collision efficiency, problem formulation, and preliminary results. *J. Atmos. Sci.*, **56**, 2585–2600.

Pointin, Y., 1985: Numerical simulation of organized convection. Part I: Model description and preliminary comparisons with squall line observations. *J. Atmos. Sci.*, **42**, 155–172.

Randall, D. A., K.-M. Xu, R. J. C. Sommerville, and S. Iacobellis, 1996: Single-column models and cloud ensemble models as links between observations and climate models. *J. Climate*, **9**, 1583–1697.

Rangno, A. L., and P. V. Hobbs, 1991: Ice particle concentrations and precipitation development in small polar maritime cumuliform clouds. *Quart. J. Roy. Meteor. Soc.*, **117**, 207–241.

——, and ——, 1994: Ice particle concentrations and precipitation development in small continental cumuliform clouds. *Quart. J. Roy. Meteor. Soc.*, **120**, 573–601.

Roach, W. T., 1976: On the effect of radiative exchange on the growth by condensation of a cloud or fog droplet. *Quart. J. Roy. Meteor. Soc.*, **102**, 361–372.

Rogers, D. C., 1982: Field and laboratory studies of ice nucleation in winter orographic clouds. Ph. D. dissertation, University of Wyoming, Laramie, 161 pp.

Ryan, B. F., and P. Lalousis, 1979: A one-dimensional time-dependent model for small cumulus. *Quart. J. Roy. Meteor. Soc.*, **105**, 615–628.

——, G. J. Tripoli, and W. R. Cotton, 1990: Convection in high based stratiform cloud bands: Some numerical experiments. *Quart. J. Roy. Meteor. Soc.*, **116**, 943–964.

Schmidt, F. H., 1947: Some speculations on the resistence to motion of cumuliform clouds. *K. Ned. Meteor. Inst. Meded. Verh.*, **3**, 1.

Schmidt, J. M., and W. R. Cotton, 1990: Interactions between upper and lower tropospheric gravity waves on squall line structure and maintenance. *J. Atmos. Sci.*, **47**, 1205–1222.

Shaw, R. A., W. C. Reade, L. R. Collins, and J. Verlinde, 1998: Preferential concentration of cloud droplets by turbulence: Effects on the early evolution of cumulus cloud droplet spectra. *J. Atmos. Sci.*, **55**, 1965–1976.

Simpson, J., 1971: On cumulus entrainment and one-dimensional models. *J. Atmos. Sci.*, **28**, 449–455.

——, 1972: Reply. *J. Atmos. Sci.*, **29**, 220–225.

——, and W.-K. Tao, 1993: Goddard cumulus ensemble model. Part II: Applications for studying cloud precipitation processes and for NASA TRMM. *Terr. Atmos. Oceanic Sci.*, **4**, 73–116.

——, and V. Wiggert, 1969: Models of precipitating cumulus towers. *Mon. Wea. Rev.*, **97**, 471–489.

Snook, J. S., and R. A. Pielke, 1995: Diagnosing a Colorado heavy snow event with a nonhydrostatic mesoscale numerical model structured for operational use. *Wea. Forecasting*, **10**, 261–285.

Soong, S. T., and Y. Ogura, 1973: A comparison between axisymmetric and slab-symmetric cumulus cloud models. *J. Atmos. Sci.*, **30**, 879–893.

Squires, P., and J. S. Turner, 1962: An entraining jet model for cumulonimbus updraughts. *Tellus*, **14** (4), 422–434.

Steiner, J. T., 1979: Comments on "Cumulus Convection in shear flow—Three-dimensional numerical experiments." *J. Atmos. Sci.*, **36**, 1609–1611.

Stephens, G. L., 1983: The influence of radiative transfer on the mass and heat budgets of ice crystals falling in the atmosphere. *J. Atmos. Sci.*, **40**, 1729–1739.

Stevens, B., G. Feingold, W. R. Cotton, and R. L. Walko, 1996: Elements of the microphysical structure of numerically simulated stratocumulus. *J. Atmos. Sci.*, **53**, 980–1006.

——, W. R. Cotton, G. Feingold, and C.-H. Moeng, 1998: Large-eddy simulations of strongly precipitating, shallow, stratocumulus-topped boundary layers. *J. Atmos. Sci.*, **55**, 3616–3638.

Stohlgren, T. J., T. N. Chase, R. A. Pielke, T. G. F. Kittel, and J. Baron, 1998: Evidence that local land use practices influence regional climate and vegetation patterns in adjacent natural areas. *Global Change Biol.*, **4**, 495–504.

Stommel, H., 1947: Entrainment of air into a cumulus cloud. *J. Meteor.*, **4**, 91–94.

Tao, W.-K., and J. Simpson, 1989: A further study of cumulus interactions and mergers: Three-dimensional simulations with trajectory analyses. *J. Atmos. Sci.*, **46**, 2974–3004.

——, and ——, 1993: Goddard cumulus ensemble model. Part I: Model description. *Terr. Atmos. Oceanic Sci.*, **4**, 35–72.

——, ——, and S.-T. Soong, 1987: Statistical properties of a cloud ensemble: A numerical study. *J. Atmos. Sci.*, **44**, 3175–3187.

Taylor, C. M., R. J. Harding, R. A. Pielke Sr., P. L. Vidale, R. L. Walko, and J. W. Pomeroy, 1998: Snow breezes in the boreal forest. *J. Geophys. Res.*, **103**, 23 087–23 101.

Telford, J. W., and S. K. Chai, 1980: A new aspect of condensation theory. *Pageoph.*, **118**, 720–742.

——, T. S. Keck, and S. K. Chai, 1984: Entrainment at cloud tops and the droplet spectra. *J. Atmos. Sci.*, **41**, 3170–3179.

Tripoli, G., and W. R. Cotton, 1989a: A numerical study of an observed orogenic mesoscale convective system. Part 1: Simulated genesis and comparison with observations. *Mon. Wea. Rev.*, **117**, 273–304.

——, and ——, 1989b: A numerical study of an observed orogenic mesoscale convective system. Part 2: Analysis of governing dynamics. *Mon. Wea. Rev.*, **117**, 305–328.

Uliasz, M., R. A. Stocker, and R. A. Pielke, 1996: Regional modeling of air pollution transport in the southwestern United States. *Environmental Modeling*, Vol. 3, P. Zannetti, Ed., Computational Mechanics Publications, 145–181.

Walko, R. L., W. R. Cotton, and R. A. Pielke, 1992: Large eddy simulations of the effects of hilly terrain on the convective boundary layer. *Bound.-Layer Meteor.*, **53**, 133–150.

——, and Coauthors, 2000: Coupled atmosphere–biophysics–hydrology models for environmental modeling. *J. Appl. Meteor.*, **39**, 931–944.

Wang, J. Y., 1983: A quasi-one-dimensional, time-dependent, and nonprecipitating cumulus cloud model: On the bimodal distribution of cumulus cloud height. *J. Atmos. Sci.*, **40**, 651–664.

Warner, J., 1970: On steady-state one-dimensional models of cumulus convection. *J. Atmos. Sci.*, **27**, 1035–1040.

Weinstein, A. I., and L. G. Davis, 1968: A parameterized numerical model of cumulus convection. Rep. 11, NSF Grant GA-777, Dept. of Meteorology, Pennsylvania State University, State College, 42 pp.

Weissbluth, M. J., and W. R. Cotton, 1993: The representation of convection in mesoscale models. Part I: Scheme fabrication and calibration. *J. Atmos. Sci.*, **50**, 3852–3872.

Woodward, E. B., 1959: The motion in and around isolated thermals. *Quart. J. Roy. Meteor. Soc.*, **85**, 144–151.

Wu, T., 1999: Numerical modeling study of the November 26, 1991 cirrus event. Ph.D. dissertation, Colorado State University, 188 pp.

——, W. R. Cotton, and W. Y. Y. Cheng, 2000: Radiative effects on the diffusional growth of ice particles in cirrus clouds. *J. Atmos. Sci.,* **57,** 2892–2904.

Zeng, X., and R. A. Pielke, 1995a: Further study on the predictability of landscape-induced atmospheric flow. *J. Atmos. Sci.,* **52,** 1680–1698.

——, and ——, 1995b: Landscape-induced atmospheric flow and its parameterization in large-scale numerical models. *J. Climate,* **8,** 1156–1177.

Ziegler, C. L., W. J. Martin, R. A. Pielke, and R. L. Walko, 1995: A modeling study of the dryline. *J. Atmos. Sci.,* **52,** 263–285.

Chapter 9

Goddard Cumulus Ensemble (GCE) Model: Application for Understanding Precipitation Processes

WEI-KUO TAO

Laboratory for Atmospheres, NASA Goddard Space Flight Center, Greenbelt, Maryland

ABSTRACT

One of the most promising methods to test the representation of cloud processes used in climate models is to use observations together with cloud resolving models (CRMs). The CRMs use more sophisticated and realistic representations of cloud microphysical processes, and they can reasonably well resolve the time evolution, structure, and life cycles of clouds and cloud systems (size about 2–200 km). The CRMs also allow explicit interaction between outgoing longwave (cooling) and incoming solar (heating) radiation with clouds. Observations can provide the initial conditions and validation for CRM results.

The Goddard Cumulus Ensemble (GCE) model, a cloud-resolving model, has been developed and improved at the National Aeronautics and Space Administration (NASA) Goddard Space Flight Center over the past two decades. Dr. Joanne Simpson played a central role in GCE modeling developments and applications. She was the lead author or coauthor on more than 40 GCE modeling papers. In this paper, a brief discussion and review of the application of the GCE model to 1) cloud interactions and mergers, 2) convective and stratiform interaction, 3) mechanisms of cloud–radiation interaction, 4) latent heating profiles and TRMM, and 5) responses of cloud systems to large-scale processes are provided. Comparisons between the GCE model's results, other cloud resolving model results, and observations are also examined.

1. Introduction

The global hydrological cycle is central to understanding climate system interactions and rainfall, and its associated processes are a key link in this cycle. Fresh water provided by tropical rainfall and its variability can exert a large impact upon the structure of the upper ocean layer. In addition, approximately two-thirds of the global rain falls in the Tropics, while the associated latent heat release accounts for about three-fourths of the total heat energy for the Earth's atmosphere (Riehl and Simpson 1979). Convective cloud systems account for a large portion of tropical heating and rainfall. Furthermore, the vertical distribution of latent heat release modulates large-scale tropical circulations (e.g., the 30–60-day intraseasonal oscillation; see Hartmann et al. 1984; Sui and Lau 1989), which, in turn, impacts midlatitude weather through teleconnection patterns such as those associated with El Niño. Shifts in these global circulations can result in prolonged periods of droughts and floods, thereby exerting a tremendous impact upon the biosphere and human habitation. And yet monthly rainfall over the tropical oceans is still not known within a factor of 2 over large (5° latitude × 5° longitude) areas (Simpson et al. 1988, 1996). The Tropical Rainfall Measuring Mission (TRMM), a joint United States–Japan space project, was established to provide a more accurate estimate of rainfall and the four-dimensional structure of diabatic heating over the Tropics. The distributions of rainfall and inferred heating can be used to advance our understanding of the global energy and water cycle. In addition, this information can be used for global circulation and climate models for testing and improving their parameterizations.

Cloud resolving models (CRMs) are one of the most important tools used to establish quantitative relationships between diabatic heating and rainfall. This is because latent heating is dominated by phase changes between water vapor and small, cloud-sized particles, which cannot be directly detected using remote sensing techniques (though some passive microwave frequencies do respond to path-integrated cloud water). The CRMs, however, explicitly simulate the conversion of cloud condensate into raindrops and various forms of precipitation ice. It is these different forms of precipitation that are most readily detected from space and that ultimately reach the surface in the form of rain. In addition, the highest science priority identified in the Global Change Research Program (GCRP) is the role of clouds in climate and hydrological systems, which have been identified as being the most problematic issue facing global change studies. For this reason, the Global Energy and Water Cycle Experiment (GEWEX) formed the GEWEX Cloud System Study (GCSS) specifically for the purpose of studying such problems. CRMs were

TABLE 9.1. Major highlights of CRM development in the past four decades. Some (by no means all) key contributors are also listed. (The author apologizes for omitting any other major contributors to CRM development.)

	Highlights	Key contributors
1960s	Loading, buoyancy, and entrainment	J. Simpson (first 1D model)
		Y. Ogura and N. Phillips (first 2D anelastic)
1970s	Slab-symmetric vs axissymmetric model	T. Clark, W. Cotton, E. Kessler, J. Klemp, M. Miller, M. Moncrieff, H. D. Orville, R. Schlesinger, G. Sommeria, S.-T. Soong, R. Wilhelmson, and others
	Cloud seeding	
	Supercell dynamics	
	Cloud dynamics and warm rain	
	Wind shear effect on cloud organization	
1980s	Ensemble of clouds—cumulus parameterization	C.-S. Chen, N. A. Crook, K. K. Droegemeier, J. Dudhia, D. Durran, R. D. Farley, R. Fovell, B. Ferrier, S. Krueger, Y.-L. Lin, J.-L. Redelsperger, R. Rotunno, W. Skamarock, W.-K. Tao, G. J. Tripoli, M. L. Weisman, M. Yoshizaki, and many others
	Cloud interactions and mergers	
	Ice processes	
	Squall line	
	Convective and stratiform	
	Wind shear and cool pool	
	Gravity wave and density current	
	Large-scale and cloud-scale interactions	
	Cloud–radiation interaction	
1990s	2D vs 3D	W. Grabowski, X. Wu, K.-M. Xu, and many other talented scientists
	Land and ocean processes	
	Multiscale interactions	
	Cloud chemistry	
	Process modeling—climate variation implications	
	GCSS	
	Coupled with microwave radiative model for TRMM	

chosen as the primary approach (GEWEX Cloud System Science Team 1993; Moncrieff et al. 1997).

The pioneering one-dimensional (1D) cloud model was developed by Dr. J. Simpson in the 1960s. The 1D cloud model was used extensively to study the cloud seeding problem. A two-dimensional anelastic model that filtered out sound waves was developed by Drs. Y. Ogura and N. Phillips (1962; Ogura and Charney 1962). The models were used to study cloud development under the influence of the surrounding environment. In the 1970s, three-dimensional cloud models were developed (Steiner 1973; Wilhelmson 1974; Miller and Pearce 1974; Sommeria 1976; Clark 1979; Klemp and Wilhelmson 1978a; Cotton and Tripoli 1978; Schlesinger 1975, 1978). The effect of model designs (i.e., slab vs axisymmetric, and 2D vs 3D) on cloud development and liquid water content were the major foci in the 1970s. Also, the dynamics of midlatitude supercells, which are usually associated with tornados, was another major focus in the 1970s (i.e., Klemp and Wilhelmson 1978b; Wilhelmson and Klemp 1978). After the Global Atmospheric Research Program Atlantic Tropical Experiment (GATE; 1974), cloud ensemble modeling[1] was developed to study the collective feedback of clouds on the large-scale tropical environment with the aim of improving cumulus parameterization in large-scale models (i.e., Tao 1978; Soong and Ogura 1980; Soong and Tao 1980; Tao and Soong 1986; Lipps and Helmer

1986; Krueger 1988). The effect of ice processes on cloud formation and development, stratiform rain processes and their relation to convective cells, and the effect of wind shear on squall-line development were the other major areas of interest for cloud resolving models in the 1980s. The impact of radiative processes on cloud development was also investigated in the late 1980s. In the 1990s, cloud resolving models were used to study multiscale interactions (i.e., Tripoli and Cotton 1989; Peng et al. 2001), cloud chemistry interaction (see a review by Thompson et al. 1997), idealized climate variations (i.e., Held et al. 1993; Lau et al. 1993, 1994; Sui et al. 1994; Tao et al. 1999), and surface processes (i.e., Wang et al. 1996, 2002). The cloud resolving model was also used for the development and improvement of satellite rainfall retrieval algorithms (see a review by Simpson et al. 1998). Table 9.1 lists the major foci and some (not all) of the key contributors to cloud resolving model development over the past four decades.

During the past 20 years, observational data on atmospheric convection has been accumulated from measurements by various means, including radars, instrumented aircraft, satellites, and rawinsondes in special field observations [e.g., GATE, Preliminary Regional Experiment for Stormscale Operational and Research Meteorology (PRESTORM), Cooperative Huntsville Meteorological Experiment (COHMEX), Taiwan Area Mesoscale Experiment (TAMEX), Equatorial Mesoscale Experiment (EMEX), Tropical Ocean Global Atmosphere Coupled Ocean–Atmosphere Response Experiment (TOGA COARE), and many others]. This has made it possible for cloud resolving modelers to test

[1] A cloud ensemble model is a specific type of cloud resolving model. Its unique feature is to allow several convective clouds to develop simultaneously inside the model domain. Typically, cyclic lateral boundary conditions are used.

TABLE 9.2. Key developments in the CRM approach for studying tropical deep convective systems over the past two decades.

	Model	Microphysics	Turbulence	Domain	Integration
Tao (1978), Soong and Ogura (1980)	2D	Water	TKE	64 km	24 h
Soong and Tao (1980)	2D	Water	TKE	64 km	24 h
Soong and Tao (1984)	2D	Water	TKE	128 km	6 h
Ogura and Jiang (1985)	2D	Water	TKE	128 km	16 h
Tao and Soong (1986)	3D	Water	TKE	32×32 km^2	6 h
Lipps and Helmer (1986)	2D	Water	K-theory	32–64 km	4 h
	3D			24×16 km^2	
Tao et al. (1987)	2D	Water	TKE	128 km	6 h
	3D			32×32 km^2	6 h
Nikajima and Matsuno (1988)	2D	Water	K-theory	512 km	50 h
Dudhia and Moncrieff (1987)	3D	Water	Prescribed fluxes	25×50 km^2	3 h
Krueger (1988)	2D	Water	3rd Moment	30 km	2 h
	2D	Water and Ice	TKE	512 km	12 h
Tao and Simpson (1989)	3D			96×96 km^2	4 h
Gregory and Miller (1989)	2D	Water	Prescribed fluxes	256 km	9 h
Xu and Krueger (1991)	2D	Water and Ice	3rd moment	512 km	120 h
McCumber et al. (1991)	2D	Water	TKE	512 km	12 h
	3D	Water and Ice		64×32 km^2	3 h
Xu et al. (1992), Xu and Arakawa (1992)	2D	Water and Ice	3rd moment	512 km	120 h
Held et al. (1993)	2D	Water and Ice	K-theory	640 km	1000 h
Sui et al. (1994)	2D	Water and Ice	TKE	768 km	1248 h
Grabowski et al. (1996, 1999), Grabowski et al. (1998)	2D	Water and Ice	K-theory	900 km	7 days
	3D			400×400 km^2	
Xu and Randall (1996), (1995)	2D	Water and Ice	3rd Moment	512 km	18 days
Donner (1999)	3D	Water and Ice	K-theory	220×200 km^2	73 h
Wu et al. (1998), (1999)	2D	Water and Ice	K-theory	900 km	39 days
Li et al. (1999)	2D	Water and Ice	TKE	768 km	7 days
Su et al. (1999)	3D	Water and Ice	Blackadar-Type	210×210 km^2	7 days
Johnson et al. (2002)	2D	Water and Ice	TKE	1024 km	7 days

their simulations against observations and thereby improve their models. In turn, the models have helped scientists to understand the complex dynamical and physical processes that interact in atmospheric convective systems and for which observations alone still cannot provide a complete and consistent picture. The past decades have also seen substantial advances in the numerical modeling of convective clouds and mesoscale convective systems (e.g., squall-type and non-squall-type convective systems), which have substantially elucidated complex dynamical cloud–environment interactions in the presence of varying vertical wind shear. With the advent of powerful scientific computers, many important and complex processes, such as ice microphysics and radiative transfer, can now be simulated to a useful (but still oversimplified) degree in these numerical cloud models. Table 9.2 lists the key modeling papers for studying *tropical* convection over the past two decades along with selected model characteristics. As shown, over the last 20 years these models have become increasingly sophisticated through the introduction of improved (bulk type) microphysical processes, radiation and boundary layer effects, and improved turbulent parameterizations for subgrid-scale processes. In addition, the availability of exponentially increasing computer capabilities has resulted in time integrations increasing from hours to days, domain grid boxes (points) increasing from less than 2000 to more than 2 500 000 grid points, and 3D models becoming

increasingly prevalent. The CRM is now at a stage where it can provide reasonably accurate statistical information of the subgrid, cloud resolving processes now poorly parameterized in climate models and numerical prediction models.

2. Goddard Cumulus Ensemble (GCE) model

The Goddard Cumulus Ensemble (GCE) model is a cloud resolving model with its main features and applications reported by Tao and Simpson (1993), Simpson and Tao (1993), and Tao et al. (1997). The model is nonhydrostatic and model variables include horizontal and vertical velocities, potential temperature, perturbation pressure, turbulent kinetic energy, and mixing ratios of all water phases (vapor, liquid, and ice). The cloud microphysics includes a parameterized Kessler-type two-category liquid water scheme (cloud water and rain), and a three-category ice-phase scheme (cloud ice, snow, and hail/graupel) mainly based on Lin et al. (1983) and Rutledge and Hobbs (1984). However, there are several minor differences between the Goddard three-category ice-phase scheme and the Lin et al. (1983) and Rutledge and Hobbs (1984) schemes.[2] The

---

[2] The Goddard three-category ice-phase scheme was coded by Tao (with some discussions with Dr. Y.-L. Lin). These modifications pertain to the original GCE model (Tao et al. 1989; Tao and Simpson 1989).

TABLE 9.3. Characteristics of the GCE model.

Parameters/processes	GCE model
Vertical coordinate	z
Explicit convective processes	Two-class water and two-moment four-class ice
Implicit convective processes	Betts and Miller, Kain and Frisch
Numerical methods	Positive definite advection for scalar variables; fourth-order for dynamic variables
Initialization	Initial condition with forcing from observations/large-scale model
Radiation	Broad-band in LW and solar explicit cloud radiation interaction
Subgrid diffusion	TKE
Two-way interactive nesting	Radiative-type (2D model only)
Surface energy budget	Force-restore method
	Seven-layer Soil Model (PLACE)
	TOGA COARE flux module

first is the option to choose either graupel or hail as the third class of ice (McCumber et al. 1991). Graupel has a low density and a high number concentration. In contrast, hail has a high density and a low number concentration. These differences can affect not only the description of the hydrometeor population, but also the relative importance of the microphysical–dynamical–radiative processes. Second, a saturation technique was implemented by Tao et al. (1989). This saturation technique is designed to ensure that supersaturation (subsaturation) cannot exist at a grid point that is clear (cloudy). This saturation technique is one of the last microphysical processes to be incorporated. It is only done prior to evaluating the evaporation of rain and snow/graupel/hail deposition or sublimation. A third difference is that all microphysical processes (transfer rates from one type of hydrometeor to another) are calculated based on one thermodynamic state. This ensures that all processes are treated equally. The opposite approach is to have one particular process calculated first and then to modify the temperature and water vapor content (i.e., through latent heat release) before the second process is computed. The fourth difference is that the sum of all the sink processes associated with one species will not exceed its mass. This ensures that the water budget will be balanced in the microphysical calculations.

The following major improvements have been made to the model during the past seven-year period. (i) A multidimensional positive definite advection transport algorithm (MPDATA; Smolarkiewicz and Grabowski 1990) was implemented. All scalar variables (potential temperature, water vapor, turbulence coefficient, and all hydrometeor classes) use forward time differencing and the MPDATA for advection. The dynamic variables u, v, and w use a fourth-order accurate advection scheme and a leapfrog time integration (kinetic energy semiconserving method). (ii) An improved four-class, multiple-moment, multiple-phase ice scheme (Ferrier 1994) was developed, which resulted in improved agreement with observed radar and hydrometeor structures for convective systems simulated in different geographic locations without the need for adjusting coefficients (Fer-

rier et al. 1995). (iii) Solar and infrared radiative transfer processes (Chou 1992; Chou and Suarez 1994) that have been used to study the impact of radiation upon the development of clouds and precipitation (Tao et al. 1991, 1996) and upon the diurnal variation of rainfall (Tao et al. 1996; Sui et al. 1998) for tropical and midlatitude squall systems were included. (iv) Land and ocean surface processes were incorporated to investigate their impact upon the intensity and development of organized convective systems (Wang et al. 1996; Lynn et al. 1998). Mesoscale circulations, which formed in response to landscape heterogeneities represented by a land surface model, were crucial in the initiation and organization of the convection. These recent GCE model improvements can be found in Tao et al. (2002).

A stretched vertical coordinate (height increments from 40 to 1150 m) is used to maximize resolution in the lowest levels of the model. Typically, a total of 1024 grid points are used in the horizontal with 500–1000-m resolution in the two-dimensional version of the GCE model. In the three-dimensional version of the GCE model, the horizontal resolution is usually 2000 m with 200×200 grid points. The time step is 5–10 s. Table 9.3 lists the characteristics of the GCE model.

3. Applications of the GCE model to the study of precipitation processes

The application of the GCE model to the study of precipitation processes can be generalized into 14 categories (Table 9.4). It has been used to provide essential insights into the interactions of clouds with each other (Tao and Simpson 1984, 1989a), with their surroundings, and their associated heat, moisture, momentum, mass, and water budgets (Tao 1978; Soong and Tao 1980, 1984; Tao and Soong 1986; Tao et al. 1987; Tao and Simpson 1989b), with radiative transfer processes (Tao et al. 1991, 1993a, 1996; Sui et al. 1998), with ocean surfaces (Tao et al. 1991; Wang et al. 1996, 2002), and with idealized climate variations (Lau et al. 1993, 1994; Sui et al. 1994; Tao et al. 1999). Cloud draft structure, trace gas transport (Scala et al. 1990; Pickering et al. 1992a,b; and a review by Thompson et al.

TABLE 9.4. Applications of the GCE model. The specific topics and their respective GCE model characteristics, major results, and references are shown.

Topics	Model characteristics	Major results	References
Cloud–cloud interactions and mergers	2D/3D warm rain	Cloud downdraft and its associated cold outflow play major role in cloud merger.	Tao and Simpson (1984, 1989a)
Q_1 and Q_2 budgets	2D/3D warm rain and ice processes	Importance of evaporative cooling in Q_1 budget. Importance of vertical transport of moisture by convection in Q_2 budget.	Tao (1978), Soong and Tao (1980), Tao and Soong (1986), Tao and Simpson (1989b), Tao et al. (1992, 1993a, 1996), Johnson et al. (2002)
Cloud characteristics	2D/3D warm rain	Active convective updrafts cover small area but major contributors in mass, Q_1 and Q_2 budgets. Excellent agreement with aircraft measurements.	Tao and Soong (1986), Tao et al. (1987)
Convective momentum transport	2D/3D smaller domain in 3D	Identify the role of horizontal pressure gradient force on upgradient transport of momentum.	Soong and Tao (1984), Tao and Soong (1986), Tao et al. (1995)
Ice processes	2D/3D	The importance of ice processes for stratiform rain formation and its associated mass, Q_1 and Q_2 budgets.	Tao and Simpson (1989b), McCumber et al. (1991), Tao et al. (1993a), Ferrier et al. (1995)
Convective and stratiform interactions	2D	The horizontal transport of hydrometeors and water vapor from convective towers to stratiform region are quantified.	Tao et al. (1993a), Sui et al. (1994), Tao (1995), Lang et al. (2002)
Cloud–radiation interactions and diurnal variation of precipitation	2D (short- and long-term integration)	Longwave cooling can enhance precipitation significantly for tropical cloud systems, but only slightly for midlatitude systems. Modulation in relative humidity by radiative processes is major reason for diurnal variation of precipitation.	Tao et al. (1993a, 1996), Sui et al. (1998)
Cloud chemistry interactions	2D/3D	Significant redistribution of trace gases by convection. Enhancement of O_3 production related to deep convection in Tropics.	Thompson et al. (1997, a review)
Air–sea interactions	2D/3D	TOGA COARE flux algorithm performs well compared with observations, better than other flux algorithms. Surface fluxes are important for precipitation processes and maintain CAPE and boundary layer structure.	Wang et al. (1996, 2002)
Precipitation efficiency (PE)	2D	Examined different definitions of PE. Identify several important atmospheric parameters for better PE.	Ferrier et al. (1996)
Land processes	2D/3D	Importance of mesoscale circulation induced by soil gradient on precipitation. Identify the atmospheric parameters for triggering convection.	Lynn et al. (1998, 2001), Lynn and Tao (2001), Baker et al. (2001)
Idealized climate variations in tropics	2D	Examined several hypotheses associated with climate variation and climate warming. Identified physical processes that cause two different statistical equilibrium states (warm/humid and cold/dry) in idealized climates.	Sui et al. (1994), Lau et al. (1993, 1994), Tao et al. (1999, 2001b), Shie et al. (2002, manuscript submitted to *J. Climate*)
TRMM rainfall retrieval	3D	Improved the performance of TRMM rainfall retrieval algorithms by providing realistic cloud profiles.	Simpson et al. (1996, a review)
Latent heating profile retrieval	2D	Developed algorithms for retrieving 4D vertical structure of latent heating profiles over global Tropics.	Tao et al. (1990, 1993b, 2000, 2001a)

1997), and precipitation efficiency (Ferrier et al. 1996) have also been investigated. Further, the GCE model has been used to convert the radiances received by cloud-observing microwave radiometers into predicted rainfall rates (Simpson et al. 1988; and a review by Simpson et al. 1996). Remote sensing of cloud-top properties by high-flying aircraft bearing microwave and other instruments is now beginning to provide powerful tests of the GCE model, particularly when such observations are augmented by simultaneous ground-based radar measurements (Adler et al. 1991; Prasad et al. 1995; Yeh et al. 1995). The GCE model has also been used to study the distribution of rainfall and inferred heating (Tao et al. 1990, 1993b, 2000, 2001b). In this paper, a brief discussion of the application of the GCE model to 1) cloud interaction and mergers, 2) convective and stratiform interaction, 3) mechanisms of cloud–radiation interaction, 4) latent heating profiles and TRMM, and 5) responses of deep cloud systems to large-scale processes will be provided. Comparisons between the GCE model's results, and other cloud resolving model results and observations will also be examined. Please note that the survey selected and discussed in the paper is from the author's point of view.

a. Cloud interactions and mergers

Field experiment data [e.g., Florida Area Cumulus Experiment (FACE), GATE, and Island Thunderstorm Experiment (ITEX)] have shown that the merging of shower clouds is a crucial factor in the development of organized convective complexes, which are the major producers of rainfall in the Tropics (Houze and Cheng 1977), in the Florida peninsula (Simpson et al. 1980), and in the Maritime Continent region north of Darwin, Australia (Simpson et al. 1993). The observational data consisted of calibrated radar and rain gauges. The mergers usually yield more than an order of magnitude more precipitation than unmerged cells. For example, Simpson et al. (1980) found that mergers were responsible for 86% of the rainfall observed, even though 90% of the cells were unmerged. Most of the increase in total rainfall comes from the increased areal extent and duration of the second-order mergers. [A first-order merger is identified as a consolidation of two or more previously independent single-cell radar echoes, while a second-order merger is the result of the juncture of two or more first-order radar merged echoes (Westcott 1984).]

However, the physical mechanisms that affect the merging process are not clearly specified through observational studies, largely because of the difficulty of measuring the air circulations in and around cumulus clouds. Westcott (1984) reviewed observational analyses of mergers in detail and also raised some key questions concerning the mechanisms involved. From observational studies, several processes have been proposed as important in merging events. These processes fall into two main categories. The first involves addition of moisture to neighboring air, thereby reducing dilution by entrainment (Byers and Braham 1949; Scorer and Ludlam 1953; Malkus 1954). Moistening of the cloud environment can be accomplished in several ways. One source of moisture is precipitation falling from an overhanging canopy, which produces a favorable environment for new convective growth. Dissipation of previous and nearby clouds also provides a moister, more favorable environment. The merging cells can be better protected from the entrainment of dry environmental air (Lopez 1978). The second category involves dynamic processes that enhance low-level convergence leading to new growth and merging. Low-level convergence can be enhanced by 1) collision of downdraft outflows (Simpson 1980; Simpson et al. 1980); 2) differential motions of cloud masses (Holle and Maier 1980; Cunning et al. 1982; LeMone 1989); and 3) hydrostatic and nonhydrostatic pressure responses within the boundary layer (Cunning and Demaria 1986; LeMone et al. 1988).

1) GCE MODEL RESULTS

A two-dimensional version of the GCE model was used with a GATE dataset to study cloud interactions and merging (Tao and Simpson 1984). Over 200 groups of cloud systems with a life history of over 60 min were generated under the influence of different combinations of stratification and large-scale forcing (through a total of 48 numerical experiments). The GCE model results demonstrated the increase in convective activity and in the amount of precipitation with increased intensity of large-scale forcing (lifting). In the GCE model, a cloud merger is defined as a joining of the surface rainfall contour of 1 mm h^{-1}. Additional criteria are also considered. The merged clouds need to join for at least 15 min and the distance between previous separate clouds must be at least four to five grid intervals initially. These conditions are a combination of the definitions of merger found in several observational studies (Changnon 1976; Houze and Cheng 1977; Simpson et al. 1980). Based on the GCE model results, the most unfavorable environmental conditions for cloud merging are 1) less unstable stratification of the atmosphere and 2) weaker large-scale forcing.

One advantage of the model simulations is that the model can be rerun in order to investigate the sensitivity of its results to various physical processes. For example, Tao and Simpson (1984) performed an additional run (sensitivity test) using identical initial conditions that produced merged clouds. The only difference was that the drag force of rainwater in the vertical equation was set to zero in the sensitivity test. The absence of the drag force could lead to a delay in either the onset or the weakening of the downdraft below the cloud. The new convective cell in the merged situation did not occur in the run with weaker downdrafts. This sensitivity test demonstrated the importance of downdrafts on mergers.

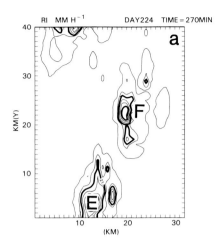

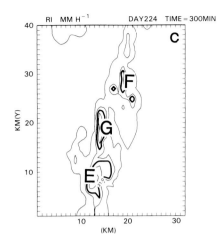

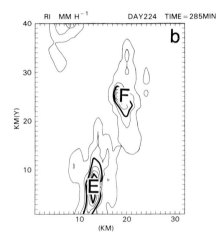

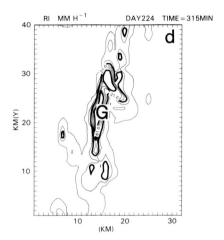

FIG. 9.1. Estimated surface rainfall intensity over part of the horizontal domain at (a) 270-, (b) 285-, (c) 300-, and (d) 315-min simulation time from a 3D GCE model. The contour interval is 10 mm h^{-1} starting at 1 mm h^{-1}. This type of merger is identified as a parallel-cells case.

Later, a total of nine three-dimensional experiments were made using the same GATE dataset (Tao and Simpson 1989a). Ten merged systems involving precipitating clouds were identified. Eight of ten mergers were from two clouds, and two involved merging of three clouds. Seven of the two-cloud mergers lie along a line roughly parallel to the initial environmental wind shear vector (called *parallel cells*; see Fig. 9.1). Only one merger lies along a line roughly perpendicular to the wind shear vector prior to the merging (called *perpendicular cells*; see Fig. 9.2). The dominance of parallel cells is consistent with observations in FACE and GATE (Simpson et al. 1980; Turpeinen 1982). The remaining two systems involve three clouds and are a combination merger of parallel and perpendicular cells. It was also found that a cloud bridge, which consists of a few low-level cumuli that develop and connect the clouds before the merger is detected on radar, occurs in most of the simulated merger cases. (This phenomenon was also well-simulated in the 2D model.) New cells (cell G in the parallel merger case and cell K in the perpendicular merger case) at the cloud bridge area developed rapidly. Both backward and forward air parcel trajectory analyses (Fig. 9.3) were performed. Forward air parcel trajectories are computed using grid points located in the merging area. Then, a backward trajectory calculation was performed to locate the origins of the high-rising parcels. These trajectory analyses show that the high-rising air parcels at the bridge area originated close to or within the regions occupied by previous separated cells (cells E and F). These air parcels were strongly affected by either one or two interacting cold outflows. Both 2D and 3D GCE model studies clearly suggest that the primary initiating mechanism for the occurrence of a precipitating cloud merger is the cloud downdrafts and their associated cold outflows as proposed by Simpson (1980). A significant difference between the simulated parallel and perpendicular cells is that the latter

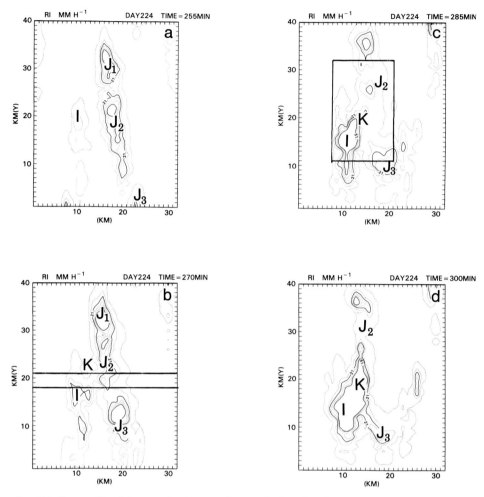

FIG. 9.2. Same as Fig. 9.1 except for the case of perpendicular cells. Time is (a) 255, (b) 270, (c) 285, and (d) 300 min.

cells are usually situated closer to each other (5–6 km) prior to merging, compared to the former (10 km or more). An explanation for this difference is that the direction of individual cell movement as well as the direction of cold outflow are predominantly directed down shear.

2) COMPARISON WITH OTHER CLOUD RESOLVING MODEL RESULTS

The causes of merging have been investigated by Hill (1974), Wilkins et al. (1976), Orville et al. (1980), Turpeinen (1982), Bennetts et al. (1982), and Kogan and Shapiro (1996) using cloud resolving models. Orville et al. (1980) investigated the effects of varying the spacing, timing, and intensity of two initial impulses in the context of a two-dimensional cloud model including warm rain and hail processes. Merging was found to result if two clouds were relatively close to each other (less than 7 km) and if the clouds were of different strength or initiated at different times (at intervals of 6

min). The mechanism of merging was attributed to the existence of a low-level (2–4-km elevation) pressure gradient directed from the weaker and younger cell toward the older and stronger one. By using a three-dimensional cloud model, Turpeinen (1982) also found that the mechanism of merging was dependent on the perturbation pressure distribution. Note that these two modeling studies used the joining of the 100% relative humidity isopleth of water vapor as a criterion for merger. The formation of a cloud bridge observed by Simpson (1980) has been simulated by both studies. But vigorous development of the new convective cell at the cloud bridge area did not occur in Orville et al. (1980) and Turpeinen (1982). Turpeinen (1982) suggested that this discrepancy might be attributed to the absence of mesoscale convergence in the model simulations. Chin and Wilhelmson (1998) used a three-dimensional cloud resolving model to simulate a GATE squall line. Their results indicated that a low-level cloud bridge was formed between a new cell and an existing squall line. Their results also indicated that this shallow cloud

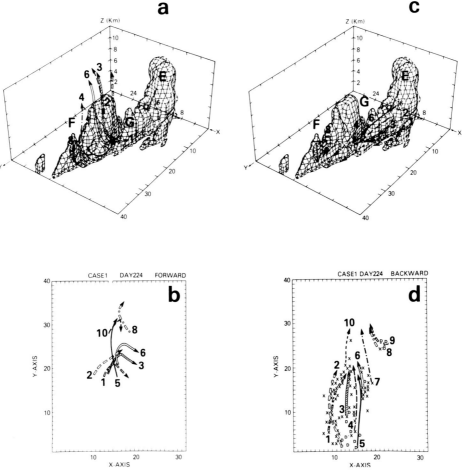

FIG. 9.3. Depiction of trajectory paths calculated from the evolving 3D model wind fields for the parallel-cell case shown in Fig. 9.1: (a) 3D depiction of upward paths as viewed from north-northwest; (b) as viewed from the overhead, computed forward from 300 to 340 min; (c),(d) same as (a),(b), respectively, except they are computed backward from 300 to 270 min; (a),(c) shaded area indicates the 3D depiction of estimated 20 dBZ isosurface at 300 min.

bridge is the response of colliding cold outflows [as suggested by Tao and Simpson (1984, 1989a)].

Kogan and Shapiro (1996) performed three-dimensional numerical simulations of mergers using explicit microphysics in a shear-free environment. Their criterion for cloud merger was based on the visual form of cloud updraft merger on a horizontal cross section. An arbitrary contour interval specified in the graphics routine (2 m s^{-1}) for coalescence of vertical velocity was used. This criterion was examined every 300 s. Kogan and Shapiro (1996) found that updraft merger occurred in four of the six simulations. They also found that after updraft merger, the maximum vertical velocity and domain-averaged kinetic energy were increased over the single bubble simulation. They hypothesize that the mergers were a consequence of mutual advection; that is, each of the clouds advected its neighbor in its radial inflow. Bennetts et al. (1982) also attributed merging in their numerical simulations to "mutual attraction." Kogan and Shapiro (1996) also found that the most favorable conditions for merger occur when the cells are closer than 4.5–6 radii apart (about 3–5 km between the centers of the temperature perturbations). No vigorous development occurred after the two updrafts merged, however. No precipitating downdraft was present in their simulations.

There is one major difference between the GCE model simulated mergers and those from others (Orville et al. 1980; Turpeinen 1982; Bennetts et al. 1982; Kogan and Shapiro 1996). The simulated mergers from other modeling studies are the consolidation of two initial independent single bubbles (the first-order merger). Their simulated mergers do not have vigorous development in contrast to the GCE model simulations. The basic design of the GCE modeling study is to generate several convective clouds randomly inside the model domain and then to observe and analyze the interactions between the simulated clouds. Neither locations nor intensities of simulated clouds are predetermined. The mergers identified in Tao and Simpson (1984, 1989a) only involve precipitating clouds (by definition). Their merged cases lasted longer and produced quite significant surface precipitation as observed by Simpson et al. (1980, 1993). Tao and Simpson (1989a) found that some of the previously distinct clouds associated with merger cases resulted from the consolidation of smaller-sized clouds. [This may explain why the mergers discussed in Tao and Simpson (1989a) are very similar to the second-order merged systems observed by Simpson et al. (1980).] These smaller-sized cells were predominantly oriented along the direction of the wind shear vector when they merged together. This result is inconsistent with the simulation performed by Turpeinen (1982). Situations for this type of merger only involve shallow clouds with little or no surface precipitation. Thus, the mechanism responsible for their merging cannot be cloud downdrafts and their associated cold outflows. The pressure distribution, as suggested by Orville et al. (1980) and Turpeinen (1982), mutual advection, as suggested by Kogan and Shapiro (1996), and the differential motions between convective elements (LeMone 1989) are probably the major mechanisms for this type of merger. All first-order simulated mergers may require two initially separated convective cells to be very close [from 7 km in Orville et al. (1982) to about 4 km in Kogan and Shapiro (1996) and Tao and Simpson (1989a)].

The definition of cloud merger is not unique in observational studies (Westcott 1984). Observational studies are based on radar-derived information. The observational studies usually define mergers in terms of coalescence of precipitation areas or radar reflectivity (at 1 mm h^{-1}, minimum detectable reflectivity signal). Additional criteria related to the distance between initially distinct convective elements and the duration of precipitation are also sometimes applied. Numerical simulations have used modeled dynamical and thermodynamical parameters (i.e., overlap of buoyancy, updraft, humidity, or circulation fields) to define mergers. Westcott (1984) pointed out that in order to perform better merger studies, it is necessary to clearly relate convective system's dynamical, thermodynamic, and microphysical structures and their radar image.

b. Convective–stratiform interaction

One of the major findings from GATE was the important contribution to rainfall from mesoscale convective systems[3] (MCSs). For example, Houze (1977) estimated that four MCSs accounted for 50% of the rainfall at one of the GATE ships during phase III. It was also estimated that the widespread stratiform rain accounted for about 32%–49% of the total rainfall from

the GATE MCSs (Houze 1977; Zipser et al. 1981; Gamache and Houze 1983). In addition, observations indicated that little stratiform rain fell during the early stages of *tropical* MCSs. As the stratiform cloud developed and expanded, the total amount of rain falling from it became equal to that generated in the convective region. The large, thick anvils were first observed to the rear of midlatitude squall lines (Newton 1950, 1963). The fraction of stratiform rainfall from midlatitude squall lines has been estimated at 29%–43% (Rutledge and Houze 1987; Johnson and Hamilton 1988). The existence of unsaturated warm mesoscale descent beneath the stratiform region was identified by Zipser (1969), modeled by Brown (1979), and conceptualized in Houze (1977) and Zipser (1977). The associated mesoscale ascent at the middle and upper layers of the stratiform region was diagnosed from indirect observations by Gamache and Houze (1983). One type of MCS is a squall line. The conceptual models of tropical and midlatitude squall lines are shown in Fig. 9.4.

The vertical distribution of heating in the stratiform region of MCSs is also considerably different from the vertical profile of heating in the convective region (Houze 1982; Johnson 1984). The convective profiles always show heating throughout the depth of the troposphere that is maximized in the lowest 2–5 km. The shapes of the heating profiles are quite similar with only slight variations in their magnitude for different MCSs from different geographic locations. The same can generally be said about the stratiform region. Heating is maximized in the upper troposphere between 5 and 9 km while cooling prevails at about 4 km. In addition, many recent studies (Adler and Negri 1988; Tao et al. 1993b) indicated that a separation of convective and stratiform clouds is necessary for a successful surface rain and latent heating profile retrieval from remote sensors.

These findings lead to an important question: *what are the origins and growth mechanisms of particles in stratiform precipitation?* Chen and Zipser (1982) suggested that both depositional growth associated with upward motion in the anvil and the horizontal flux of hydrometeors from the convective region are important in the maintenance of anvil precipitation. In a kinematic model study of a GATE squall line, Gamache and Houze (1983) showed quantitatively that 25%–40% of the stratiform condensate was created by mesoscale ascent at mid-to-upper levels in the stratiform region itself. Gallus and Johnson (1991) found that the contribution to surface rainfall from condensation in the mesoscale updraft was comparable in magnitude to the transport of condensate rearward from the convective line during a rapidly weakening stage of a midlatitude squall line. Using a kinematic (steady state) cloud model, Rutledge (1986) suggested that the condensate produced by mesoscale ascent is largely responsible for the large horizontal extent of light stratiform precipitation to the rear of the same GATE squall line analyzed by Gamache

---

[3] Houze (1997) defined a mesoscale convective system (MCS) as "a cloud system that occurs in connection with an ensemble of thunderstorms and produces a contiguous precipitation area ~100 km or more in horizontal scale in at least one direction."

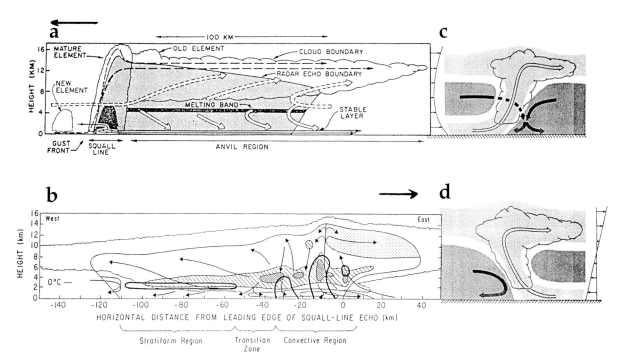

FIG. 9.4. (a),(b) Schematic cross sections through mature squall lines observed in the Tropics and midlatitudes, respectively; they are adopted from Gamache and Houze (1983) and Smull and Houze (1987). (c),(d) Schematic diagrams of major conceptual models of tropical and midlatitude squall lines derived from case studies, respectively. Areas of light and dark stippling indicate areas of high- and low-θ_e air, respectively. They are originally shown in Zipser (1969) and Newton (1963, 1966), but are adopted from Rotunno et al. (1988).

and Houze (1983). Using higher-resolution, Doppler-derived air motions associated with a midlatitude squall line as input in their two-dimensional kinematic model, Rutledge and Houze (1987) found that deposition in the mesoscale updraft accounted for 80% of the stratiform precipitation. They also conducted a series of sensitivity tests and found that almost no rain reached the surface in the stratiform region without the influx of hydrometeors from the convective cells, while only about one-fourth as much stratiform rain reached the surface in the absence of mesoscale ascent.

1) GCE MODEL RESULTS

Observational studies have had to use a steady-state assumption to estimate the transfer of hydrometeors from the convective region to its associated stratiform region as well as a relatively simple 1D cloud model to estimate the microphysical processes within the convective and stratiform regions. The time-dependent cloud resolving models (Tao et al. 1993a; Chin 1994; and Caniaux et al. 1994; Tao 1995; and others) have been used to explicitly quantify the origins and growth mechanisms of particles in stratiform precipitation by calculating the water budgets (microphysical processes and transfer processes of hydrometeors between convective and stratiform regions).

Several organized convective systems (EMEX, TOGA COARE, TAMEX, and PRESTORM), which occurred in different large-scale environments, have been

simulated by the GCE model and the associated water budgets were analyzed (Tao et al. 1993a; Tao 1995). Table 9.5 compares several characteristics of the large-scale flow (i.e., stability, Richardson number, and precipitable water) in which these convective systems were embedded. The propagation speed of these systems and the references for the GCE model simulations are also listed. The convective available potential energy (CAPE) associated with the tropical convective systems is moderate (from 1400 to 1660 $m^2\ s^{-2}$) and smaller than that of the midlatitude system (PRESTORM). The vertical integrated water vapor contents are much higher for the TOGA COARE and EMEX cases compared to the PRESTORM case.

The water budgets in the convective precipitating, stratiform precipitating (hereafter the convective and stratiform stand for convective and stratiform precipitating, respectively), and nonprecipitating regions associated with the TOGA COARE, EMEX, TAMEX, and PRESTORM convective systems are shown in Fig. 9.5. The water budgets are separated into three different layers: lower (surface to 10°C level), middle (from 10° to −10°C), and upper (−10°C to −70°C). The horizontal transfer of hydrometeors from the convective to the stratiform region occurs mainly in the middle troposphere for the EMEX and TOGA COARE convective systems. By contrast, two-thirds of the horizontal transfer of hydrometeors is accomplished in the upper troposphere for the PRESTORM case. This is caused by the strong convective updrafts associated with the

TABLE 9.5. Initial environmental conditions expressed in terms of CAPE, precipitable water, and Richardson number for the TAMEX, EMEX, TOGA COARE, and PRESTORM MCSs.

	CAPE $m^2 s^{-2}$	Precipitable water $(g\ cm^{-2})$	Richardson number	$\Delta x,\ \Delta z$ (m) $(L_x,\ L_z)$ (km)	References
TAMEX	1450	5.275	29	750, 240–1150 (1906, 22)	Tao et al. (1991)
EMEX	1484	6.175	555	750, 240–1150 (1906, 22)	Tao et al. (1993a)
PRESTORM	2300	4.385	43	1000, 225–1000 (2542, 20)	Tao et al. (1993a)
TOGA COARE	1776	6.334	74	750, 40–1100 (1906, 22)	Wang et al. (1996)

PRESTORM case. Also, a more vigorous transfer of hydrometeors in the lower troposphere from the stratiform region back into the convective region occurs for the PRESTORM case. This is a consequence of the strong rear inflow simulated for this midlatitude case. For the TAMEX case, the horizontal transfer of hydrometeors can occur in both the middle and upper troposphere. A downward transfer of hydrometeors from the middle to the lower troposphere is a dominant process in the stratiform regions for all four cases. The interaction between the stratiform and non-surface-rain-

ing region is less significant than that between the convective and stratiform region.

The contribution to stratiform rain by the convective region has to be quantified by estimating a ratio (R), $R = C_T/(C_T + C_m)$, where C_T is the horizontal transfer of hydrometeors from the convective region into the stratiform region above the 10°C level, and C_m is the sum of the net condensation in the stratiform region and in the nonprecipitating region above the 10°C level. A small ratio indicates that the horizontal transfer of hydrometeors from the convective region is a small source

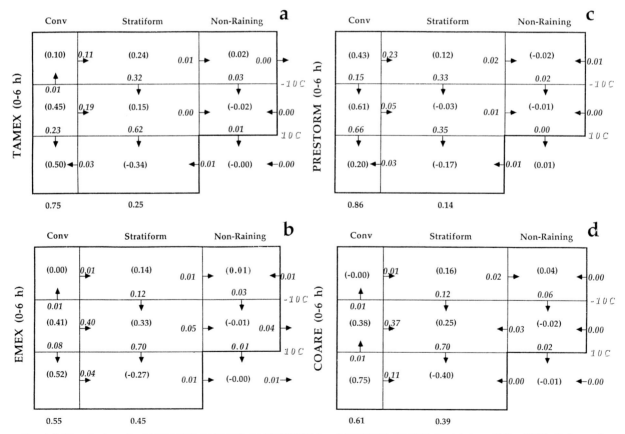

FIG. 9.5. Water budgets for (a) TAMEX, (b) EMEX, (c) PRESTORM, and (d) TOGA COARE simulated squall-line MCSs. Italic numbers indicate the amount of condensate transfer between various regions and layers while quantities in parentheses are the net condensation generated through microphysical processes.

TABLE 9.6. Values of the ratio R [$R = C_T/(C_T - C_m)$] for the GCE model simulations of TAMEX, EMEX, TOGA COARE, and PRE-STORM convective systems, as well as for the 6-h periods corresponding to the initial and mature stages. Their respective stratiform rain components are shown in the first column.

	Stratiform (0–12 h)	Ratio (0–12 h)	Initial stage (0–6 h)	Mature stage (6–12 h)
COARE	42.3%	0.40	0.46	0.36
EMEX	49.1%	0.41	0.47	0.35
TAMEX	29.6%	0.37	0.43	0.33
PRE-STORM	22.5%	0.54	0.82	0.43

of condensate for the stratiform anvil, whereas a ratio near unity indicates that nearly all of the condensate in the stratiform region was transported from the convective region. All four GCE-modeled cases showed large ratios, from 0.33 to 0.82, implying the role of the convective region in the generation of stratiform rainfall cannot be neglected[4] (Table 9.6). The relative importance of the horizontal transfer processes to the stratiform water budget is similar between the initial and the mature stages of the TAMEX, TOGA COARE, and EMEX systems, and this is likely due to the fact that stratiform precipitating clouds developed rapidly. In contrast, during the initial stage of the PRESTORM simulation, nearly all of the condensate in the stratiform region was a result of the horizontal transport from the convective region. As the PRESTORM system matured, the contribution made by the horizontal transport of hydrometeors from the convective region (i.e., the ratio R) decreased, such that the sources of condensate in the stratiform water budget were similar for all of the mature storms. It is hypothesized that during the initial stage

of the PRESTORM simulation, much of the condensate transported from the convective region is used to moisten and modify the dry environment at middle and upper levels. Condensation and deposition become increasingly more important with time in the stratiform water budget once the larger-scale environment reaches saturation. This evolution in the stratiform water budget is less obvious in the TAMEX, TOGA COARE, and EMEX cases because the environment is much more moist. (Note that the TAMEX, EMEX, and TOGA COARE cases have more stratiform rainfall than their PRESTORM counterpart.) The GCE model results also indicated that the similarity in R at the mature stages of all systems is likely to result from large stratiform regions.

2) COMPARISON WITH OBSERVATIONAL AND OTHER CLOUD-RESOLVING MODEL RESULTS

Table 9.7 lists the ratio (R) from observational studies using composite wind and thermodynamic fields for five different GATE MCSs (Leary and Houze 1980; Gamache and Houze 1983), a midlatitude squall line (Gallus and Johnson 1991), and a tropical–continental squall line (Chong and Hauser 1989). For six out of the seven observed cases the ratio is very close to or above 0.50. This implies that the convective region plays a very

---

[4] The convective region can also transport water vapor originally from the low troposphere into the stratiform region. Sui et al. (1994) indicated that this water vapor transport is the source for stratiform formation (deposition and condensation). Dynamic triggering of the stratiform formation can be gravity waves excited by strong and deep convective cells associated with the convective region.

TABLE 9.7. The same ratio defined in Table 9.6 except for different MCS cases, A, B, and C of Leary and Houze (1980), cases I and II of Gamache and Houze (1983), the 10–11 Jun squall line of Gallus and Johnson (1991), and the COPT squall line of Chong and Hauser (1989). Ratios from other CRM studies of convective–stratiform interaction are also shown.

	Ratio	Stratifrom amount (%)	Case
Leary and Houze (1980)	1.00	40	Tropics (GATE)
Leary and Houze (1980)	1.00	40	Tropics (GATE)
Leary and Houze (1980)	0.50	40	Tropics (GATE)
Gamache and Houze (1983)	0.55	49	Tropics (GATE)
Gamache and Houze (1983)	0.64	49	Tropics (GATE)
Gallus and Johnson (1991)	0.37	30	Midlatitude (PRESTORM)
Chong and Hauser (1989)	0.47	40	Tropics (COPT 81, 22 Jun)
Chin (1994)	*0.90*	*10*	Midlatitude
Chin et al. (1995)	*0.37*	*39*	Tropics (GATE)
Caniaux et al. (1994)	*0.22 (4–5 h)*	*17*	Tropics (COPT 81,
	0.09 (7–8 h)	*44*	23 Jun)

important role in the generation of stratiform rain. Very good agreement is evident between the ratio at the mature stage of the GCE modeled PRESTORM squall system (Table 9.6) and that estimated by Gallus and Johnson (1991). The GCE-modeled EMEX and TOGA COARE cases indicate a relatively small contribution (~0.40) to stratiform formation from the convective region compared to those determined from the kinematic studies.

Table 9.7 also shows the ratios determined from other CRM results (Chin 1994; Chin et al. 1995; Caniaux et al. 1994). Very good agreement is evident between the ratio at the mature stage of our modeled EMEX and TOGA COARE squall systems (shown in Table 9.6) and a tropical (GATE) squall case simulated by Chin et al. (1995). The comparison between our simulated PRESTORM and other CRM simulated midlatitude cases (Chin 1994; Caniaux et al. 1994), however, is quite different. Caniaux et al. (1994) suggested that the smaller contribution from the convective region to stratiform formation compared to observational studies (Chong and Hauser 1989) was due to the inability of convective updrafts to transport condensate to high levels in their two-dimensional simulation, the slower propagation speed, and the existence of a transition zone. The much smaller stratiform portion (10%) in the midlatitude case simulated by Chin (1994) is the reason for the higher R. The GCE-modeled PRESTORM case has a higher R (0.8) and smaller stratiform portion (14%) at the initial stage.

A direct comparison between these studies and the GCE model studies should be made with caution, because a different spatial resolution and a different definition for the convective-stratiform regions was used. For example, Caniaux et al. (1994) had a fixed number of model grid points (50) designated as the convective region. Remaining grid points with surface precipitation comprised the stratiform region. In the GCE model, the convective and stratiform regions are identified using information from surface rain rates first (i.e., Churchill and Houze 1984). Additional criteria are applied that have been included to identify convective regions where convection may be quite active *aloft* though there is little or no precipitation yet at the surface, such as areas associated with tilted updrafts and *new cells* initiated ahead of organized squall lines (Tao et al. 1993a; Lang et al. 2002, manuscript submitted to *J. Appl. Meteor.*). The GCE method was adopted by Chin (1994) and Chin et al. (1995). The comparison between the GCE simulated PRESTORM, TOGA COARE, TAMEX, and EMEX cases, however, is consistent because the same type of dataset and the same criteria for partitioning the convective and stratiform regions were being used.

3) THE CONVECTIVE AND STRATIFORM PROCESSES IN LARGE-SCALE MODELS

Molinari and Dudek (1992) and Frank (1993) suggested that the best approach to cumulus parameterization in large-scale models (30–120-km horizontal resolution, 150–300-s time steps) appears to be "to use a scheme that operates simultaneously with and interacts explicitly with the explicit scheme (grid scale microphysical processes)." They termed such schemes "hybrid schemes." The hybrid approach (by separating out the forcing mechanism for the mesoscale component) resolves the "mesoscale" circulations and microphysical processes that directly influence the development of the "stratiform clouds." Cumulus parameterization makes use of steady-state cloud models that interact with grid-scale variables and provide net heating, drying, and condensate associated with "convective cells." The interaction between parameterized and explicitly resolved cloud processes is through the *detrainment* of water vapor and condensate generated from the steady-state cloud model into the "resolved" stratiform clouds.

Recently, some global models [i.e., Colorado State University (CSU) GCM and Goddard Institute for Space Studies (GISS) GCM] have allowed both a cumulus parameterization scheme and an explicit moisture scheme to be activated simultaneously in the model simulations. The cumulus parameterization scheme is generally used to represent convective precipitation (10-km spatial scale) and the explicit moisture scheme to represent grid-resolvable precipitation such as stratiform/cirrus clouds (100–200-km spatial scale). The CSU GCM has implemented an explicit microphysical scheme with *five* prognostic variables for the mass of water vapor, cloud water, cloud ice, rain, and snow (Fowler and Randall 1996). The GISS global climate model has added an efficient prognostic cloud water (one species only). Stratiform clouds can be coupled with parameterized convection through detrainment of cloud water and/or cloud ice from the "tops" of cumulus towers or at any level above 550 mb (Del Genio et al. 1996).

The explicit interaction between cumulus parameterized and grid-scale resolved microphysics is only one-way in the current large-scale models. Note that some water condensate generated by the stratiform region can be transported into the convective region. In addition, how much (all or part) and where the (cloud tops or above melting layer) parameterized water condensate should detrain into the explicitly resolved microphysical scheme needs to be addressed. CRM results can and should be used for improving the cumulus parameterization schemes as well as for understanding the interaction between the cumulus parameterization schemes and the explicit moisture schemes. In the future, CRMs can be used to study the time evolution of each of the water budget terms associated with MCSs in different geographical regions as well as to determine whether any important variations in the evolution of the water budget can be explained in terms of differences in the wind and thermodynamic characteristics of the large-scale environments.

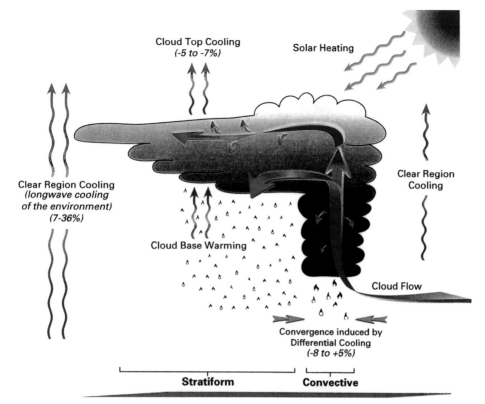

FIG. 9.6. Schematic diagram demonstrating the effects by different cloud-radiation mechanisms (cloud-top cooling and cloud-base warming—alters the thermal stratification of the stratiform cloud layer; differential cooling between clear and cloudy regions—enhances dynamic convergence into the cloud system; and the large-scale radiative cooling—destabilizes the large-scale environment).

c. Mechanisms of cloud–radiation interaction

The interaction between clouds and radiation is two-way. On the one hand, clouds can reflect incoming solar and outgoing longwave radiation. On the other hand, radiation can enhance or reduce the cloud activity. Gray and Jacobsen (1977) suggested that differential cooling between cloudy and clear regions can enhance cloud activity in the cloudy region. Longwave radiation cools the stratiform cloud top but warms the stratiform cloud base (Cox and Griffith 1979). As a result, longwave radiation can destabilize the stratiform cloud layer. Webster and Stephens (1980), also suggested that this destabilization was quite an important process in the light precipitation region during the Winter Monsoon Experiment (WMONEX). Stephens (1983) further suggested that the effects of radiation on the growth and sublimation rates of ice particles are significant. Particle growth (sublimation) is enhanced (suppressed) in a radiatively cooled (heated) environment. Radiative cooling could also destabilize the large-scale environment (Dudhia 1989). Cloud–radiation interaction can also have a major impact on the diurnal variation of precipitation processes over the Tropics. For example, two different (thermodynamic and dynamic) mechanisms responsible for the diurnal variation of precipitation over

tropical oceans were proposed. Kraus (1963) and Randall et al. (1991) suggested that the thermodynamic response of clouds to radiative heating (cloud development is reduced by solar heating and enhanced by IR cooling) is the main mechanism responsible for the diurnal variation of precipitation. On the other hand, Gray and Jacobson (1977) indicated that the large-scale dynamic response to the radiational differences between cloudy and clear regions was the main mechanism.

1) GCE MODEL RESULTS

A two-dimensional version of the GCE model has been used to perform a series of sensitivity tests to identify which is the dominant cloud-radiative forcing mechanism with respect to the organization, structure and precipitation processes for both a tropical (EMEX) and a midlatitude (PRESTORM) mesoscale convective system (Tao et al. 1996). Figure 9.6 shows a schematic diagram demonstrating the impact of cloud–radiation mechanisms on surface precipitation for the EMEX and PRESTORM cases. The GCE model results indicated that the dominant process for enhancing the surface precipitation in both the PRESTORM and EMEX squall cases was the large-scale radiative cooling (in the en-

TABLE 9.8. Summary of previous cloud-radiation modeling study results. The percentage increase or decrease in surface precipitation due to longwave (LW) and shortwave (SW) effects are given along with the mesoscale lifting, if used, for each case.

	LW radiative processes	Constant LW	LW and SW radiative processes	Imposed lifting
Chen and Cotton (1988)	0%	No	No	No
Chin (1994)	11%	No	−7%	No
Tripoli and Cotton (1989)	n/a	No	n/a	No
Tao et al. (1996)	8%	8%	−6%	No
Chin et al. (1995)	15%	No	−18%	2 cm s^{-1} continuous
Fu et al. (1995)	5%	15%	−1%	8–14 cm s^{-1} not continuous
Xu and Randall (1995)	n/a	n/a	n/a	8–14 cm s^{-1} continuous
Tao et al. (1991)	20%	No	No	4 cm s^{-1} not continuous
Tao et al. (1996)	36%	2%	−7%	7 cm s^{-1} not continuous
Dharssi et al. (1997)	30%	No	No	7 cm s^{-1} not continuous
Dudhia (1989)	36%	No	No	No
Churchill and Houze (1991)	0%	No	0%	Strong/ continuous
Miller and Frank (1993)	No	34%	18–21%	Strong/ continuous

vironment). However, the overall effect is really to increase the relative humidity and not the CAPE. Because of the high moisture in the Tropics, the increase in relative humidity by radiative cooling can have more of an impact on precipitation in the tropical case than in the midlatitude case. The large-scale radiative cooling led to a 36% increase in rainfall for the tropical case. The midlatitude squall line with a higher CAPE and lower humidity environment was only slightly affected (7%) by any of the longwave mechanisms. The GCE model results also indicated that the squall systems' overall (convective and stratiform) precipitation is increased by turning off the cloud-top cooling and cloud-base warming. Therefore, the cloud-top cooling–cloud-base warming mechanism was not the responsible cloud-radiative mechanism for enhancing the surface precipitation. However, the circulation as well as the microphysical processes were indeed (slightly) enhanced in the stratiform region by the cloud-top cooling and cloud-base warming mechanism for the midlatitude squall case. For both cases, the model results show that the mechanism associated with differential cooling between the clear and cloudy regions may or may not enhance precipitation processes (+5% to −7%, respectively, for the EMEX and PRESTORM cases). However, this mechanism is definitely less important than the large-scale longwave radiative cooling.

Solar heating was run from 9 AM to 1 PM local time in both environments and was found to decrease the precipitation by 7% in each case, compared to the runs with longwave radiation only. This result suggests that solar heating may play a significant role in the daytime minimum/nighttime maximum precipitation cycle found over most oceans, as noted in the observational study

of Kraus (1963). Sui et al. (1998) used the GCE model and performed a 15-day integration to simulate TOGA COARE convective systems. Their simulated diurnal variation of surface rainfall is in reasonable agreement with that determined from radar observations. They also found that modulation of convection by the diurnal change in available water as a function of temperature was responsible for a maximum in rainfall after midnight. This simply implies that the increase (decrease) in surface precipitation associated with IR cooling (solar heating) was mainly due to an increase (decrease) in relative humidity. The GCE model results also showed that the diurnal variation of sea surface temperature only plays a secondary role in the diurnal variation of precipitation processes.

2) COMPARISON WITH OTHER CLOUD RESOLVING MODEL RESULTS

Table 9.8 lists the previous modeling studies that have investigated the impact of cloud–radiation interactive processes on various cloud systems. The increments in surface precipitation in Table 9.8 are relative to the run without radiative processes. The conclusions associated with cloud–radiation mechanisms for our GCE-modeled tropical (EMEX) and midlatitude (PRESTORM) squall cases are in good agreement with many of these previous modeling studies. For example, Xu and Randall (1995), Miller and Frank (1993), and Fu et al. (1995) indicated that the differential cooling between cloudy and clear regions plays only a secondary role in enhancing precipitation processes. Xu and Randall (1995) and Fu et al. (1995) suggested that the cloud-top cooling and cloud-base warming destabilization mechanism could be important for prolonging the lifespan of high anvil clouds (around 10 km). Xu and Randall (1995) showed that this direct cloud destibilization does not have any impact on surface precipitation. The modeling studies (Fu et al. 1995; Miller and Frank 1993) also indicated that more surface precipitation can be generated in runs with constant clear-air radiative cooling than without. In addition, previous modeling results (Chin 1994; Chin et al. 1995; Miller and Frank 1993) indicated that solar radiative processes can reduce precipitation processes. However, the amount of increase or decrease in surface precipitation varies quite significantly among these different modeling studies, but only in regard to the tropical convective systems and not the midlatitude systems. One possible explanation is that large-scale forcing (lifting) was needed in some of these different tropical convective system studies. The imposed lifting varied from 2 cm s^{-1} to 14 cm s^{-1} in magnitude and was applied continuously or discontinuously in time among the different studies (see Table 9.8). Using an earlier version of the GCE model (Tao and Simpson 1989b) that included a superimposed large-scale vertical velocity as the main forcing, sensitivity tests using two different large-scale vertical velocities were performed. The results show that the radiative effects on the clouds are quite sensitive to the imposed background ascent (or lifting). The larger the imposed vertical velocity (9–12 cm s^{-1}), the less the impact of longwave cooling on surface precipitation processes (over 24 h of simulation time). Miller and Frank (1993) also obtained a similar conclusion using a regional-scale model. Further, note that the larger the imposed vertical velocity, the larger the cloud coverage that was generated.

The physical processes responsible for the diurnal variation of precipitation were found to be quite different between the GCE model and other CRM studies. For example, Xu and Randall (1995) found that nocturnal convection is basically a direct result of cloud–radiation interactions, in which solar absorption by clouds stabilized the atmosphere. However, their simulated rainfall for both noninteractive and interactive radiation were quite similar. Liu and Moncrieff (1998) performed a 15-day integration, as did Sui et al. (1998). They showed that direct interaction of radiation with organized convection was the major process that determined the diurnal variability of rainfall. Their results also indicated that well- (less) organized cloud systems can have strong (weak) diurnal variations of rainfall. In addition, they suggested that ice processes are needed. The model setups between Sui et al. (1998) and Liu and Moncrieff (1998) are quite different, however. In Liu and Moncrieff (1998), the horizontal momentum was relaxed to its initial value, which had a strong vertical shear in horizontal wind. On the other hand, the horizontal wind was nudged to time-varying observed values in Sui et al. (1998). Consequently, only long-lived squall lines (or fast-moving convective systems) were simulated in Liu and Moncrieff (1998) over the entire

simulation. In Sui et al. (1998), however, their simulated cloud systems had many different sizes and various life cycles. A more rigorous cloud resolving model intercomparison involving mechanisms associated with diurnal variation is needed in the future. A good quality-controlled long-term observational dataset that can provide large-scale initial conditions is also required.

d. Latent heating profiles and TRMM

1) GODDARD CONVECTIVE–STRATIFORM HEATING ALGORITHM

The GCE model has been used to develop a convective-stratiform heating (CSH) algorithm. The CSH algorithm uses surface precipitation rates, amount of stratiform rain, and information on the type and location of observed cloud systems as input. The CSH algorithm also utilizes a lookup table that consists of convective and stratiform diabatic heating profiles for various types of cloud systems in different geographic locations. These profiles stored in the lookup table are obtained from GCE model simulations by temporally and spatially averaging the heating distributions in the convective and stratiform regions of the systems, which are then normalized by their total surface rainfall (i.e., Figs. 9.7a,b).

Tao et al. (2000) evaluated the CSH algorithm's performance by retrieving the latent heating profiles associated with three TOGA COARE convective episodes (10–17 Dec 1992; 19–27 Dec 1992; and 9–13 Feb 1993). The inputs for the CSH algorithm were Special Sensor Microwave Imager–[SSM/I; similar to the TRMM Microwave Imager (TMI)] and radar- [similar to the TRMM Precipitation Radar (PR)] derived rainfall and stratiform amount. Diagnostically determined latent heating profiles calculated using 6-hourly soundings were used for validation. The temporal variability of retrieved latent heating profiles using radar-estimated rainfall and stratiform amount was in good agreement with that diagnostically determined for all three periods. However, less rainfall and a smaller stratiform percentage estimated by radar resulted in weaker (underestimated) latent heating profiles and lower maximum latent heating levels compared to those determined diagnostically. Rainfall information from SSM/I cannot retrieve individual convective events due to the limited temporal sampling of polar-orbiting satellites.

The four-dimensional latent heating structure over the global Tropics for February 1998 was obtained using TRMM rain products[5] in Tao et al. (2001b). Figure 9.8 shows monthly (Feb 1998) mean latent heating at three different altitudes (2, 5, and 8 km) over the global Tropics from the CSH algorithm. The horizontal distributions

---

[5] TRMM rainfall products are the daily 0.5° gridded rainfall and percentage of rainfall classified as convective from the TMI, SSM/I, and the PR, respectively, for February 1998.

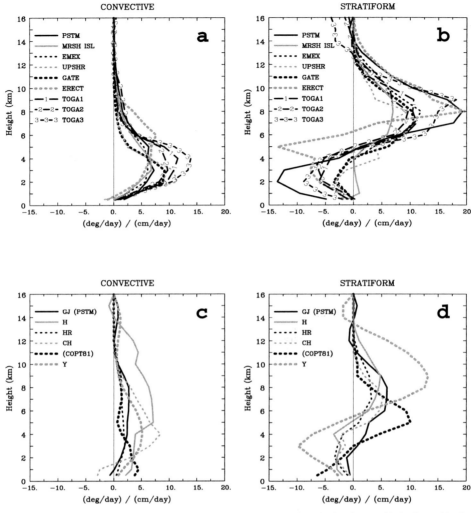

FIG. 9.7. (a) Convective and (b) stratiform heating profiles stored in the heating profile lookup table for the CSH algorithm. The profiles were obtained from GCE model simulations for cloud systems in various geographic locations [the Pacific warm pool region (TOGA1, TOGA2, TOGA3, ERECT—a squall system with erect updrafts; UPSHR—a squall system with upshear tilted updrafts; MRSH ISL—Marshall Island), the East Atlantic region (GATE), midlatitude United States (PRESTORM—PSTM), and Australia (EMEX)]. (c),(d) Same as (a),(b) except that these profiles are from Gallus and Johnson (1991; curve GJ) and Yanai et al. (1973) but partitioned into convective and stratiform components by Johnson (1984; curve Y), Houze (1989; curve H), Houze and Rappaport (1984; curve HR), Chong and Hauser (1990; curve CH), and an African squall line simulated by Caniaux et al. (1994; curve COPT81).

or patterns of latent heat release identify the areas of major convective activity [i.e., a well defined intertropical convergence zone (ITCZ) in the Pacific, a distinct South Pacific convergence zone (SPCZ)] in the global Tropics. A well defined ITCZ in the east and central Pacific and in the Atlantic Ocean, a distinct SPCZ, and broad areas of precipitation events spread over the continental regions are all present. Also, stronger latent heat release (10 K day^{-1} or greater) in the middle and upper troposphere is always associated with heavier surface precipitation. Heating in the upper troposphere over the Pacific and Indian Oceans is much stronger than the heating over Africa, South America, and the Atlantic

Ocean. The difference in retrieved convective and stratiform properties between the various geographic locations is the major reason for the difference in the heights of the maximum latent heating level. Higher stratiform amounts always contribute to higher maximum latent heating levels. Whether the higher stratiform proportions and more frequent vigorous convective events in the Pacific are related to the warmer SSTs needs to be studied using multiseason and multiyear retrieved latent heating profiles. Note that differential heating between land and ocean in the upper troposphere could generate strong horizontal gradients in the thermodynamic fields and interact with the global circulation.

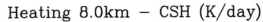

Heating 8.0km – CSH (K/day)

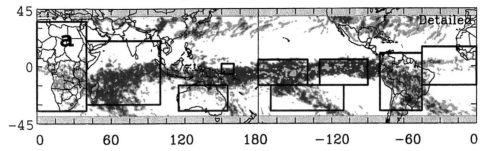

Heating 5.0km – CSH (K/day)

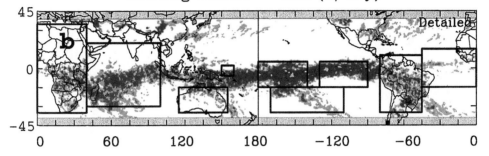

Heating 2.0km – CSH (K/day)

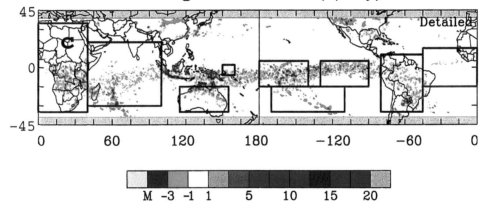

FIG. 9.8. Monthly (Feb 1998) mean latent heating at (a) 8, (b) 5, and (c) 2 km over the global Tropics derived from the CSH algorithm.

One interesting result from Fig. 9.8 is the relatively strong cooling (-1 to -2 K day^{-1}) at 2 km over the (East, Central, and South) Pacific and Indian Oceans but not Africa and South America. This result is due to the fact that the TMI observations had less stratiform precipitation over the continental regions, which is not conducive to retrieving stronger low-level cooling over the continental regions relative to the tropical oceans. However, it is still not an expected result because the moisture content is higher over oceans. Cooling by evaporation of raindrops in the lower troposphere should be stronger over dry areas. Several previous observational studies were performed to analyze the heating budget obtained from sounding networks over the Pacific warm pool region and the Amazon region. For example, Lin and Johnson (1996) found weak cooling at low levels, probably induced by mesoscale downdrafts or evaporation by shallow cumuli, in the mean heating profile over the TOGA COARE region for the month of February 1993. In addition, LeMone et al. (1998) found that the relative humidity tended to be lower in the lower troposphere in TOGA COARE. This can also induce more evaporative cooling in the lower troposphere. Greco et al. (1994) calculated latent heating profiles from the Amazon Boundary Layer Experiment (ABLE) network. Their results indicated that the distribution of heating is quite similar to the studies of those of West African squall lines (Chong and Hauser 1990). Peak heating occurs between 500 and 550 hPa (about 5–6 km). Their results did not exhibit low-level diabatic cooling for the ABLE case. They suggested that the lowermost 2–3 km over the Amazon rain forest canopy is characterized by a strong diurnal cycle of evapotranspiration and upward convective fluxes of moisture producing very large mixing ratios (Fitzjarrald et al. 1990). Model results (Scala et al. 1990) also suggested that dry tropospheric air is not present for the production and maintenance of evaporatively cooled downdrafts. The high moisture content during the wet season in the lower troposphere of the Amazon basin may prevent or severely limit cooling below cloud base. Thus, more low-level cooling over the Pacific than over South America as estimated by the CSH heating algorithm is, perhaps, reasonable.

2) COMPARISON WITH OTHER LATENT HEATING ALGORITHMS' RESULTS

Two other latent heating retrieval algorithms, the Goddard profiling (GPROF) heating and the hydrometeor heating (HH) were also used to estimate the latent heating for February 1998, and their results were compared to the those estimated by the CSH algorithm. The horizontal distribution or patterns of latent heat release from the three different heating retrieval methods are quite similar. They all can identify the areas of major convective activity (i.e., a well-defined ITCZ in the Pacific, a distinct SPCZ) in the global Tropics. The magnitude of their estimated latent heating is also in good agreement with each other. However, the major difference among these three heating retrieval algorithms is the altitude of the maximum heating level. The CSH algorithm–estimated heating profiles show one maximum heating level, and the level varies with convective activity and geographic location. These features are in good agreement with the heating profiles obtained from the results of diagnostic studies over a broad range of geographic locations (Yanai et al. 1973; Johnson 1984, 1992; Thompson et al. 1979; Houze 1989; Frank and McBride 1989; Greco et al. 1994; Frank et al. 1996; Lin and Johnson 1996; and many others). In contrast, a broader heating maximum, often with two embedded peaks, is generally derived from applications of the GPROF heating and HH algorithms, and the response of the heating profiles to convective activity is less pronounced. Also, the GPROF and HH algorithms generally yield heating profiles with a maximum at somewhat lower altitudes than the CSH algorithm.

3) COMPARISON WITH DIAGNOSTIC BUDGET STUDIES

The GCE model simulated heating profiles stored in the lookup table shown in Figs. 9.7a,b all have a char-

acteristic shape for the convective and stratiform regions (e.g., Houze 1982, 1997). These include maximum convective heating in the lower to middle troposphere, maximum stratiform (anvil) heating in the upper troposphere, and regions of stratiform cooling prevailing in the lower troposphere. Also, larger heating aloft in the stratiform region is associated with larger cooling in the lower troposphere. However, some notable differences do exist. For example, GCE-modeled latent heating profiles are sometimes twice as large as those determined from diagnostic budgets (Figs. 9.7c,d). The level separating the heating and cooling in the stratiform region (indicating the freezing or melting level) is different for the convective systems simulated by the GCE model and determined by diagnostic budgets. The differences in the height of the stratiform region cooling probably reflect differences in melting-layer height or the type of convective systems (a system having erect updrafts has a higher height). The cooling is quite strong near the surface for the African convective system due to a dry boundary layer (Caniaux et al. 1994). The latent heating profiles modeled by the GCE and determined kinematically[6] are quite different for the GATE convective system. Nevertheless, there is, perhaps, more similarity than difference in these profiles shown in Fig. 9.7. This may imply that the lookup table may not need a significant number of heating profiles.

Heating profiles for the TRMM field campaign sites [i.e., the South China Sea Monsoon Experiment (SCSMEX), May–Jun 1998; TRMM Large-scale Biosphere–Atmosphere Experiment (LBA) in Brazil, Jan–Feb 1999; and the Kwajalein Experiment (KWAJEX), Jul–Sep 1999] as well as other major field campaigns such as the Department of Energy Atmospheric Radiation Measurement (DOE ARM) will be produced using the three different heating algorithms, and these will be compared to profiles determined from the field campaign sounding networks. This future comparison can provide an assessment of the absolute and relative errors of the heating retrieval algorithms. In addition, global analyses will be used to identify/compare the large-scale circulation patterns for the retrieved periods and for periods during previous field campaigns (i.e., TOGA COARE and GATE). It is reasonable to assume that the latent heating structures for westerly wind bursts (WWBs) and supercloud clusters (SCCs) occurring in similar large-scale circulations and with similar SSTs may not be very different.

e. Response of tropical deep cloud systems to large-scale processes

The role of clouds/cloud systems in global energy and hydrological balance is very complex. On the one

---

[6] The convective and stratiform heating profiles were derived using composite "kinematic and thermodynamic" fields from radar, upper-air soundings, and aircraft-measured winds.

hand, clouds owe their origin to large-scale dynamical forcing, radiative cooling in the atmosphere, and turbulent transfer processes between the ground and the atmosphere (e.g., the transfer of heat and moisture from the ocean to the atmosphere). On the other hand, the latent heat from precipitating clouds provides most of the energy received by the atmosphere. Clouds also serve as important mechanisms for the vertical redistribution of momentum, trace gases (including the greenhouse gas, CO_2), and sensible and latent heat on the large scale. They also influence the coupling between the atmosphere and the earth's surface as well as the radiative and dynamical–hydrological balance.

The use of CRMs in the study of tropical convection and its relation to the large-scale environment can be generally categorized into two groups. The first approach is so-called *cloud ensemble modeling*. In this approach, many clouds of different sizes in various stages of their lifecycles can be present at any model simulation time. The large-scale effects that are derived from observations are imposed into the models as the main forcing, however. In addition, the cloud ensemble models use cyclic lateral boundary conditions (to avoid reflection of gravity waves) and require a large horizontal domain (to allow for the existence of an ensemble of clouds). The clouds simulated from this approach could be termed *continuous large-scale forced convection*. This approach is mainly applied to simulations associated with tropical deep convection. On the other hand, the second approach for cloud resolving models usually requires initial temperature and water vapor profiles, which have a medium to large CAPE, and an open lateral boundary condition is used. The modeled clouds, then, are initialized with either a cool pool, warm bubble, or surface processes (i.e., land/ocean fluxes). These modeled clouds could be termed *self-forced convection*. The key developments in cloud ensemble modeling using the continuous large-scale forced convection approach over the past two decades are listed in Table 9.2.

1) SIMULATED RESULTS FROM THE GCE MODEL

Tao et al. (1987) used 2D and 3D versions of the GCE model to study the statistical properties of cloud ensembles for a well-organized ITCZ rainband that occurred during GATE. The statistical properties of clouds, such as mass flux by cloud drafts and vertical velocity as well as condensation and evaporation, were examined. Figure 9.9 shows heating rates by condensation (c) and evaporation (e) in the 2D model and the 3D model. Using Q_1, the apparent heat source budget defined in Yanai et al. (1973), the heating rate estimated from large-scale observations, $Q_1 - Q_R$, and the total cloud heating rate in the 3D model are also included. The rate of condensation in the 3D case is slightly larger than its 2D counterpart, but so is the rate of evaporation. As a consequence, the net heating effect of clouds in the 3D model is nearly equal to the 2D counterpart, and

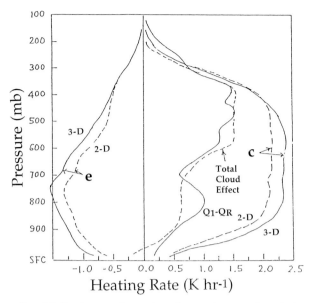

FIG. 9.9. Heating rates for condensation c and evaporation e in the 2D model (dashed line) and the 3D model (solid line). The heating rate estimated from the large-scale observations, $Q_1 - Q_R$, and the total cloud heating rate in the 3D model (marked as total cloud effect) are also included.

both agree with $Q_1 - Q_R$ as estimated from the large-scale heat budget. The GCE model also found that the 3D modeled surface rainfall rates have a smaller standard deviation in time than their 2D counterparts. Overall, the GCE model results indicated that collective thermodynamic feedback effects and vertical transports of mass, sensible heat, and moisture by the convective cells show profound similarities between the two- and three-dimensional GCE model simulations.

Zipser and LeMone (1980) and LeMone and Zipser (1980) presented the results of statistical analyses of convective updrafts and downdrafts. Their analyses were based on aircraft data gathered from cumulonimbus cloud penetrations for six days during GATE. In order to facilitate a comparison between our model results and their analysis results, we subdivided the updrafts and downdrafts into active or inactive updrafts and downdrafts (Table 9.9). For example, a grid point in the model is designated as an active updraft region if (a) the total liquid water content exceeds 0.01 g kg^{-1} and (b) the vertical velocity is larger than 1 m s^{-1} (or 2 m s^{-1}, depending upon how we define "active") at that grid point and at that integration time. The total cloud draft coverages measured by the aircraft are larger than either two- or three-dimensional model predictions. The degree of agreement of the active cloud fractional area coverage between the aircraft measurements and model results is fairly good. In addition, the ratios between the active cloud updrafts and downdrafts (with absolute vertical velocity of 1 m s^{-1}) indicated an excellent agreement among results between the two- and three-dimensional models as well as the cores as mea-

TABLE 9.9. Ratio of fractional cloud coverage [R = cloud updraft coverage (%)/cloud downdraft coverage (%)]. Fractional coverage occupied by cloud drafts and active cloud drafts over the domain are also shown within the parentheses. This table is from Tao et al. (1987).

Altitude range (m)	Zipser and LeMone		2D			3D		
	Draft	Core	Cloudy	Active 1 m s⁻¹	Active 2 m s⁻¹	Cloudy	Active 1 m s⁻¹	Active 2 m s⁻¹
9500			1.36 (34.6/25.4)	2.92 (2.98/1.02)	8.00 (1.60/0.20)	1.60 (45.9/28.8)	1.98 (2.67/1.35)	8.60 (1.54/0.18)
4300–8100	0.57 (16.9/29.9)	2.56 (4.6/1.8)	1.04 (11.3/10.9)	2.80 (4.2/1.5)	12.70 (2.8/0.22)	0.81 (11.7/14.4)	2.87 (4.3/1.5)	22.20 (2.89/0.13)
2500–4300	0.60 (18.3/30.3)	1.30 (2.1/1.1)	0.75 (8.4/11.2)	1.05 (4.1/3.9)	2.45 (2.7/1.1)	0.58 (8.3/14.3)	1.23 (3.8/3.1)	7.21 (2.81/0.39)
700–2500	0.65 (16.3/25.2)	1.91 (2.1/1.1)	1.07 (13.9/12.9)	1.37 (4.1/3.0)	2.63 (2.1/0.8)	0.76 (13.4/17.0)	1.72 (4.3/2.5)	4.90 (2.45/0.50)
300–700	0.88 (16.6/18.8)	1.88 (1.5/0.8)	1.13 (15.3/13.5)	1.24 (2.47/2.00)	2.03 (0.73/0.36)	0.95 (17.2/18.2)	1.68 (3.2/1.9)	3.40 (0.85/0.25)
0–300	1.01 (15.9/15.7)	1.50 (0.3/0.2)	1.11 (15.6/14.0)	1.36 (0.87/0.64)	1.78 (0.16/0.09)	0.99 (17.6/17.8)	1.38 (1.1/0.8)	1.20 (0.12/0.10)

sured by Zipser and LeMone (1980). They all indicate a local minimum ratio near 2500–4300-m altitude. The three-dimensional results are in better agreement with observations in the middle troposphere than the two-dimensional ones. The mean intensities in the numerical experiments and in Zipser and LeMone's (1980) analyzed results are both about 2 to 4 m s⁻¹ in the active updrafts and about −2 to −3 m s⁻¹ in the active downdrafts (see Fig. 9.10). Figure 9.10 shows that the modeled mean intensity for strong and median cores in GATE systems exhibit a *triangular* distribution (defined as sharply peaked extrema in the middle troposphere and, by steady, approximately linear increase and de-

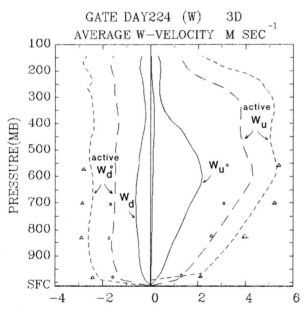

FIG. 9.10. Vertical profiles of the upward and downward mean velocities inside all clouds and mean velocities inside the active updraft and downdraft areas. Aircraft measurements from Zipser and LeMone (1980) are also indicated by circles for the mean (50%) and by triangles for the strong (10%) intensity cores.

crease) in the updrafts and a nearly uniform distribution in the downdrafts. Zipser and LeMone (1980), however, noted that both updrafts and downdrafts are triangle shaped.

Both 2D and 3D model results also showed a similar feature in that the active updrafts account for approximately 75% of the upward mass flux due to clouds and yet they only cover about 12%–14% of the total area (see Fig. 9.11). This result is consistent with the concept, first proposed by Riehl and Malkus (1958; see also Riehl and Simpson 1979), that hot towers play a critical role in the heat and moisture budgets in the Tropics, even though they occupy a small fraction of the area. Overall, our comparison study has indicated that the statistical properties of the clouds obtained in the 2D model are essentially the same as the 3D counterpart given an identical large-scale environment (see Tao 1983). The explanation for this similarity between the 2D and 3D simulations is that the same large-scale advective forcing in temperature and water was superimposed into the GCE model as the main forcing. The cyclic lateral boundary condition used in the GCE model does not allow for additional forcing in the model domain. A two-dimensional simulation should, therefore, give a good approximation of the continuous large-scale forced convection.

Large-scale models (i.e., general circulation and climate models) require not only the global surface rainfall pattern but also the associated vertical distribution within the Q_1 and Q_2 budgets, where Q_2 is the apparent moisture sink budget defined in Yanai et al. (1973). The GCE model can help to identify which processes should be parameterized by the large-scale model, as well as provide information on the vertical profiles of the Q_1 and Q_2 budgets (Tao 1978; Soong and Tao 1980; Tao and Soong 1986; Tao et al. 1993a; and many others). The GCE model was used to examine the Q_1 and Q_2 budgets of various cloud systems that developed in different geographic locations (GATE, EMEX, PRES-

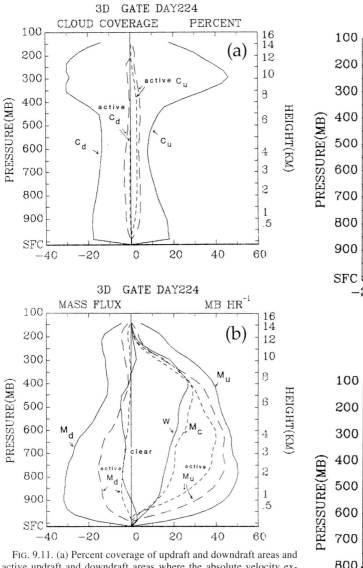

FIG. 9.11. (a) Percent coverage of updraft and downdraft areas and active updraft and downdraft areas where the absolute velocity exceeds 1 (long-dashed line) or 2 m s^{-1} (short-dashed line). Here C_u represents cloud updrafts and C_d cloud downdrafts. (b) as in (a) except for mass fluxes. Here M_u denotes upward mass flux inside clouds and M_d downward mass flux inside clouds; M_c ($=M_u + M_d$) is the total cloud mass flux; ω is the domain-averaged vertical velocity.

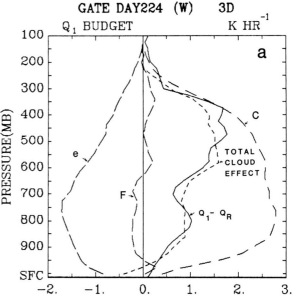

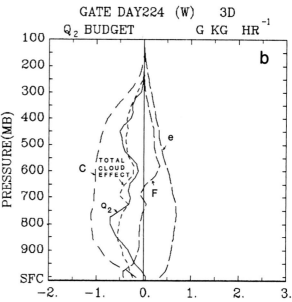

FIG. 9.12. (a) Vertical profiles of the heating rate by condensation of moisture, c; evaporation of liquid water drops, e; net vertical flux of sensible heat, F; the total heating rate by clouds and the heating rate estimated from large-scale observations, $Q_1 - Q_R$. (b) Vertical profiles of the moistening rate by condensation of moisture, evaporation from liquid water drops, net vertical moisture flux, the total moistening rate by clouds, and the moistening rate estimated from the large-scale observations, $-(c_p/L) Q_2$. These results were from a 3D GCE model simulation of a GATE convective system (Tao and Soong 1986).

TORM, TOGA COARE, ABLE, TAMEX, and others). In all of these simulations, the heating due to the vertical eddy convergence/divergence term in sensible heat by convective clouds is always one order of magnitude smaller than that produced by condensation at most levels (Fig. 9.12a). On the other hand, the maximum value of the cooling rate by evaporation is more than half of the heating rate by condensation. This finding implies that the sum of the condensation and evaporation would provide a good approximation to the total cloud heating rate. The cloud heating effect would be considerably overestimated if heating by condensation alone is considered, ignoring cooling by evaporation. For the Q_2

budget, the GCE model results indicated that the net vertical eddy convergence/divergence of moisture by clouds is generally smaller than the rate of condensation or evaporation, but it is not negligible (Fig. 9.12b). The different roles of the vertical eddy convergence/divergence term in the Q_1 (temperature) and Q_2 (water vapor) budgets is the major reason for Q_1 and Q_2 decoupling (the level of maximum values in the Q_1 and Q_2 profiles is not at the same level). The GCE model-generated heating and drying effects agree well with those estimated from observations.

GATE (1974, in the East Atlantic) and TOGA COARE (1992–93, in the West Pacific warm pool region) are perhaps two of the best-planned and well-coordinated field campaigns for understanding tropical convective systems and their interactions with the large-scale environments within which they are embedded. The GCE model was integrated for 7 and 8 days, respectively, for GATE (1–8 Sep 1974) and TOGA COARE (19–27 Dec 1992). The large-scale environments associated with these organized cloud systems that occurred in TOGA COARE and GATE were quite different. The large-scale advective forcing in temperature and water vapor as well as the large-scale vertical velocity are stronger for TOGA COARE than for GATE. The large-scale vertical velocity shows a diurnal signature in TOGA COARE but not in GATE. The (spatial and temporal) mean CAPE is larger in GATE than in TOGA COARE. Note that LeMone et al. (1998) examined the wind shear and thermodynamic conditions in determining the structure and evolution of (20 individual) mesoscale convective systems during TOGA COARE. Their results indicated that the environmental sounding in TOGA COARE tended to have higher CAPEs than GATE. Their results are more representative of preconvective conditions than the mean (average over large area and several days). The SST is higher for TOGA COARE (about 29°C vs 27.4°C for GATE). The vertically integrated water vapor content (precipitable water) is much drier for GATE (2.47 g cm^{-2}) than TOGA COARE (5.15 g cm^{-2}). The mean vertical shear from the surface to 700 mb of the large-scale horizontal wind is slightly larger for GATE than TOGA COARE during the GCE model simulation periods. However, the shear is much stronger from over the entire depth of the troposphere in TOGA COARE. The low-level wind shear can determine the organization of convective systems. An improved 2D GCE model (ice microphysics, cloud–radiation interaction, dynamics, and surface fluxes) was used to study the response of convective systems to the large-scale environment. Both (TOGA COARE and GATE) runs used 1024 horizontal grid points with a 1-km resolution and 41 vertical grid points with varying resolution (40 m near the surface to 1000 m at the top level). The time step was 7.5 s.

Figures 9.13a,b show the temporal variation of the GCE model simulated domain mean surface rain rate for TOGA COARE and GATE, respectively. There are more convective systems simulated by the GCE model for TOGA COARE than for GATE. This is due to the stronger large-scale forcing imposed in the TOGA COARE simulation. The model-simulated surface precipitation showed a very complex structure for TOGA COARE compared with GATE. Overall, the GCE model-simulated cloud systems propagated in one direction while the individual cells embedded within the systems propagated in the opposite direction. In addition, the cloud tops propagate in the opposite direction of the associated surface precipitation. These two hierarchies of convective organization are in good agreement with other modeling studies (Wu et al. 1998). In the GATE simulation, only shallow convective systems developed during the first day. Then, deep convective clouds and nonsquall (slow moving) cloud systems developed and propagated westward with the mean wind. Squall-line type (fast moving) cloud systems developed after 4 September. After 6 September, the systems simulated by the GCE model were less organized and produced less surface precipitation compared to the nonsquall and squall systems. The GCE model simulated GATE features are in good agreement with other modeling studies (Grabowski et al. 1996; Xu and Randall 1996) and observations.

The GCE model simulated domain-averaged surface rainfall (mm), and stratiform amount (percentage) for both TOGA COARE and GATE are shown in Table 9.10. The ratios between evaporation and condensation, sublimation and deposition, and deposition and condensation were examined for both cases. These ratios illustrate the relative importance of warm versus ice processes and source and sink terms associated with water vapor over the course of the TOGA COARE and GATE simulations. The microphysical processes are broken down according to convective organization (i.e., slow-moving, fast-moving, less-organized convective episodes from GATE, vigorous deep convection, and weaker convective events during the westerly wind burst period) in Table 9.10. As expected, more surface rainfall was simulated by the GCE model for TOGA COARE (153.9 mm) than for GATE (91.46 mm). Also, a higher stratiform component was simulated for TOGA COARE (45%) than for GATE (32%). The surface rainfall and stratiform component simulated by the GCE model for TOGA COARE are in reasonable agreement with the rainfall determined from soundings and the stratiform amount measured by radar [see Tao et al. (2000) for a detailed comparison]. This close agreement is mainly caused by the fact that the GCE model was forced by large-scale tendencies in temperature and water vapor that were derived from the sounding network. However, the GCE model simulated surface rainfall is almost twice that estimated by radar. Johnson and Ciesielski (2000) indicated that the ship radars were located within a relatively dry region of the intensive flux array (IFA). The lower rainfall estimates from the ship radars could also be caused by the specific Z–R relationship applied in Short et al. (1997). Based on radar observations (Houze 1997), the GCE model may have underestimated the stratiform rain for GATE fast-moving squall systems. The dominance of warm rain processes in the

(a)   (b)

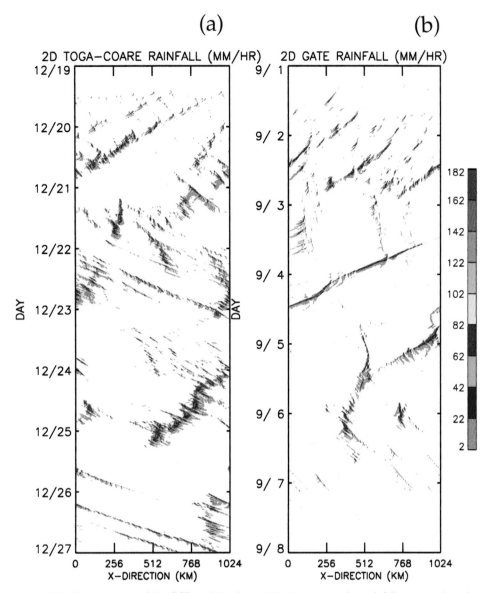

FIG. 9.13. Time sequence of the GCE model estimated domain mean surface rainfall rate (mm h^{-1}) for (a) TOGA COARE and (b) GATE. This type of CRM diagnostic and graphical presentation has been very popular and was first presented in Tao and Simpson (1984).

GATE squall and nonsquall convective systems may explain the smaller stratiform rain amounts simulated by the GCE model. Very little ice processes on 6 and 8 September are an indication of shallower convection. In contrast, ice processes are quite important for both active and relatively inactive convective periods during TOGA COARE. The GCE model results also indicated that evaporation was only 54% of the condensation in GATE compared to 71% in TOGA COARE. This may be a reflection of the lower troposphere being drier in TOGA COARE than in GATE (LeMone et al. 1998). Weak convective episodes in both GATE and TOGA COARE had high ratios between evaporation and con-

densation compared to more intense convective periods. The ratio of sublimation to deposition was smaller in the GATE simulation.

2) COMPARISON WITH OTHER CLOUD RESOLVING MODEL RESULTS

A two-dimensional cloud model with a third-moment turbulence closure for simulating an ensemble of cumulus clouds was developed by Krueger (1988). He simulated the response of cumulus clouds to large-scale forcing, under large-scale conditions observed during GATE. Krueger (1988) found that cloud-scale vertical

TABLE 9.10. The GCE model simulated domain-averaged surface rainfall (mm), stratiform amount (%), and microphysical processes (ratios between evaporation and condensation, sublimation and deposition, and deposition and condensation): (a) TOGA COARE and (b) GATE. For TOGA COARE, the GCE model results are also separated into subperiods, deep strong convection during 20–23 and 24–25 Dec and weaker convection prior to, in between, and after the deep convection (19–20, 23–24, and 25–26 Dec 1992). Slow-moving (nonsquall, 2–4 Sep), fast-moving (squall, 4–6 Sep) and less organized (6–8 Sep) periods for the GCE model simulated GATE results are also shown.

(a) TOGA COARE (19–26 Dec 1992)

	19–26 Dec	19–20, 23–24, and 25–26 Dec	20–23 and 24–25 Dec
Total surface rainfall (mm)	153.9	29.06	124.84
Stratiform amount (%)	45%	42%	55%
Evaporation/condensation	71%	80%	69%
Sublimation/deposition	50%	56%	48%
Deposition/condensation	39%	39%	39%

(b) GATE (2–8 Sep 1974)

	2–8 Sep	Slow moving 2–4 Sep	Fast moving 4–6 Sep	Random 6–8 Sep
Total surface rainfall (mm)	91.46	43.34	39.62	8.50
Stratiform amount (%)	32%	27%	26%	44%
Evaporation/condensation	54%	58%	44%	68%
Sublimation/deposition	32%	36%	27%	24%
Deposition/condensation	22%	23%	25%	9.5%

transport of moisture and evaporative cooling are significant in the Q_2 and Q_1 budgets, respectively. The cloud-scale vertical advection of heat is only important in the subcloud layer in the Q_1 budget. These results are consistent with our GCE model simulations. Lafore and Redelsperger (1991) applied a two-dimensional cloud model to simulate a fast-moving tropical squall line observed during Convection Profonde Tropical (COPT81) and a frontal system observed during Europe's FRONTS87 experiment. Their results also indicated the importance of evaporative cooling and cloud transport of moisture for these two cases. Furthermore, their results showed a relatively small effect by cloud transport of heat on the Q_1 budget except near the subcloud layer. The different roles of the vertical eddy convergence/divergence term in the Q_1 (temperature) and Q_2 (water vapor) budgets are also the major reason for Q_1 and Q_2 decoupling in both systems as indicated by Lafore and Redelsperger (1991).

Grabowski et al. (1998) examined the effects of resolution and the third spatial dimension for cloud systems observed during phase III of GATE (1–7 Sep 1974). Xu and Randall (1996) used the two-dimensional model developed by Krueger (1988) to simulate cloud systems observed during phase III of GATE (1–18 Sep 1974). Wu et al. (1998) also used a two-dimensional model to examine the cloud properties associated with cloud systems observed during TOGA COARE (5 Dec 1992–12 January 1993). Donner et al. (1999) used a three-dimensional model developed by Lipps and Helmer (1986) to simulate several GATE convective systems. The major difference for these modeling studies (and the improved GCE model simulation shown in Fig. 9.13) from the previous CRM simulations (i.e., Tao and Soong 1986; Tao et al. 1987; Krueger 1988) is that they performed long-term integrations. All these studies sim-

ulated Q_1 and Q_2 budgets that are in good agreement with observations. This is due to the fact that the observed large-scale advective forcing in temperature and water vapor was imposed as suggested by Soong and Tao (1980). Cloud organization in all these studies also agreed well with observations due to the fact that the modeled simulated horizontal wind was relaxed to the observed time-varying large-scale horizontal wind. The importance of vertical shear of the large-scale horizontal wind on the organization of tropical and midlatitude convective systems was recognized in observational works (Chisholm and Renick 1972; LeMone and Moncrieff 1994; LeMone et al. 1998), theoretical studies (i.e., Moncrieff 1992; LeMone and Moncrieff 1994), and numerical simulations (Tao 1983; Weisman and Klemp 1984; Tao and Soong 1986; Dudhia et al. 1987).

Larger temporal variability in the two-dimensional integration than in the three-dimensional integration was found in Grabowski et al. (1998) and Donner et al. (1999). Donner et al. (1999) suggested that this is probably related to the different behavior of the CAPE and convective inhibition (CIN) in two and three dimensions. Grabowski et al. (1998), however, concluded that, as long as high-frequency temporal variability is not of primary importance, low-resolution two-dimensional simulations can be used as realizations of tropical cloud systems for addressing the climate problem and for improving and testing cloud parameterizations for large-scale models. This conclusion is only valid for CRMs using large-scale advective forcing and applied with periodic lateral boundary conditions. A similar conclusion was also obtained using the GCE model.

However, there are several notable differences between two- and three-dimensional CRM simulations. For example, a weaker convective updraft and a stronger convective downdraft velocity were simulated for a

GATE fast-moving system in the GCE two-dimensional model compared with in the three-dimensional model. Yet the total upward and downward mass fluxes are almost identical between the two- and three-dimensional GCE model simulations. Lipps and Helmer (1986), however, found that their two-dimensional model had stronger upward and downward mass fluxes than their three-dimensional counterpart for the same GATE simulation as Tao and Soong (1986). They also found more evaporation of cloud water in the two-dimensional simulation and, consequently, less cloud water was present. These results are very different from those of Wu and Moncrieff (1997) for simulations of TOGA COARE convective systems. More ice water and liquid water were simulated in the two-dimensional model than the three-dimensional model in Wu and Moncrieff (1997). The different cases simulated between Lipps and Helmer (1986) and Wu and Moncrieff (1997) is, perhaps, one of the major reasons for the differences. The microphysical schemes used in these two studies are also different. More detailed comparisons are needed in the future. The GCSS model intercomparison project and field campaigns [(ARM, TRMM LBA, KWAJEX), and NASA Convection and Moisture Experiment (CA-MEX)] can provide good quality observational datasets for CRM initialization as well as for its validation.

4. Future developments and works

There is much more work to be done comparing simulated cloud systems over various types of land and vegetation environments, ranging from arid to jungle. Recently completed field programs (TOGA COARE, ARM, TRMM LBA, TRMM KWAJEX, and NASA CA-MEX) should provide a good opportunity to orchestrate combined observational and numerical studies of convective systems. These large-scale field campaigns can provide some of the desperately needed observations for key locations. These observations can guide and correct existing microphysical schemes used in the CRMs.

Recently, physical processes represented in the spectral bin-microphysical scheme have been implemented into the two-dimensional version of the GCE model. The formulation of the microphysical processes is based on solving stochastic kinetic equations for the size distribution functions of water droplets (cloud droplets and raindrops), and six types of ice particles: ice crystals (columnar, platelike, and dendrites), snowflakes, graupel, and frozen drops. Each type is described by a special size distribution function containing 43 categories (bins). The bulk density is equal to 0.9 g cm^{-3} for ice crystals. Snowflakes, graupel, and frozen drops are assumed to be spheres and their densities range from 0.01 to 0.9 g cm^{-3}. The terminal fall velocities used are those applied by Khain and Sednev (1996), List and Schemenauer (1971), and Cotton et al. (1986). Nucleation (activation) processes are based on the size distribution function for cloud condensation nuclei (43 size categories). The GCE model using the spectral bin microphysics can be used to study cloud–aerosol interactions and nucleation scavenging of aerosols, as well as the impact of different concentrations and size distributions of aerosol particles upon cloud formation. These findings will, in turn, be used to improve the bulk parameterizations. With the improved GCE model, it is expected to lead to a better understanding of the mechanisms that determine the intensity and the formation of precipitation for a wide spectrum of atmospheric phenomenon (i.e., clean or dirty environment) related to clouds.

In addition, cloud microphysical processes, heat fluxes from the warm ocean and land, and radiative transfer processes should interact with each other. How these processes interact under different environmental conditions should be a main focus of modeling studies in the future. Also, a major area of needed development involves scale interactions and how cloud processes must be included in simulations of mesoscale to global-scale circulation models. Specifically, Moncrieff and Tao (1999) suggested that improved CRMs can be used to address the following aspects in the near future:

1) derive physically based parameterizations for numerical weather prediction models and climate models;
2) test single-column representations of physical processes (i.e., the processes that trigger convection, cloudiness, and convective momentum transport);
3) complement large-scale field experiments that would otherwise be subcritical in terms of cloud-scale measurements;
4) add value to datasets in situations where standard soundings are the only measurement available;
5) improve the physical basis of surface– (land– and ocean–)-atmosphere interaction in coupled climate models;
6) help in the design of space-based and earth-based remote sensing and in the interpretation of the datasets; and
7) understand the vortex formation that may be important for initial tropical cyclone (hurricane) development.

Almost all cloud systems presented in this paper organized to form a line structure (squall system). A two-dimensional simulation, therefore, gives a good approximation to such a line of convective clouds. Since the real atmosphere is three-dimensional, three-dimensional cloud resolution model simulations are also needed to address the above scientific problems.

Acknowledgments. The work presented in this paper was done collectively by many members of the mesoscale modeling group in the Mesoscale Atmospheric Processes Branch, Dr. W. Lau's group in the Climate and Radiation Branch, and Drs. A. Thompson and K.

Pickering in the Atmospheric Chemistry and Dynamics Branch at NASA/Goddard Space Flight Center.

The author appreciates the inspiring and enthusiastic support by his mentor, Dr. Joanne Simpson, over the past 18 years. Dr. Simpson has also been a good example for the author to follow due to her high expectations in doing quality research and publications. She also taught the author to appreciate the art of observational work. The author would like to also thank his Ph.D. advisor Dr. S.-T. Soong for his guidance and advice over the past 25 years. The author also thanks Mr. S. Lang for reading the manuscript and Drs. R. Adler and F. Einaudi of NASA/Goddard Space Flight Center for their support.

The author is grateful to Dr. R. Kakar at NASA headquarters for his continuous support of this research. In addition, Drs. J. Theon and J. Dodge are acknowledged for their early support of Goddard Cumulus Ensemble model improvements and applications. The work is mainly supported by the NASA headquarters Physical Climate Program and the NASA Tropical Rainfall Measuring Mission (TRMM). The author thanks the members of the Goddard Cumulus Ensemble Modeling Group (Mr. S. Lang, Drs. Y. Wang, C.-L. Shie, B. Ferrier, M. McCumber, and D. Johnson) for their excellent teamwork. The author also thanks Dr. M. LeMone, Dr. R. Wilhelmson, and one anonymous reviewer for their constructive comments that improved this paper considerably.

Acknowledgment is also made to the NASA/Goddard Space Flight Center for computer time used in this research.

REFERENCES

Adler, R. F., and A. J. Negri, 1988: A satellite infrared technique to estimate tropical convective and stratiform rainfall. *J. Appl. Meteor., 27,* 30–51.

——, H.-Y. Yeh, N. Prasad, W.-K. Tao, and J. Simpson, 1991: Microwave rainfall simulations of a tropical convective system with a three-dimensional cloud model. *J. Appl. Meteor., 30,* 924–953.

Baker, R. D., B. H. Lynn, A. Boone, and W.-K. Tao, 2001: The influence of soil moisture, coastline curvature, and the land-breeze circulation on sea-breeze initiated precipitation. *J. Hydrometeor., 2,* 193–211.

Bennetts, D. A., M. J. Bader, and R. H. Marles, 1982: Convective cloud merging and its effect on rainfall. *Nature, 300,* 42–45.

Brown, J. M., 1979: Mesoscale unsaturated downdrafts driven by rainfall evaporation: A numerical study. *J. Atmos. Sci., 36,* 313–338.

Byers, H. R., and R. R. Braham Jr., 1949: *The Thunderstorm: Report of the Thunderstorm Project.* U.S. Government Printing Office, 287 pp.

Caniaux, G., J.-L. Redelsperger, and J.-P. Lafore, 1994: A numerical study of the stratiform region of a fast-moving squall line. Part I: General description of water and heat budgets. *J. Atmos. Sci., 51,* 2046–2074.

Changnon, S. A., Jr., 1976: Effects of urban areas and echo merging on radar echo behavior. *J. Appl. Meteor., 15,* 561–570.

Chen, S., and W. R. Cotton, 1988: The sensitivity of a simulated extratropical mesoscale convective system to longwave radiation and ice-phase microphysics. *J. Atmos. Sci., 45,* 3897–3910.

Chen, Y.-L., and E. J. Zipser, 1982: The role of horizontal advection

of hydrometeors in the water budget of a large squall system. Preprints, *12th Conf. on Severe Local Storms,* San Antonio, TX, Amer. Meteor. Soc., 355–358.

Chin, H.-N. S., 1994: The impact of the ice phase and radiation on a midlatitude squall line system. *J. Atmos. Sci., 51,* 3320–3343.

——, and R. B. Wilhelmson, 1998: Evolution and structure of tropical squall line elements within a moderate CAPE and strong low-level jet environment. *J. Atmos. Sci., 55,* 3089–3113.

——, Q. Fu, M. M. Bradley, and C. R. Molenkamp, 1995: Modeling of a tropical squall line in two dimensions and its sensitivity to environmental winds and radiation. *J. Atmos. Sci., 52,* 3172–3193.

Chisholm, A. J., and J. H. Renick, 1972: The kinematics of multicell and supercell Alberta hailstorms. Hail Studies Report 72-2, Alberta Hail Studies, Research Council of Alberta, 53 pp.

Chong, M., and D. Hauser, 1989: A tropical squall line observed during the COPT 81 experiment in West Africa. Part II: Water budget. *Mon. Wea. Rev., 117,* 728–744.

——, and ——, 1990: A tropical squall line observed during the COPT 81 experiment in West Africa. Part III: Heat and moisture budgets. *Mon. Wea. Rev., 118,* 1696–1706.

Chou, M.-D., 1992: A solar radiation model for use in climate studies. *J. Atmos. Sci., 49,* 762–772.

——, and M. J. Suarez, 1994: An efficient thermal infrared radiation parameterization for use in general circulation models. NASA Tech. Memo. 104606, 85 pp.

Churchill, D. D., and R. A. Houze Jr., 1984: Development and structure of winter monsoon cloud clusters on 10 December 1978. *J. Atmos. Sci., 41,* 933–960.

——, and ——, 1991: Effects of radiation and turbulence on the diabatic heating and water budget of the stratiform region of a tropical cloud cluster. *J. Atmos. Sci., 48,* 903–922.

Clark, T. L., 1979: Numerical simulations with a three-dimensional cloud model: Lateral boundary condition experiments and multicellular severe storm simulations. *J. Atmos. Sci., 36,* 2191–2215.

Cotton, W. R., and G. J. Tripoli, 1978: Cumulus convection in shear flow—Three-dimensional numerical experiments. *J. Atmos. Sci., 35,* 1503–1521.

——, ——, R. M. Rauber, and E. A. Mulvihill, 1986: Numerical simulation of the effect of varying ice crystal nucleation rates and aggregation processes on orographic snowfall. *J. Appl. Meteor., 25,* 1658–1679.

Cox, S. K., and K. T. Griffith, 1979: Estimates of radiative divergence during Phase III of the GARP Atlantic Tropical Experiment: Part II. Analysis of Phase III results. *J. Atmos. Sci., 36,* 586–601.

Cunning, J. B., and M. DeMaria, 1986: An investigation of the development of cumulonimbus systems over South Florida. Part I: Boundary layer interactions. *Mon. Wea. Rev., 114,* 5–24.

——, R. L. Holle, P. T. Gannnon, and A. I. Watson, 1982: Convective evolution and merger in the FACE experiment area: Mesoscale convection and boundary layer interaction. *J. Appl. Meteor., 21,* 953–977.

Del Genio, A. D., M.-S. Yao, W. Kovari, and K. K.-W. Lo, 1996: A prognostic cloud water parameterization for global climate models. *J. Climate, 9,* 270–304.

Dharssi, I., R. Kershaw, and W.-K. Tao, 1997: Longwave radiative forcing of a simulated tropical squall line. *Quart. J. Roy. Meteor. Soc., 123,* 187–206.

Donner, L. J., C. J. Seman, and R. S. Hemler, 1999: Three-dimensional cloud-system modeling of GATE convection. *J. Atmos. Sci., 56,* 1885–1912.

Dudhia, J., 1989: Numerical study of convection observed during the Winter Monsoon Experiment using a mesoscale two-dimensional model. *J. Atmos. Sci., 46,* 3077–3107.

——, M. W. Moncrieff, and D. W. K. So, 1987: The two-dimensional dynamics of west African squall lines. *Quart. J. Roy. Meteor. Soc., 113,* 567–582.

——, and M. W. Moncrieff, 1987: A numerical simulation of quasi-stationary tropical convective bands. *Quart. J. Roy. Meteor. Soc.,* **113,** 929–967.

Ferrier, B. S., 1994: A double-moment multiple-phase four-class bulk ice scheme. Part I: Description. *J. Atmos. Sci.,* **51,** 249–280.

——, W.-K. Tao, and J. Simpson, 1995: A double-moment multiple-phase four-class bulk ice scheme. Part II: Simulations of convective storms in different large-scale environments and comparisons with other bulk parameterizations. *J. Atmos. Sci.,* **52,** 1001–1033.

——, J. Simpson, and W.-K. Tao, 1996: Factors responsible for different precipitation efficiencies between midlatitude and tropical squall simulations. *Mon. Wea. Rev.,* **124,** 2100–2125.

Fitzjarrad, D. R., K. E. Moore, O. M. R. Cabral, J. Scala, A. O. Manzi, and L. D. de Abreu, 1990: Daytime turbulent exchange between the Amazon forest and the atmosphere. *J. Geophys. Res.,* **95,** 16 825–16 838.

Fowler, L. D., and D. A. Randall, 1996: Liquid and ice cloud microphysics in the CSU general circulation model. Part I: Model description and simulated microphysical processes. *J. Climate,* **9,** 489–529.

Frank, W. M., 1993: A hybrid parameterization with multiple closures. *The Representation of Cumulus Convection in Numerical Models,* K. Emanuel and D. Raymond, Eds., Amer. Meteor. Soc., 151–154.

——, and J. L. M. Bride, 1989: The vertical distribution of heating in AMEX and GATE cloud clusters. *J. Atmos. Sci.,* **46,** 3464–3478.

——, H. Wang, and J. L. McBride, 1996: Rawinsonde budget analysis during the TOGA COARE IOP. *J. Atmos. Sci.,* **53,** 1761–1780.

Fu, Q., S. K. Krueger, and K. N. Liou, 1995: Interactions of radiation and convection in simulated tropical cloud clusters. *J. Atmos. Sci.,* **52,** 1310–1328.

Gallus, W. A., Jr., and R. H. Johnson, 1991: Heat and moisture budgets of an intense midlatitude squall line. *J. Atmos. Sci.,* **48,** 122–146.

Gamache, J. F., and R. A. Houze Jr., 1983: Water budget of a mesoscale convective system in the Tropics. *J. Atmos. Sci.,* **40,** 1835–1850.

GEWEX Cloud System Science Team, 1993: The GEWEX cloud system study (GCSS). *Bull. Amer. Meteor. Soc.,* **74,** 387–400.

Grabowski, W. W., X. Wu, and M. W. Moncrieff, 1996: Cloud resolving modeling of tropical cloud systems during PHASE III of GATE. Part I: Two-dimensional experiments. *J. Atmos. Sci.,* **53,** 3684–3709.

——, ——, ——, and W. D. Hall, 1998: Cloud resolving modeling of tropical cloud systems during PHASE III of GATE. Part II: Effects of resolution and the third dimension. *J. Atmos. Sci.,* **55,** 3264–3282.

——, ——, and ——, 1999: Cloud resolving modeling of tropical cloud systems during PHASE III of GATE. Part III: Effects of cloud microphysics. *J. Atmos. Sci.,* **56,** 2384–2402.

Gray, W. M., and R. W. Jacobsen, 1977: Diurnal variation of deep cumulus convection. *Mon. Wea. Rev.,* **105,** 1171–1188.

Greco, S., J. Scala, J. Halverson, H. L. Massie, W.-K. Tao, and M. Garstang, 1994: Amazon coastal squall lines. Part II: Heat and moisture transports. *Mon. Wea. Rev.,* **122,** 623–635.

Gregory, D., and M. J. Miller, 1989: A numerical study of the parameterization of deep tropical convection. *Quart. J. Roy. Meteor. Soc.,* **115,** 1209–1241.

Hartmann, D. L., H. H. Hendon, and R. A. Houze Jr., 1984: Some implications of the mesoscale circulations in tropical cloud clusters for large-scale dynamics and climate. *J. Atmos. Sci.,* **41,** 113–121.

Held, I. M., R. S. Hemler, and V. Ramaswamy, 1993: Radiative-convective equilibrium with explicit two-dimensional moist convection. *J. Atmos. Sci.,* **50,** 3909–3927.

Hill, G., 1974: Factors controlling the size and spacing of cumulus clouds as revealed by numerical experiments. *J. Atmos. Sci.,* **31,** 1934–1941.

Holle, R. L., and M. W. Maier, 1980: Tornado formation from down-draft interaction in the FACE mesonetwork. *Mon. Wea. Rev.,* **108,** 991–1009.

Houze, R. A., Jr., 1977: Structure and dynamics of a tropical squall-line system. *Mon. Wea. Rev.,* **105,** 1540–1567.

——, 1982: Cloud clusters and large-scale vertical motions in the tropics. *J. Meteor. Soc. Japan,* **60,** 396–409.

——, 1989: Observed structure of mesoscale convective systems and implications for large-scale heating. *Quart. J. Roy. Meteor. Soc.,* **115,** 425–461.

——, 1997: Stratiform precipitation in regions of convection: A meteorological paradox. *Bull. Amer. Meteor. Soc.,* **78,** 2179–2196.

——, and C.-P. Cheng, 1977: Radar characteristics of tropical convection observed during GATE: Mean properties and trends over the summer season. *Mon. Wea. Rev.,* **105,** 964–980.

——, and E. N. Rappaport, 1984: Air motions and precipitation structure of an early summer squall line over the eastern tropical Atlantic. *J. Atmos. Sci.,* **41,** 553–574.

Johnson, D., W.-K. Tao, J. Simpson, and C.-H. Sui, 2002: A study of the response of deep tropical clouds to large-scale processes. Part I: Model setup strategy and comparison with observation. *J. Atmos. Sci.,* in press.

Johnson, R. H., 1984: Partitioning tropical heat and moisture budgets into cumulus and mesoscale components: Implication for cumulus parameterization. *Mon. Wea. Rev.,* **112,** 1656–1665.

——, 1992: Heat and moisture sources and sinks of Asian Monsoon precipitating systems. *J. Meteor. Soc. Japan,* **70,** 353–371.

——, and P. J. Hamilton, 1988: The relationship of surface pressure features to the precipitation and airflow structure of an intense midlatitude squall line. *Mon. Wea. Rev.,* **116,** 1444–1471.

——, and P. E. Ciesielski, 2000: Rainfall and radiative heating estimates from TOGA COARE atmospheric budgets. *J. Atmos. Sci.,* **57,** 1497–1514.

Khain, A. P., and I. Sednev, 1996: Simulation of precipitation formation in the Eastern Mediterranean coastal zone using a spectral microphysics cloud ensemble model. *Atmos. Res.,* **43,** 77–110.

Klemp, J. B., and R. Wilhelmson, 1978a: The simulation of three-dimensional convective storm dynamics. *J. Atmos. Sci.,* **35,** 1070–1096.

——, and ——, 1978b: Simulations of right and left moving storms through storm splitting. *J. Atmos. Sci.,* **35,** 1097–1110.

Kogan, Y. L., and A. Shapiro, 1996: The simulation of a convective cloud in a 3D model with explicit microphysics. Part II: Dynamical and microphysical aspects of cloud merger. *J. Atmos. Sci.,* **53,** 2525–2545.

Kraus, E. B., 1963: The diurnal precipitation change over the sea. *J. Atmos. Sci.,* **20,** 546–551.

Krueger, S. K., 1988: Numerical simulation of tropical cumulus clouds and their interaction with the subcloud layer. *J. Atmos. Sci.,* **45,** 2221–2250.

Lafore, J.-P., and J.-L. Redelsperger, 1991: Effects of convection on mass and momentum fields as seen from cloud-scale simulations of precipitating systems. *ECMWF Workshop on Fine-scale Modelling and the Development of Parameterization Schemes,* Reading, U.K., 165–197.

Lau, K. M., C. H. Sui, and W.-K. Tao, 1993: A preliminary study of the tropical water cycle and its sensitivity to surface warming. *Bull. Amer. Meteor. Soc.,* **74,** 1313–1321.

——, ——, M.-D. Chou, and W.-K. Tao, 1994: An enquiry into the cirrus-cloud thermostat effect for tropical sea surface temperature. *Geophys. Res. Lett.,* **21,** 1157–1160.

Leary, C. A., and R. A. Houze Jr., 1980: The contribution of mesocale motions to the mass and heat fluxes of an intense tropical convective system. *J. Atmos. Sci.,* **37,** 784–796.

LeMone, M. A., 1989: The influence of vertical wind shear on the diameter of cumulus clouds in CCOPE. *Mon. Wea. Rev.,* **117,** 1480–1491.

——, and E. J. Zipser, 1980: Cumulonimbus vertical velocity events in GATE. Part I: Diameter, intensity and mass flux. *J. Atmos. Sci.,* **37,** 2444–2457.

——, and M. W. Moncrieff, 1994: Momentum and mass transport by convective bands: Comparisons of highly idealized dynamical models to observations. *J. Atmos. Sci.,* **51**, 281–305.

——, G. M. Barnes, J. C. Fankhauser, and L. F. Tarleton, 1988: Perturbation pressure fields measured by aircraft around the cloud-base updraft of deep convective clouds. *Mon. Wea. Rev.,* **116**, 313–327.

——, E. Zipser, and S. B. Trier, 1998: The role of environmental shear and thermodynamic conditions in determining the structure and evolution of mesoscale convective systems during TOGA COARE. *J. Atmos. Sci.,* **55**, 3493–3518.

Li, X., C.-H. Sui, K.-M. Lau, and M.-D. Chou, 1999: Large-scale forcing and cloud–radiation interaction in the tropical deep convective regime. *J. Atmos. Sci.,* **56**, 3028–3042.

Lin, X., and R. H. Johnson, 1996: Heating, moistening, and rainfall over the western Pacific during TOGA COARE. *J. Atmos. Sci.,* **53**, 3367–3383.

Lin, Y.-L., R. D. Farley, and H. D. Orville, 1983: Bulk parameterization of the snow field in a cloud model. *J. Climate Appl. Meteor.,* **22**, 1065–1092.

Lipps, F. B., and R. S. Helmer, 1986: Numerical simulation of deep tropical convection associated with large-scale convergence. *J. Atmos. Sci.,* **43**, 1796–1816.

List, R., and R. S. Schemenauer, 1971: Free-fall behavior of planar snow crystals, conical graupel and small hail. *J. Atmos. Sci.,* **28**, 110–115.

Liu, C., and M. W. Moncrieff, 1998: A numerical study of the diurnal cycle of tropical oceanic convection. *J. Atmos. Sci.,* **55**, 2329–2344.

Lopez, R. E., 1978: Internal structure and development processes of c-scale aggregates of cumulus clouds. *Mon. Wea. Rev.,* **106**, 1488–1494.

Lynn, B. H., and W.-K. Tao, 2001: A parameterization for the triggering of landscape-generated moist convection. Part II: Zero-order and first-order closure. *J. Atmos. Sci.,* **58**, 593–607.

——, ——, and P. Wetzel, 1998: A study of landscape-generated deep moist convection. *Mon. Wea. Rev.,* **126**, 928–942.

——, ——, and F. Abramopoulos, 2001: A parameterization for the triggering of landscape-generated moist convection. Part I: Analyses of high-resolution model results. *J. Atmos. Sci.,* **58**, 575–592.

Malkus, J. S., 1954: Some results of a trade cumulus clouds investigation. *J. Meteor.,* **11**, 220–237.

McCumber, M., W.-K. Tao, J. Simpson, R. Penc, and S.-T. Soong, 1991: Comparison of ice-phase microphysical parameterization schemes using numerical simulations of convection. *J. Appl. Meteor.,* **30**, 985–1004.

Miller, M. J., and R. P. Pearce, 1974: A three-dimensional primitive equation model of cumulonimbus and squall lines. *Quart. J. Roy. Meteor. Soc.,* **100**, 133–154.

Miller, R. A., and W. M. Frank, 1993: Radiative forcing of simulated tropical cloud clusters. *Mon. Wea. Rev.,* **121**, 482–498.

Molinari, J., and M. Dudek, 1992: Parameterization of convective precipitation in mesoscale numerical models: A critical review. *Mon. Wea. Rev.,* **120**, 326–344.

Moncrieff, M. W., 1992: Organized convective systems: Archetypal dynamical models, momentum flux theory, and parameterization. *Quart. J. Roy. Meteor. Soc.,* **118**, 819–850.

——, and W.-K. Tao, 1999: Cloud-resolving models. *Global Water and Energy Cycles,* K. Browing and R. J. Gurney, Eds., Cambridge University Press, 200–209.

——, S. K. Krueger, D. Gregory, J.-L. Redelsperger, and W.-K. Tao, 1997: GEWEX Cloud System Study (GCSS) Working Group 4: Precipitating convective cloud systems. *Bull. Amer. Meteor. Soc.,* **78**, 831–845.

Newton, C. W., 1950: Structure and mechanism of the prefrontal squall line. *J. Meteor.,* **7**, 210–222.

——, 1963: Dynamics of severe convective storms. *Severe Local Storms, Meteor. Monogr.,* No. 5, Amer. Meteor. Soc., 33–58.

——, 1966: Circulations in large sheared cumulonimbus. *Tellus,* **18**, 699–712.

Nikajima, K., and T. Matsuno, 1988: Numerical experiments concerning the origin of cloud cluster in tropical atmosphere. *J. Meteor. Soc. Japan,* **66**, 309–329.

Ogura, Y., and J. G. Charney, 1962: A numerical model of thermal convection in the atmosphere. *Meteor. Soc. Japan,* **40**, 431–451.

——, and N. A. Phillips, 1962: Scale analysis of deep and shallow convection in the atmosphere. *J. Atmos. Sci.,* **19**, 173–179.

——, and J.-Y. Jiang, 1985: A modeling study of heating and drying effects of convective clouds in an extratropical mesoscale system. *J. Atmos. Sci.,* **42**, 2478–2492.

Orville, H. D., Y.-H. Kuo, R. D. Farley, and C. S. Hwang, 1980: Numerical simulation of cloud interactions. *J. Rech. Atmos.,* **14**, 499–516.

Peng, L., C.-H. Sui, K.-M. Lau, and W.-K. Tao, 2001: Genesis and evolution of super cloud clusters in a 2D numerical cloud resolving model. *J. Atmos. Sci.,* **58**, 877–895.

Pickering, K. E., A. M. Thompson, J. R. Scala, W.-K. Tao, R. R. Dickerson, and J. Simpson, 1992a: Free tropospheric ozone production following entrainment of urban plumes into deep convection. *J. Geophys. Res.,* **97**, 17 985–18 000.

——, J. R. Scala, A. M. Thompson, W.-K. Tao, and J. Simpson, 1992b: A regional estimate of convective transport of CO from biomass burning. *Geophys. Res. Lett.,* **19**, 289–292.

Prasad, N., H.-Y. M. Yeh, R. F. Adler, and W.-K. Tao, 1995: Infrared and microwave simulations of an intense convective system and comparison with aircraft observations. *J. Appl. Meteor.,* **34**, 153–174.

Randall, D. A., Harshvardhan, and D. A., Dazlich, 1991: Diurnal variability of the hydrologic cycle in a general circulation model. *J. Atmos. Sci.,* **48**, 40–62.

Riehl, H., and J. S. Malkus, 1958: On the heat balance in the equatorial trough zone. *Geophysica,* **6**, 503–535.

——, and J. Simpson, 1979: The heat balance of the equatorial trough zone, revisited. *Beitr. Phys. Atmos.,* **52**, 287–305.

Rotunno, R., J. B. Klemp, and M. L. Weisman, 1988: A theory for strong, long-lived squall lines. *J. Atmos. Sci.,* **45**, 463–485.

Rutledge, S. A., 1986: A diagnostic modeling study of the stratiform region associated with a tropical squall line. *J. Atmos. Sci.,* **43**, 1356–1377.

——, and P. V. Hobbs, 1984: The mesoscale and microscale structure and organization of clouds and precipitation in midlatitude clouds. Part XII: A diagnostic modeling study of precipitation development in narrow cold frontal rainbands. *J. Atmos. Sci.,* **41**, 2949–2972.

——, and R. A. Houze Jr., 1987: A diagnostic modeling study of the trailing stratiform rain of a midlatitude squall line. *J. Atmos. Sci.,* **44**, 2640–2656.

Scala, J. R., and Coauthors, 1990: Cloud draft structure and trace gas transport. *J. Geophys. Res.,* **95**, 17 015–17 030.

Schlesinger, R. E., 1975: A three-dimensional numerical model of an isolated deep convective cloud. Preliminary results. *J. Atmos. Sci.,* **32**, 934–957.

——, 1978: A three-dimensional numerical model of an isolated thunderstorm. Part I: Comparative experiments for variable ambient wind shear. *J. Atmos. Sci.,* **35**, 690–713.

Scorer, R. S., and F. H. Ludlam, 1953: Bubble theory of penetrative convection. *Quart. J. Roy. Meteor. Soc.,* **79**, 94–103.

Short, D. A., B. S. Ferrier, J. C. Gerlach, S. A. Rutledge, and O. W. Thiele, 1997: Shipboard radar rainfall patterns within the TOGA COARE IFA. *Bull. Amer. Meteor. Soc.,* **78**, 2817–2836.

Simpson, J., 1980: Downdrafts as linkages in dynamic cumulus seeding effects. *J. Appl. Meteor.,* **19**, 477–487.

——, and W.-K. Tao, 1993: The Goddard Cumulus Ensemble Model. Part II: Applications for studying cloud precipitating processes and for NASA TRMM. *Terr. Atmos. Oceanic Sci.,* **4**, 73–116.

——, N. E. Westcott, R. J. Clerman, and R. A. Pielke, 1980: On cumulus mergers. *Arch. Meteor. Geophys. Bioklim.,* **A29**, 1–40.

——, R. F. Adler, and G. R. North, 1988: A proposed Tropical Rainfall Measuring Mission (TRMM) satellite. *Bull. Amer. Meteor. Soc.,* **69**, 278–295.

——, T. D. Keenan, B. Ferrier, R. H. Simpson, and G. J. Holland, 1993: Cumulus mergers in the Maritime continent region. *Meteor. Atmos. Phys.*, **51**, 73–99.

——, C. Kummerow, W.-K. Tao, and R. Adler, 1996: On the Tropical Rainfall Measuring Mission (TRMM). *Meteor. Atmos. Phys.*, **60**, 19–36.

Smolarkiewicz, P. K., and W. W. Grabowski, 1990: The multidimensional positive advection transport algorithm: Nonoscillatory option. *J. Comput. Phys.*, **86**, 355–375.

Smull, B. F., and R. A. Houze Jr., 1987: Rear inflow in squall lines with trailing stratiform precipitation. *Mon. Wea. Rev.*, **115**, 2869–2889.

Sommeria, G., 1976: Three-dimensional simulation of turbulent processes in an undisturbed trade wind boundary layer. *J. Atmos. Sci.*, **33**, 216–241.

Soong, S.-T., and Y. Ogura, 1980: Response of trade wind cumuli to large-scale processes. *J. Atmos. Sci.*, **37**, 2035–2050.

——, and W.-K. Tao, 1980: Response of deep tropical clouds to mesoscale processes. *J. Atmos. Sci.*, **37**, 2016–2036.

——, and ——, 1984: A numerical study of the vertical transport of momentum in a tropical rainband. *J. Atmos. Sci.*, **41**, 1049–1061.

Steiner, J. T., 1973: A three-dimensional model of cumulus cloud development. *J. Atmos. Sci.*, **30**, 414–435.

Stephens, G. L., 1983: The influence of radiative transfer on the mass and heat budget of ice crystals falling in the atmosphere. *J. Atmos. Sci.*, **40**, 1729–1739.

Su, H., S. S. Chen, and C. S. Bretherton, 1999: Three-dimensional week-long simulations of TOGA-COARE convective systems using the MM5 mesoscale model. *J. Atmos. Sci.*, **56**, 2326–2344.

Sui, C.-H., and K.-M. Lau, 1989: Origin of low-frequency (intraseasonal) oscillations in the tropical atmosphere. Part II: Structure and propagation of mobile wave–CISK modes and their modification by lower boundary forcings. *J. Atmos. Sci.*, **46**, 37–56.

——, ——, W.-K. Tao, and J. Simpson, 1994: The tropical water and energy cycles in a cumulus ensemble model. Part I: Equilibrium climate. *J. Atmos. Sci.*, **51**, 711–728.

——, ——, and X. Li, 1998: Convective–radiative interaction in simulated diurnal variations of tropical cumulus ensemble. *J. Atmos. Sci.*, **55**, 2345–2357.

Tao, W.-K., 1978: A numerical simulation of deep convection in the tropics. M.S. thesis, Department of Atmospheric Science, University of Illinois, 66 pp.

——, 1983: A numerical study of the structure and vertical transport properties of a tropical convective system. Ph.D. dissertation, University of Illinois, 228 pp.

——, 1995: Interaction of parameterized convection and explicit stratiform cloud microphysics. *Cloud Microphysics Parameterizations in Global Atmospheric General Circulation Models*, D. Randall, Ed., WCRP, 199–210.

——, and J. Simpson, 1984: Cloud interactions and merging: Numerical simulations. *J. Atmos. Sci.*, **41**, 2901–2917.

——, and S.-T. Soong, 1986: A study of the response of deep tropical clouds to mesoscale processes: Three-dimensional numerical experiments. *J. Atmos. Sci.*, **43**, 2653–2676.

——, and J. Simpson, 1989a: A further study of cumulus interaction and mergers: Three-dimensional simulations with trajectory analyses. *J. Atmos. Sci.*, **46**, 2974–3004.

——, ——, 1989b: Modeling study of a tropical squall-type convective line. *J. Atmos. Sci.*, **46**, 177–202.

——, and ——, 1993: The Goddard Cumulus Ensemble Model. Part I: Model description. *Terr. Atmos. Oceanic Sci.*, **4**, 19–54.

——, ——, and S.-T. Soong, 1987: Statistical properties of a cloud ensemble: A numerical study. *J. Atmos. Sci.*, **44**, 3175–3187.

——, ——, and M. McCumber, 1989: An ice-water saturation adjustment. *Mon. Wea. Rev.*, **117**, 231–235.

——, ——, S. Lang, M. McCumber, R. Adler, and R. Penc, 1990: An algorithm to estimate the heating budget from vertical hydrometeor profiles. *J. Appl. Meteor.*, **29**, 1232–1244.

——, ——, and S.-T. Soong, 1991: Numerical simulation of a sub-

tropical squall line over the Taiwan Strait. *Mon. Wea. Rev.*, **119**, 2699–2723.

——, ——, C.-H. Sui, B. Ferrier, S. Lang, J. Scala, M.-D. Chou, and K. Pickering, 1993a: Heating, moisture, and water budgets of tropical and midlatitude squall lines: Comparisons and sensitivity to longwave radiation. *J. Atmos. Sci.*, **50**, 673–690.

——, S. Lang, J. Simpson, and R. Adler, 1993b: Retrieval algorithms for estimating the vertical profiles of latent heat release: Their applications for TRMM. *J. Meteor. Soc. Japan*, **71**, 685–700.

——, J. Scala, B. Ferrier, and J. Simpson, 1995: The effect of melting processes on the development of a tropical and a midlatitude squall line. *J. Atmos. Sci.*, **52**, 1934–1948.

——, S. Lang, J. Simpson, C.-H. Sui, B. Ferrier, and M.-D. Chou, 1996: Mechanisms of cloud–radiation interaction in the Tropics and midlatitudes. *J. Atmos. Sci.*, **53**, 2624–2651.

——, J. Simpson, and B. Ferrier, 1997: Cloud resolving model simulations of mesoscale convective systems. *New Insights and Approaches to Convective Parameterization*, D. Gregory, Ed., ECMWF, 77–112.

——, ——, C.-H. Sui, C.-L. Shie, B. Zhou, K. M. Lau, and M. Moncrieff, 1999: On equilibrium states simulated by cloud-resolving models. *J. Atmos. Sci.*, **56**, 3128–3139.

——, S. Lang, J. Simpson, W. S. Olson, D. Johnson, B. Ferrier, C. Kummerow, and R. Adler, 2000: Vertical profiles of latent heat release and their retrieval in TOGA COARE convective systems using a cloud resolving model, SSM/I, and radar data. *J. Meteor. Soc. Japan*, **78**, 333–355.

——, C.-L. Shie, and J. Simpson, 2001a: Comments on the sensitivity study of radiative–convective equilibrium in the Tropics with a convective resolving model. *J. Atmos. Sci.*, **58**, 1328–1333.

——, and Coauthors, 2001b: Retrieved vertical profiles of latent heat release using TRMM products for February 1998. *J. Appl. Meteor.*, **40**, 957–982.

——, and Coauthors, 2002: Microphysics, radiation and surface processes in a non-hydrostatic model. *Meteor. Atmos. Phys.*, in press.

Thompson, R. M., Jr., S. W. Payne, E. E. Recker, and R. J. Reed, 1979: Structure and properties of synoptic-scale wave disturbances in the intertropical convergence zone of the eastern Atlantic. *J. Atmos. Sci.*, **36**, 53–72.

Thompson, A. M., W.-K. Tao, K. E. Pickering, J. Scala, and J. Simpson, 1997: Tropical deep convection and ozone formation. *Bull. Amer. Meteor. Soc.*, **78**, 1043–1054.

Tripoli, G. J., and W. R. Cotton, 1989: Numerical study of an observed orogenic mesoscale convective system. Part II: Analysis of governing dynamics. *Mon. Wea. Rev.*, **117**, 305–328.

Turpeinen, O., 1982: Cloud interactions and merging on day 261 of GATE. *Mon. Wea. Rev.*, **110**, 1238–1254.

Wang, Y., W.-K. Tao, and J. Simpson, 1996: The impact of a surface layer on a TOGA COARE cloud system development. *Mon. Wea. Rev.*, **124**, 2753–2763.

——, ——, ——, and S. Lang, 2002: The sensitivity of tropical squall lines (GATE and TOGA COARE) to surface fluxes: 3D Cloud resolving model simulations. *Quart. J. Roy. Meteor. Soc.*, in press.

Webster, P. J., and G. L. Stephens, 1980: Tropical upper-troposphere extended clouds: Inferences from winter MONEX. *J. Atmos. Sci.*, **37**, 1521–1541.

Weisman, M. L., and J. B. Klemp, 1984: The structure and classification of numerically simulated convective storms in directionally varying wind shears. *Mon. Wea. Rev.*, **112**, 2479–2498.

Westcott, N., 1984: A historical perspective on cloud mergers. *Bull. Amer. Meteor. Soc.*, **65**, 219–226.

Wilhelmson, R. B., 1974: The life cycle of a thunderstorm in three dimensions. *J. Atmos. Sci.*, **31**, 735–743.

——, and J. B. Klemp, 1978: A numerical study of storm splitting that leads to long-lived storms. *J. Atmos. Sci.*, **35**, 1975–1986.

Wilkins, E. M., Y. K. Sasaki, G. E. Gerber, and W. H. Chaplin Jr.,

1976: Numerical simulation of the lateral interactions between buoyant clouds. *J. Atmos. Sci.,* **33,** 1321–1329.

Wu, X., and M. W. Moncrieff, 1997: Recent progress on cloud-resolving modeling of TOGA COARE and GATE cloud systems. *New Insights and Approaches to Convective Parameterization,* D. Gregory, Ed., ECMWF, 128–156.

——, W. W. Grabowski, and M. W. Moncrieff, 1998: Long-term behavior of cloud systems in TOGA COARE and their interactions with radiative and surface processes. Part I: Two-dimensional modeling study. *J. Atmos. Sci.,* **55,** 2693–2714.

——, W. W. D. Hall, W. Grabowski, and M. W. Moncrieff, 1999: Long-term behavior of cloud systems in TOGA COARE and their interactions with radiative and surface processes. Part II: Effects of ice microphysics on cloud–radiation interaction. *J. Atmos. Sci.,* **56,** 3177–3195.

Xu, K.-M., A. Arakawa, 1992: Semiprognostic tests of the Arakawa-Schubert cumulus parameterization using simulated data. *J. Atmos. Sci.,* **49,** 2421–2436.

——, and D. A. Randall, 1995: Impact of interactive radiative transfer on the microscopic behavior of cumulus ensembles. Part II: Mechanisms for cloud–radiation interactions. *J. Atmos. Sci.,* **52,** 800–817.

——, and ——, 1996a: A semiempirical cloudiness parameterization for use in climate models. *J. Atmos. Sci.,* **53,** 3084–3102.

——, and ——, 1996b: Explicit simulation of cumulus ensembles with the GATE Phase III data: Comparison with observations. *J. Atmos. Sci.,* **53,** 3709–3736.

——, and S. K. Krueger, 1991: Evaluation of cloudiness parameterizations using a cumulus ensemble model. *Mon. Wea. Rev.,* **119,** 342–367.

Yanai, M., S. Esbensen, and J. Chu, 1973: Determination of average bulk properties of tropical cloud clusters from large-scale heat and moisture budgets. *J. Atmos. Sci.,* **30,** 611–627.

Yeh, H.-Y., M. N. Prasad, R. Meneghini, W.-K. Tao, and R. F. Adler, 1995: Model-based simulation of TRMM spaceborne radar observations. *J. Appl Meteor.,* **34,** 175–197.

Zipser, E. J., 1969: The role of organized unsaturated convective downdrafts in the structure and rapid decay of an equatorial disturbance. *J. Appl. Meteor.,* **8,** 799–814.

——, 1977: Mesoscale and convective-scale downdrafts as distinct components of squall-line structure. *Mon. Wea. Rev.,* **105,** 1568–1589.

——, and M. A. LeMone, 1980: Cumulonimbus vertical velocity events in GATE. Part II: Synthesis and model core structure. *J. Atmos. Sci.,* **37,** 2458–2469.

——, R. J. Meitin, and M. A. LeMone, 1981: Mesoscale motion fields associated with a slowly moving GATE convective band. *J. Atmos. Sci.,* **38,** 1725–1750.

Chapter 10

Hot Towers and Hurricanes: Early Observations, Theories, and Models

RICHARD A. ANTHES

University Corporation for Atmospheric Research, Boulder, Colorado

ABSTRACT

This paper describes the exciting period of discovery in the 1950s and 1960s in tropical meteorology, and the important role played by Joanne Malkus (Simpson) in her studies of cumulus convection and tropical cyclones. A key concept developed by Joanne, with Herbert Riehl, was that of the "hot tower." Hot towers were deep tropical cumulonimbus clouds whose cores were undiluted by entrainment and thus carried heat and water vapor from the boundary layer to high in the troposphere. Joanne's observational work led to a major effort by a number of theoreticians and modelers in the 1960s and 1970s to incorporate the effects of the relatively small-scale but energetically important cumulus clouds in numerical models of tropical cyclones.

The important theory of *conditional instability of the second kind,* or CISK, and its contribution to tropical cyclone theory and modeling, is summarized. The CISK theory envisioned a cooperation between the tropical cyclone–scale circulation and the much smaller-scale convective clouds, including hot towers, that caused tropical cyclones to form and intensify. Although the CISK and hot tower theories were misunderstood and misused by some, they both contributed much to the development of tropical cyclone models and scientific understanding of these violent storms, and their general concepts and importance remain valid today.

1. Introduction

Prior to the 1950s, the tropical oceans were largely a meteorologically unexplored mystery. Little was known about the characteristics of tropical clouds and cloud systems, how hurricanes formed, or about how the Tropics interacted with the rest of the global climate system. For example, Herbert Riehl introduced his 1954 book *Tropical Meteorology* with a story of his first visit to the Tropics (Puerto Rico) in 1943 when he observed torrential rain falling from clouds with tops no higher than 8000 feet, far below the freezing level. At that time ice processes were thought to be essential in the precipitation process and Riehl and others schooled in the dominant culture of extratropical meteorology were unprepared for what they found in the Tropics.

The decade of the 1950s saw an enormous amount of intellectual excitement and energy directed at observing and understanding tropical meteorology. Following the devastating hurricanes of 1954 and 1955 along the Atlantic coast of the United States, the U.S. Congress in 1955 appropriated funds for the investigation of hurricanes and in 1956 the National Hurricane Research Project (NHRP) was formed. Research aircraft from NHRP enabled for the first time detailed in situ observations of tropical clouds and cyclones from the surface of the ocean to the upper troposphere. Leading the observational assault on the mysteries of the Tropics were Herbert Riehl and Joanne Malkus.

Increasing observations stimulated and challenged theoreticians and early numerical modelers (e.g., J. Charney, A. Eliassen, S. M. A. Haque, D. Lilly, A. Kasahara, H. L. Kuo, Y. Ogura, V. Ooyama, S. Syōno, and M. Yamasaki) to explain the formation of tropical cyclones using relatively simple linear mathematical models and more complex, nonlinear numerical models.

This paper discusses the observational origin of the *hot tower* concept and the role that this important conceptual model played in early theories and models of tropical cyclone formation. Although not a widely used term today, hot towers dominated the thinking about the interactions of cumulus convection with the large-scale tropical environment for more than 20 years. In later years, based on field programs such as the Line Islands Experiment (1967), the Barbados Oceanographic and Meteorological Experiment (BOMEX, 1969), and the Global Atmospheric Research Program (GARP) Atlantic Tropical Experiment (GATE, 1974), scientists discovered that real tropical convective systems are much more complex than the simple hot tower concept suggests. However, the key aspects of the hot tower theory remain valid today. Appendix A contains photographs that illustrate hot towers from several different perspectives.

2. Origin of the hot tower hypothesis

The latter part of the 1950s and early 1960s saw an intense dialogue among the leading atmospheric scientists of the day about the formation of tropical cy-

On the heat balance in the equatorial trough zone

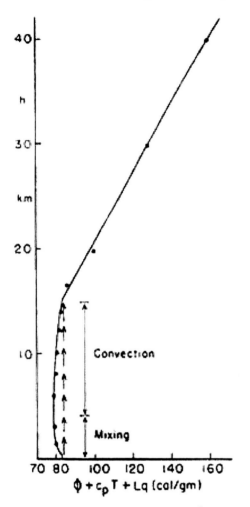

FIG. 10.1 Model of mean vertical distribution of $\Phi + CpT + Lq$ in the troposphere and stratosphere and the mechanism of counter-gradient heat flux in the troposphere by hot towers (Riehl and Malkus 1958).

clones and the role of cumulus clouds in their formation. As a graduate student in 1967 attending my first tropical meteorology conference in Caracas, Venezuela (AMS 1967), I recall vigorous and sometimes heated arguments on this topic. The debate centered on how the latent heat of condensation is actually released in tropical cyclones and how this heat produces the warm inner region of the cyclone and drives its transverse and horizontal circulations. A central part of this discussion was the hot tower. The concept of hot towers as protected, undilute (no entrainment) cumulonimbus clouds was introduced by Riehl and Malkus (1958) in their classic paper on the heat balance in the equatorial trough zone. They showed (Fig. 10.1) how "undilute cloud towers" could transport heat upward from the surface of the ocean into the upper troposphere against the mean gradient of potential energy (Φ), enthalpy (CpT), and latent

energy (Lq). In this paper Riehl and Malkus referred to "narrow warm towers" rather than hot towers, but by 1960 they were calling it the hot tower hypothesis and applying the conceptual model of moist adiabatic ascent in cumulus clouds to the tropical cyclone problem (Malkus and Riehl 1960).

> These temperature deficiencies suggest, however, that the moist adiabatic ascent does not take place by means of uniform and gradual ascent of the whole mass of the hurricane but, as postulated by Riehl and Malkus (1958) for the equatorial trough zone, it is largely concentrated in regions of rapidly ascending buoyant hot towers.

In August 1958, NHRP research aircraft observed on four separate days the formation, deepening, and intensification of Hurricane Daisy. Although she was not able to fly on any of the research flights,[1] Joanne Malkus eagerly began analysis of the fresh data upon return of the aircraft. Her observations and photographic analysis of the cloud structure of the hurricane during the different stages of its life cycle are remarkable in their care and completeness, especially considering that her analyses were done in an era before Global Positioning System (GPS) and without computer graphics or mapping routines. In "Cloud patterns in Hurricane Daisy, 1958" (Malkus et al. 1961), Joanne and her colleagues used time-lapse kodachrome movies from cameras mounted in the nose of each aircraft to map the positions and types of clouds relative to the moving hurricane. She superimposed on the cloud maps the 3-cm radar precipitation echoes, providing an unprecedented view of the relationship of visible clouds and precipitation and radar echoes to the hurricane circulation. In the introduction of this paper, she describes in her characteristic exciting, highly descriptive words the role of cumulus convection in hurricanes.

> The furious winds are driven by latent heat released by giant clouds; their concentrated and highly buoyant character is also apparently essential in the heat and mass transports of the storm. The models envisage that the ascent in the core occurs in extremely restricted regions of undilute, rapidly rising cumulonimbus towers, which only come to occupy a significant fraction of the rain area as the eye wall is approached (Malkus 1958). This latter concept has been informally christened the 'hot tower hypothesis.'

And indeed the aircraft studies of Hurricane Daisy confirmed that the convective activity was concentrated into only "a few lines of cumulonimbus towers

---

[1] I recently asked Joanne if she flew in any of the Hurricane Daisy research flights. Here is her answer: "In the 1958 period NO WOMEN were allowed on the NHRP aircraft. The first hurricane mission I was allowed on was not until 1963. I flew as Chief Scientist in Beulah and Betsy (1965). It really felt weird working on the data virtually as soon as they came in and still not being able to fly. However, in the early days the aircraft were all military and they could make whatever rules they wanted to." (Joanne Simpson, 16 March 2000)

which even on the mature day occupied only about 4% of the rain area (radius within about 200 nautical miles)."

Following the descriptive paper on the cloud patterns in Hurricane Daisy, Riehl and Malkus (1961) used the research flight data to compute the budgets of heat, moisture, kinetic energy, and momentum for two different days. Their paper confirmed the hot tower hypothesis and perhaps represents the peak of the use of the hot tower concept in the literature describing the intensification and maintenance of the hurricane.

> Heat budget computations permitted assessing the fraction of mass flow reaching upper levels in undilute 'hot' cumulonimbus towers and estimation of the number of towers required, to compare with previous photographic studies. Hot towers proved to be the dominant mechanism of raising warm air to upper levels, particularly in the inner core... The hot cores thus play an ever-increasing role as the center is approached, and they furnish the primary mechanism for heat flux to the high troposphere.

In spite of her focus on hot towers, Joanne was well aware of the complexities in the cloud structure of tropical cyclones and of the importance of cloud physics and cloud dynamics that went far beyond the simple concept of undilute ascent of moist air in a single cloud tower. The photographs of Hurricane Daisy show a rich mosaic of multilayered stratiform clouds that are anything but hot towers. And, in her remarkable paper "On the structure and maintenance of the mature hurricane eye" (Malkus 1958), Joanne discusses the role of momentum transport by convective clouds in the eyewall and the role of entrainment and evaporation of cloud water and cloud ice in maintaining the observed temperature and humidity structure of the eye. Nevertheless, like any great concept, the notion of hot towers conveyed in just two words a key process in tropical cyclones in a simple and intuitive way that captured the imagination of scientists for many years.

In their definitive 1961 paper on the structure and budgets of Hurricane Daisy, Malkus and Riehl sum up the role of hot towers and pose, and then answer, a key question regarding the future prospects for the numerical modeling of tropical cyclones (Riehl and Malkus 1961, 206–207).

> The Daisy data have in fact confirmed the importance of the hot tower mechanism and the concentration of the storm's heat release into 100–200 individual buoyant elements. If the details of the activity on these small scales are decisive for storm growth and maintenance, then the prospect for solving the large-scale hurricane problems and achieving predictive models recedes into the dim future. On the contrary, we shall conclude by offering hope that the dynamic and thermodynamic effects of these small scales can be introduced or parameterized effectively in a relatively simple framework, without

dwelling on their complex details, which may differ from case to case and from time to time.

The hope they offer is a remarkably prescient vision, based on their detailed budget calculations of momentum, vorticity and heat and on their observations of the hot towers (Riehl and Malkus 1961, p. 210).

> In conclusion, it appears hopeful to attempt two-dimensional models of the hurricane core, provided the vital convective-scale processes are adequately parameterized and entered in the thermal budgets where their transports and releases are vital in the basic machinery. It appears likely that scales of motion intermediate between convection and the hurricane itself, such as lateral eddy transports, are not essential in the core; thus a 'separation of scales of motion' may provide a useful attack upon this problem....

These ideas were crucial in the thinking of the theoreticians and modelers as they struggled with developing a theory of tropical cyclogenesis and a way of parameterizing the effects of cumulus convection in numerical models.

3. Toward the parameterization of cumulus convection in hurricane models

The late 1950s and early 1960s saw the beginning of serious attempts to model tropical cyclones. Kasahara (2000) provides an interesting review of the efforts at this time. (He also adds some interesting personal recollection in the appendix of this paper.) Very early on scientists realized that there were two distinct horizontal scales of motion associated with tropical cyclones: the scale of individual cumulus clouds ($\sim$1–10 km) and the much larger scale of the cyclone itself ($\sim$100–1000 km). Computers at this time were barely powerful enough to resolve the largest scales of the hurricane circulation, even with the assumption of axisymmetry that reduced the modeling problem to two dimensions. However, the observational studies of Riehl and Malkus showed that the cumulus clouds were essential components of the hurricane's energetics, and so modelers realized that somehow the effects of the cumulus clouds had to be considered in the hurricane models. They tried to do this by relating the effects of the clouds to large-scale variable, or *parameters* that could be calculated from the resolvable scales of motion in the models, a process called *parameterization.*

Although parameterization in numerical models refers generally to the representation of any physical process that occurs on scales smaller than the resolution of the model (subgrid-scale processes), such as turbulence, radiation, or sensible heat exchange, *cumulus* parameterization so dominated the research in the 1950s and 1960s that many people simply referred to the topic as "parameterization." But what gave people the idea that cumulus parameterization was even possible?

a. Observational basis for cumulus parameterization in hurricanes

As described above, the extensive diagnostic studies of Hurricane Daisy using research aircraft data by Malkus and Riehl provided strong empirical evidence that cumulus clouds were intimately related to their larger-scale environment, and thus there was hope for parameterizing their effects in large-scale models. The paper by Malkus and Williams (1963) describes the relationship beautifully.

> Thus, to grow a hurricane, we must have conditions permissive to large penetrative towers. But on the other hand, it has also been learned that we must have present some of the essential features of the storm in order to grow the giant clouds! In a photographic mapping of cloud conditions over the tropical Pacific. . . . we found that penetrative cumulonimbus (tops ≥ 35 000 ft.) occurred only where low-level, synoptic-scale convergence prevailed and were snuffed out instantly when divergence took over. . . In fact, dynamic factors have proved far more critical criteria for cloud growth in the tropics than has static instability, which in fact shows an inverse relation to convection: cumulonimbus conditions are characterized by marginal instability and low-level convergence, while very fair conditions are characterized by much stronger instability and divergence.

An earlier important observational study related to the parameterization of cumulus convection in tropical cyclones was carried out by Syōno et al. (1951). They showed that the observed rainfall at three stations in Japan during the passage of a typhoon was closely related to the large-scale tangential wind field through frictional mass convergence associated with "Ekman pumping."

These observational studies provided encouragement for modelers to try to relate the existence and effects of deep cumulus convection in tropical cyclone models to the resolvable-scale, low-level convergence in the model. The somewhat paradoxical fact that regions of strong cumulus convection tend to be less unstable (closer to a moist adiabat) than fair areas was another important observation. A few years later, H. L. Kuo (Kuo 1965) based his popular and quite successful cumulus parameterization scheme on these two observational facts.

b. Linear theories of tropical cyclogenesis

There were basically three different approaches toward trying to understand the problem of tropical cyclogenesis in the 1950s and 1960s: the observational approach led by Malkus and Riehl, the theoretical mathematical approach using linear theory (e.g., Haque 1952; Syōno 1953; Lilly 1960; Kuo 1961; Charney and Eliassen 1964; Syōno and Yamasaki 1966), and the non-

linear computer simulation approach using axisymmetric (two-dimensional) numerical models (Kasahara 1961; Syōno 1962; Ogura 1964; Ooyama 1964) and later three-dimensional models (Anthes 1972). It is beyond the scope of this paper to discuss the evolution and contributions of the nonlinear models, which have recently begun to show great promise in predicting the movement and structure of real tropical cyclones (e.g., Krishnamurti et al. 1989; Liu et al. 1997; Karyampudi et al. 1998), but let us consider the linear theory approach and the parameterization of cumulus convection in nonlinear models, which are both relevant to the concept of hot towers.

The linear theories of tropical cyclogenesis generally considered the growth of small, random perturbations in a moist, conditionally unstable and rotating atmosphere at rest. The linear models usually included a simple parameterization of surface friction and horizontal mixing, or accounted for frictional convergence in the boundary layer through a lower boundary condition on vertical velocity. The results of the linear studies showed that, unless physically unrealistic mathematical terms were introduced to suppress the smallest scales of motion, the fastest growing modes were those with horizontal scales characteristic of cumulus clouds, not the tropical cyclone. In an axisymmetric framework, Kuo (1965) summed it up as follows: "the results from linear analysis show that under such conditions, the energy tends to go into establishing strong, narrow, ring-shaped convection cells instead of cyclone-scale perturbations."

This result was simply the manifestation of *conditional instability* and was well known from earlier linear theories (Bjerknes 1938). However, it raised a key scientific question of the day; how can the tropical cyclone–scale disturbance grow when the dominant growth occurred at the cumulus cloud (hot tower) scale? Considerable effort was devoted toward trying to suppress the growth of the cumulus scales and so shift the wavelength of maximum perturbation growth to larger scales. The dilemma was phrased nicely by Charney (1971): "Of course, friction and entrainment of dry air will limit the width of the ascending branch, but it is known from the theory of Benard convection that these limitations will not be effective until the width approaches that of the depth of the troposphere, i.e., the width of a cumulus cloud. It follows that conditional instability will by itself give rise to cumulus clouds, not depressions or hurricanes."

In judging the impact of the linear analysis approach to explain tropical cyclogenesis, it can be argued that the concept itself was so seriously flawed that the many studies actually delayed progress in the understanding of the tropical cyclogenesis. In retrospect, the preoccupation with the more rapid growth of cumulus clouds compared to the larger cyclone circulation was missing the point. An analogy would be to question why a child could grow and prosper over many years in spite of the much more rapid growth of individual cells in the body!

Significant flaws in the linear theory approach include the following.

1) Tropical cyclones do not grow through linear processes from small random perturbations on a stagnant, conditionally unstable base state. Instead, they grow from finite-amplitude, large-scale perturbations in the Tropics, and nonlinear effects are important throughout the development (Anthes 1982; Gray 1998).
2) The assumption of a constant static stability is erroneous; in the real tropical cyclone environment, the effect of cumulus convection is to gradually modify the environmental lapse rate toward a moist adiabatic (neutral) state.
3) In linear models, all unstable modes grow exponentially. Thus, the fastest-growing modes dominate the solutions forever. In the real atmosphere, fast-growing modes like cumulus convection quickly reach a maximum amplitude and decay, through nonlinear effects and modification of the original environmental state.

The emphasis on linear theory and the frustration with the dominance of the fastest-growing linear modes were not broken until the idea of *cooperation* rather than *competition* of the two scales of motion was conceived. Under the concept of cooperation, individual convective cells grow and decay rapidly on short timescales, but their cumulative effect on the larger scale through latent heat release and vertical transports of heat, water vapor, and momentum gradually modify and intensify the large-scale circulation. In return, frictionally induced convergence at the surface, the addition of heat and water vapor from the warm ocean to the atmosphere,[2] and the latent heat–driven vertical–transverse circulation associated with the growing cyclone-scale circulation support the convective clouds. With cooperation between the two scales of motion, a positive feedback mechanism was envisioned that led to much progress in the understanding of tropical storms and their embedded hot towers.

c. Conditional instability of the second kind and the H. L. Kuo cumulus parameterization

Intimately related to the idea of cooperation rather than competition between the hot towers and the hurricane was the introduction of a new kind of instability, dubbed *conditional instability of the second kind,* or

CISK. The term was first used by Charney and Eliassen (1964).

> This suggests we should look upon the pre-hurricane depression and the cumulus cell as not competing for the same energy, for in this competition the cumulus cell must win; rather we should consider the two as supporting one another—the cumulus cell by supplying the heat energy for driving the depression, and the depression by providing the low-level convergence of moisture into the cumulus cell. The primary purpose of this paper is to show that this type of interaction does lead to a large-scale self-amplification, which we may call conditional instability of the second kind to contrast it with the conditional instability responsible for small-scale cumulus convection.

As discussed by Ooyama (1982) and Kasahara (2000), the concept of CISK as a cooperative intensification theory is quite valid; however, the term itself has been misunderstood and misused in many ways since it was introduced. CISK is not a cumulus parameterization scheme, as many have called it, nor does it describe or explain the scale selection and growth of a tropical cyclone from small random perturbations according to linear theory. Nevertheless, the concept of cooperation of the nonhydrostatic cumulus clouds and the quasi-balanced, hydrostatic tropical cyclone circulation is valid. In fact, the CISK concept applies as well to modern numerical models of tropical cyclones that permit nonhydrostatic convection explicitly in the model along with the larger scales of motion as it does to hydrostatic models in which cumulus convection is parameterized.

I end this section with a brief summary of H. L. Kuo's (1965) cumulus parameterization scheme because it was one of the most successful and widely used schemes for including the effects of deep precipitating cumulus convection in nonlinear computer models of tropical cyclones, as well as models of extratropical atmospheric phenomena. By successful, I mean that the use of the "Kuo scheme" permitted reasonably realistic development and evolution of model tropical cyclones, and when used in forecast models, showed some skill in predicting precipitation.

Kuo was undoubtedly influenced by the hot tower concept when he developed his parameterization of deep cumulus convection.

> The cumulus clouds exist momentarily. They dissolve by mixing with the environmental air at the same level, so that the heat and moisture carried up by the cloud are imparted to the environmental air. Thus, our model is similar to the deep cumulus towers described by Riehl and Malkus (1961).

Kuo envisioned a sudden creation of a hot tower in which the cloud temperature and specific humidity were that of a moist adiabat from the surface. Thus, in a conditionally unstable environment, the cloud was

---

[2] The development of the tropical cyclone through the self-induced fluxes of sensible and latent heat from the ocean to the atmosphere and the redistribution of this heat upward throughout the troposphere by cumulus convection is emphasized in the models of Emanuel (1986) and Rotunno and Emanuel (1987).

warmer and moister than the environment. The warm moist cloud was then assumed to mix horizontally with the environment, resulting in a new environmental state that was slightly warmer, moister and less unstable than the previous state. The number of cumulus clouds, or equivalently the percentage area covered by cumulus clouds, was related to the resolvable-scale water vapor convergence in the model.

The concept of how the clouds gradually altered the environment through horizontal mixing was appealing, but it was fundamentally incorrect, or at best incomplete. In reality, the environment is warmed and moistened through a combination of adiabatic warming associated with compensating subsidence in response to the upward mass flux in the clouds and moistening by detrainment of moist cloud air and evaporation of water in the cloud. Stratiform clouds also play a significant role in the complex interactions that gradually produce an overall warming and moistening of the inner regions of the tropical cyclone. However, the Kuo scheme worked reasonably well because it "drove" the large-scale atmosphere toward a moist adiabatic state (the empirically "correct" structure). It produced a reasonably accurate vertical distribution of heating (which is important in allowing the large-scale circulation to develop and then stabilize), and it conserved energy (because the total diabatic heating was equal to the total water vapor convergence).

4. Summary: Are hot towers relevant to hurricanes?

The hot tower concept introduced by Riehl and Malkus had a profound effect on tropical meteorology and hurricane studies in the late 1950s and throughout the 1960s.[3] The use of the term, however, faded in the 1970s and is heard less often today.[4] One might ask whether the concept is still valid and whether hot towers are relevant to hurricanes today.

In my opinion, the term "hot tower" became less used after the 1960s for two reasons. First, the concept or hypothesis was essentially validated in the Hurricane Daisy and later studies; hence, the concept became noncontroversial and therefore less interesting. Second, sci-

entific attention was diverted to a wealth of other exciting and important features and processes in tropical meteorology and the theory of hurricanes. These included the role of three-dimensional preexisting disturbances; large-scale environmental factors such as static stability and horizontal and vertical wind shear; ocean–air interactions and boundary layer processes; detailed cloud–environment interactions such as compensating subsidence, entrainment, and detrainment; stratiform clouds and precipitation, including ice processes and evaporation of cloud and rainwater; and radiation. However, as even the most casual inspection of modern time-lapse movies from geostationary satellites of the overshooting tops associated with real hot towers in hurricanes show, hot towers remain critical to the formation, intensification, and maintenance of hurricanes. Without hot towers, hurricanes would not exist; hence, the hot tower concept remains as valid today as it was 50 years ago.

Acknowledgments. I acknowledge and thank Joanne Simpson for her inspiration, support, and friendship over the years. I thank Akira Kasahara for his stimulating research, cooperation, and friendship, including his thoughtful review and comments on this paper (see appendix B).

APPENDIX A

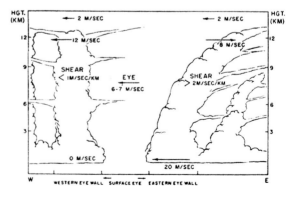

---

[3] Interestingly, in spite of the influence of the hot tower concept on science, direct measurements of the thermodynamic structure of a hot tower confirming the undilute ascent of air from near the surface were not made until 1998. In the 1998 Convection and Moisture Experiment (CAMEX-3) experiment several sondes were deliberately dropped in rising cumulus towers from aircraft flying at 11 or 12 km and these confirmed that equivalent potential temperature in the cloud was constant in the vertical for all practical purposes (J. Simpson 2000, personal communication).

[4] However, a recent paper by Joanne and her colleagues (Simpson et al. 1998) revisits the role of hot towers in Hurricane Daisy (1958) and Tropical Cyclone Oliver (1993) and shows their importance in producing high values of equivalent potential temperature in the middle troposphere where subsidence at the edge of the towers could produce warming and the development of the tropical cyclone eye.

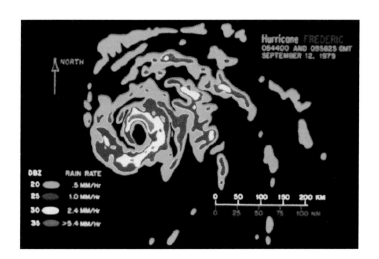

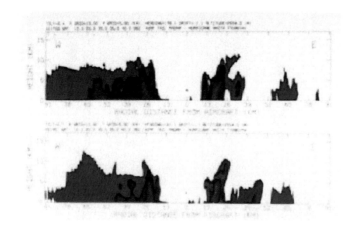

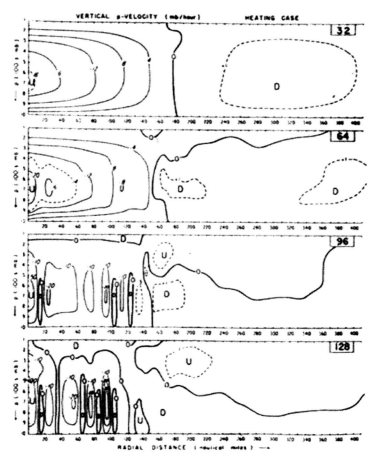

APPENDIX B

Comments by Akira Kasahara

March 22, 2000
To: Rick Anthes
From: Akira Kasahara

I enjoyed reading your article on hot towers and hurricanes. Actually, it brings back my painful memories on my struggle to simulate the development of tropical cyclones while working at the University of Chicago in 1956–61. Your article describes the importance of hot tower concept in pointing to the necessity of parameterizing the process of latent heat release in deep cumulus convection for modeling of tropical cyclone development. Your article nicely complements my article on the origin of cumulus parameterization. You are looking at the root of cumulus parameterization from the physical point of view as the means to represent ensemble effects of subgrid-scale cumulus convection in a large-scale model, while I was considering the same from the computational point of view to run stably a primitive equation model with moist physics.

Of course, while working at the University of Chicago, listening to the discussions between Riehl and Joanne, and reading many articles written by Riehl and Joanne, I was fully aware of the importance of the hot tower concept and the need to parameterize the process of heat source in my hurricane model. (One of my favorite of Joanne's articles is "Recent developments in studies of penetrative convection and an application to hurricane cumulonimbus towers" which appeared in a book entitled *Cumulus Dynamics,* Pergamon Press, 65–84, 1960. It was a very inspiring article for me in those days.)

During my Chicago days, I was running my hurricane development model at the General Circulation Research Section of the then U.S. Weather Bureau in Suitland, MD, using the IBM 701 and then IBM 704 and 709. Actually, those computers were used to make operational forecasts at the Joint Numerical Weather Prediction Unit. George Platzman made an arrangement with Joe Smagorinsky so that I was able to use the machine as a visitor from midnight to 6 A.M. In 1958, Joe hired Suki Manabe from Japan to work with him on their general circulation modeling. Obviously, Suki and I talked often and he was watching my struggle to succeed in a long-term integration of the hurricane development model. Suki and Joe were working on GCM developments and were using a convective adjustment scheme successfully. They were working on a primitive equation model with moist physics and I was too. One day Suki suggested to me to use the convective adjustment scheme in my hurricane model. I should be able to run my model stably, he said. I dismissed his suggestion, because the idea of convective adjustment was not kosher in a hurricane model. I was still hoping to find a

way to incorporate the hot tower concept to parameterize cumulus convection in the hurricane model.[B1]

I left Chicago in January 1962 to join the Courant Institute of Mathematical Sciences, NYU, to work with Eugene Isaacson and Bob Richtmyer on numerical methods for fluid dynamics. James Stoker, who was then Director of the Institute, gave me the task to study the occlusion process of frontal cyclones and I spent most of my time on this work, which was later published. I stayed in New York only a year and half and I joined NCAR in June 1963. I essentially stopped working on the hurricane problem after joining NCAR, since I felt that the development of general circulation model with Warren Washington had a higher priority. Of course, as you all know, Vic Ooyama continued to work on his quest of solving the hurricane formation problem and Akio Arakawa made an innovative work on cumulus parameterization.

Often I wonder now what would had happened on the history of hurricane formation study and of cumulus parameterization, IF I had succeeded in the development of tropical cyclone by following Suki's suggestion for using his convective adjustment scheme in my hurricane model. Of course, this is a big if.

In your summary, you asked a question: are hot towers relevant to hurricanes? The reason that I talked about my personal history here is to tell you how significant was the impact of the hot tower concept on my earlier work. Including H. L. Kuo, as discussed in your article, the hot tower concept was a guiding principle among many modelers to devise a cumulus parameterization as reminisced by Vic Ooyama and Akio Arakawa in their talks at the 22d Conference on Hurricanes and Tropical Meteorology, 19–23 May 1997. (Their papers are included in the preprint volume.) In some sense, however, the job of realizing the hot tower concept has not been finished in my view. We must set up a scheme of predicting the population of hot towers just like the prediction of cloudiness of various types in a large-scale prediction model. Of course, people are now more interested in a direct simulation of deep convective system by using a high-resolution model and explicit cloud physics.

Thank you for giving me an opportunity to read your engaging article.

Akira

[B1] However, I was not and am not against the use of the convective adjustment scheme. As a matter of fact, I have used the convective adjustment scheme in our general circulation models working with Warren Washington. The convective adjustment scheme has a respectable root as I described in my article of cumulus parameterization. Suki came up with the idea of convective adjustment to deal with the ensemble effects of moist convection in the Tropics in his study on the radiative–convective equilibrium. The point I want to stress is that I was so preoccupied in my desire to incorporate the hot tower concept in the formulation of cumulus parameterization that it didn't occur to me to try the convective adjustment scheme in my hurricane model.

REFERENCES

AMS, 1967: Program of the Fifth Technical Conference on Hurricanes and Tropical Meteorology. *Bull. Amer. Meteor. Soc.,* **48,** 604–639.

Anthes, R. A., 1972: The development of asymmetries in a three-dimensional model of a tropical cyclone. *Mon. Wea. Rev.,* **100,** 461–476.

——, 1982: *Tropical Cyclones: Their Evolution, Structure and Effects, Meteor. Monogr.,* No. 41, Amer. Meteor. Soc., 208 pp.

Bjerknes, J., 1938: Saturated-adiabatic ascent through dry-adiabatically descending environment. *Quart. J. Roy. Meteor. Soc.,* **64,** 325–330.

Charney, J. G., 1971: Tropical cyclogenesis and the formation of the intertropical convergence zone. *Lect. Appl. Math.,* **13,** 355–368.

——, and A. Eliassen, 1964: On the growth of the hurricane depression. *J. Atmos. Sci.,* **21,** 68–75.

Emanuel, K. A., 1986: An air–sea interaction theory for tropical cyclones. Part I: Steady-state maintenance. *J. Atmos. Sci.,* **43,** 585–604.

Gray, W. M., 1998: The formation of tropical cyclones. *Meteor. Atmos. Phys.,* **67,** 37–69.

Haque, S. M. A., 1952: The initiation of cyclonic circulation in a vertically unstable stagnant air mass. *Quart. J. Roy. Meteor. Soc.,* **78,** 394–406.

Karyampudi, V. Mohan, G. S. Lau, and J. Manobianco, 1998: Impact of initial conditions, rainfall assimilation, and cumulus parameterization on simulations of Hurricane Florence (1988). *Mon. Wea. Rev.,* **126,** 3077–3101.

Kasahara, A., 1961: A numerical experiment on the development of a tropical cyclone. *J. Meteor.,* **18,** 259–282.

——, 2000: On the origin of cumulus parameterization for numerical prediction models. *General Circulation Model Development: Past, Present and Future.* D. Randall, Ed., Academic Press, 199–224.

Krishnamurti, T. N., D. Oosterhof, and N. Dignon, 1989: Hurricane predictions with a high-resolution global model. *Mon. Wea. Rev.,* **117,** 631–669.

Kuo, H. L., 1961: Convection in conditionally unstable atmosphere. *Tellus,* **13,** 441–459.

——, 1965: On the formation and intensification of tropical cyclones through latent heat release by cumulus convection. *J. Atmos. Sci.,* **22,** 40–63.

Lilly, D. K., 1960: On the theory of disturbances in a conditionally unstable atmosphere. *Mon. Wea. Rev.,* **88,** 1–17.

Liu, Y., D.-L. Zhang, and M. K. Yau, 1997: A multiple numerical study of Hurricane Andrew (1992). Part I: Explicit simulation and verification. *Mon. Wea. Rev.,* **125,** 3073–3093.

Malkus, J. S., 1958: On the structure and maintenance of the mature hurricane eye. *J. Meteor.,* **15,** 337–349.

——, and H. Riehl, 1960: On the dynamics and energy transformations in steady-state hurricanes. *Tellus,* **12,** 1–20.

——, and R. T. Williams, 1963: On the interaction between severe storms and large cumulus clouds. *Severe Local Storms, Meteor. Monogr.,* No. 5, Amer. Meteor. Soc., 59–64.

——, C. Ronne, and M. Chaffee, 1961: Cloud patterns in Hurricane Daisy, 1958. *Tellus,* **13A,** 8–30.

Ogura, Y., 1964: Frictionally controlled, thermally driven circulations in a circular vortex with application to tropical cyclones. *J. Atmos. Sci.,* **21,** 610–621.

Ooyama, K., 1964: A dynamical model for the study of tropical cyclone development. *Geofis. Int.,* **4,** 187–198.

——, 1982: Conceptual evolution of the theory and modeling of the tropical cyclone. *J. Meteor. Soc. Japan,* **60,** 369–380.

Riehl, H., and J. S. Malkus, 1958: On the heat balance in the equatorial trough zone. *Geophysica,* **6,** 503–538.

——, and ——, 1961: Some aspects of Hurricane Daisy, 1958. *Tellus,* 181–213.

Rotunno, R., and K. A. Emanuel, 1987: An air–sea interaction theory for tropical cyclones. Part II: Evolutionary study using a nonhydrostatic axisymmetric numerical model. *J. Atmos. Sci.,* **44,** 542–561.

Simpson, J., J. B. Halverson, B. S. Ferrier, W. A. Petersen, R. H. Simpson, R. Blakeslee, and S. L. Durden, 1998: On the role of "hot towers" in tropical cyclone formation. *Meteor. Atmos. Phys.* **67,** 15–35.

Syōno, S., 1953: On the formation of tropical cyclones. *Tellus,* **5,** 179–195.

——, 1962: A numerical experiment of the formation of tropical cyclone. *Proc. Int. Symp. on Numerical Weather Prediction in Tokyo, Nov. 1960,* Meteor. Soc. Japan, 405–418.

——, and M. Yamasaki, 1966: Stability of symmetrical motions by latent heat released by cumulus convection under the existence of surface friction. *J. Meteor. Soc. Japan,* **44,** 353–375.

——, Y. Ogura, K. Gambo, and A. Kasahara, 1951: On the negative vorticity in a typhoon. *J. Meteor. Soc. Japan,* **29,** 397–415.

Chapter 11

On the Transverse Circulation of the Hurricane

WILLIAM M. GRAY

Department of Atmospheric Science, Colorado State University, Fort Collins, Colorado

PROLOGUE

Joanne Simpson tells the story that in the mid-1940s, when she was a young (and precocious!) graduate student at the University of Chicago, she told Carl Rossby that she wanted to study clouds and that he responded by saying that that was a good subject for a girl. We now more fully appreciate the role of clouds as the fundamental component of the hydrologic cycle. Most of us would agree that understanding the physics behind cumulus convection is a fundamental challenge for all, girl or boy. Joanne's choice of cloud studies as a career endeavor was a wiser choice than most meteorologists of that day (and many of this day) realized. Attention in the 1940s and 1950s had been focused more on the requirements of wind for the transfer of energy from the tropical to the polar regions. There is no doubt that horizontal transport of energy is a fundamental ingredient of the general circulation. But vertical energy transport to balance the troposphere's continuous radiational cooling of ~1°C per day is more important. Globally averaged, the required vertical transport of energy from the surface up into the troposphere is about four times larger than the required horizontal transport. It is this vertical energy transport that is so messy and so difficult to understand, and so hard to treat in a realistic and quantitative fashion. Many modelers and theoreticians have chosen to neglect the many hydrologic cycle complications (by assuming that the troposphere's radiational cooling is balanced by condensation warming) and to concentrate only on the horizontal energy imbalances. This has been the approach of the dishpan or annulus experiments. But this is not satisfactory for a full understanding of how the troposphere really functions. We have to face up to the need for the development of a realistic quantitative treatment of the globe's hydrologic cycle. The cumulus convection schemes in current GCMs are still inadequate. It is this continuing need to better understand the full range of cloud processes that has made Joanne's decision in the mid-1940s to concentrate on clouds such a wise one. She has since made many contributions to the understanding of the role of clouds. The paper she wrote with Herbert Riehl in 1958 (Riehl and Malkus) had much influence on the thinking of the important role of cumulus convection. Her recent work with the Tropical Rainfall Measuring Mission (TRMM) experiment is an example of her continuing drive to better understand clouds and the hydrologic cycle.

I first met and worked with Joanne in the late 1950s when I was a graduate student of Herbert Riehl's at the University of Chicago. I participated in the study she was directing on the variations of tropical Pacific cloudiness from aircraft time-lapse photography. This was before the satellite and the computer. We had more time to think and to speculate in those days. I have been most grateful to both Joanne and Bob Simpson for their interest and encouragement of my research efforts since that time.

It is a pleasure to make a contribution to this symposium honoring Joanne. The paper to follow has many similarities to the early and original paper of Joanne in 1958 titled "The Structure and Maintenance of the Mature Hurricane Eye."

ABSTRACT

This paper uses extensive aircraft, composited rawinsonde data, and an idealized hurricane structure model to analyze the physical processes that maintain the transverse circulation of the steady-state hurricane. It is shown that convective available potential energy (CAPE) or processes other than frictional forcing plays an important role in maintaining the hurricane's inner-core (radius < 60 km) in-up-and-out radial circulation. But this is not true at outer radii (60–250 km or 250–700 km) where surface friction forcing is dominant and larger than the resulting upward vertical motion.

Overall, there is less vertical motion within the hurricane's 0–250-km area than that specified by frictional forcing and, overall, CAPE or buoyancy plays a negative role in enhancing vertical motion. But this is not true of the inner-core eye-wall cloud region where nonfrictionally driven eye-wall vertical motion has an important buoyant contribution and a strong ocean-to-air energy flux is present. Frictionally forced vertical motion resulting from low-level relative vorticity is typically not balanced locally. Quasi balance between frictional forcing and vertical motion is observed only for the larger-scale vortex (approximately 0°–3° radius) as a whole.

1. Introduction

Using aircraft, composited rawinsonde data, and a steady-state model it is possible to construct reliable, quantitative estimates of the processes that maintain the hurricane's inner-core tangential wind field against frictional dissipation and upper-level tangential momentum outflow. This paper makes such calculations for cyclones of different hurricane maximum tangential winds

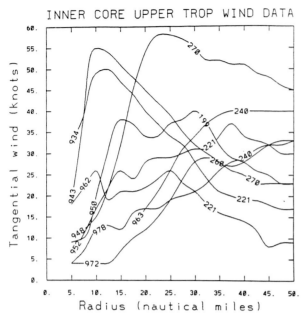

FIG. 11.1. Inner-core upper-tropospheric tangential winds as measured by National Hurricane Research Laboratory reconnaissance flights. Numbers on left side of the curves are minimum central pressure, whereas those on the right side are flight-level pressure for each curve. Storm motion has been removed (from Gray and Shea 1976). Winds are in knots.

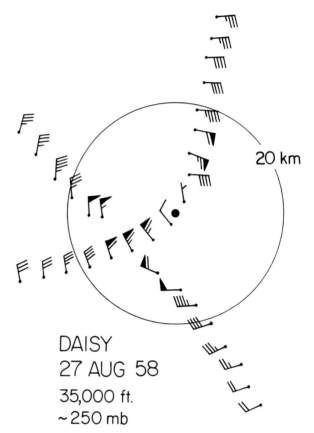

FIG. 11.2. Plan view of inner-core winds in Hurricane Daisy at 1800 UTC 27 Aug 1958. Flight-level pressure is 270 mb. Winds are in knots.

and $(V_{T\max})$, for different radial distributions of V_T, for different radius of maximum winds (RMW), and for variations of the frictional drag of the wind power profile. The relative percentage of the hurricane's inner-core transverse circulation that results directly from local forcing friction versus that which results from outer radius nonfrictional and other processes is solved for. Buoyancy is here used for all vertical motion processed not directly related to frictional forcing.

2. The inner-core upper-level winds of tropical cyclones

In general, the most intense hurricanes have the strongest upper-level inner-core tangential winds V_T. Winds are strongest around the cyclone's wall cloud outflow level. These strong upper-level tangential winds are a result of upward advection of high velocity winds from lower levels.

The more intense the cyclone, generally the greater its upper-tropospheric V_T wind and the lower the surface pressure. In fact, a common characteristic of tropical cyclones that intensify rapidly as compared to storms that intensify at moderate rates is that the ratio of upper to lower inner-core tangential winds (V_T) is higher for the rapidly intensifying systems.

Examples of upper-tropospheric tangential winds

Figure 11.1 shows upper-tropospheric tangential winds within 9–93-km radius (5–50 n mi) for nine hur-

ricanes of widely varying intensity. These measurements were taken by upper-tropospheric aircraft penetrations along constant pressure surfaces during the late 1950s and 1960s. In situations where the vertical shear of V_T is weak, indicating that the cyclonic vortex is not being disrupted by the environment, inner-core upper-tropospheric V_T wind velocities become quite strong.

One of the first measurements of inner-core V_T evolution was that of Hurricane Daisy (Riehl and Malkus 1961; Colón 1961; Shea and Gray 1972). The circulation of Daisy has been documented during its deepening stage for 1200 UTC 25 August and 1800 UTC 27 August 1958. On 25 August, Daisy was a minimal hurricane (985-mb central pressure). It approached its peak intensity of 922 mb on 27 August. Intensification of Daisy showed a striking increase of V_T in the upper troposphere. On 27 August, cyclonic winds at the 637-mb level were 57 m s^{-1} (not shown); at 10.5 km ($\approx$250 mb) they still remain high at about 35 m s^{-1} (Fig. 11.2). This is a good example of wind preservation with height in an intense tropical cyclone. Radar data confirmed that the eyewall system frequently extended above 100 mb. Clearly, inner-core convection transports cyclonic momentum to very high levels.

Another example of V_T preservation with height is

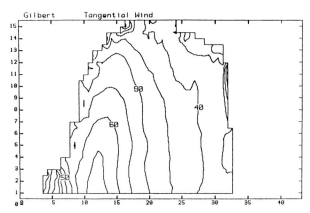

FIG. 11.3. Cross section of the inner-core winds in Hurricane Gilbert at 0600 UTC, 14 Sep 1988. Central pressure is 895 mb. Wind analysis is obtained from pseudo-dual-Doppler analysis (Marks et al. 1992). Contour intervals are 5 m s^{-1}. Figure is courtesy of Peter Dodge, Bob Burpee, and Frank Marks of the NOAA Hurricane Research Division. Ordinate and abscissa in km.

Hurricane Gilbert (1988; Fig. 11.3). Note that Gilbert has strong tangential winds extending into the upper troposphere. Hurricane Gilbert was near its record minimum central pressure of 888 mb at this time. Figure 11.4 is a plan view of the 200-mb (12 km) tangential wind of Supertyphoon Flo (1990) in the northwestern Pacific; central pressure was 890 mb and strongest winds at 200 mb (12 km) were 55 m s^{-1}; once again the conservation of V_T with height is striking. These strong

inner-core upper-level winds cannot be a consequence of horizontal inward convergence of upper-level momentum but are a result of upward advection of strong low-level V_T into the upper layers.

The vertical structure of V_T for Hurricane Gilbert (1988) and three other Atlantic hurricanes of different intensities is shown in Fig. 11.5. In agreement with our rawinsonde composite analysis of tropical cyclones, the weaker the tropical cyclone, the larger the inner-core V_T wind decrease with height. The most intense cyclones have the smallest V_T decrease with height and export the strongest tangential winds in their outflow layers.

Other papers on the vertical structure of the hurricane, as measured by Doppler radar and aircraft data, are given by Shea and Gray (1973), Marks (1985), Marks and Houze (1984), Marks et al. (1992), Black et al. (1994, 1996), Kossin and Eastin (2001), Eastin (1999), and Eastin et al. (2001a, b).

3. Structural assumptions

We will first study the transverse circulation necessary for the maintenance of a standard hurricane with a RMW at 30 km, $x = 0.5$ (for the wind law $V_T r^x =$ constant), a frictional drag coefficient $C_D = 1.5 \times 10^{-3}$, and a square wind relationship for surface stress. Tangential frictional dissipation (F_θ) is calculated from the formula

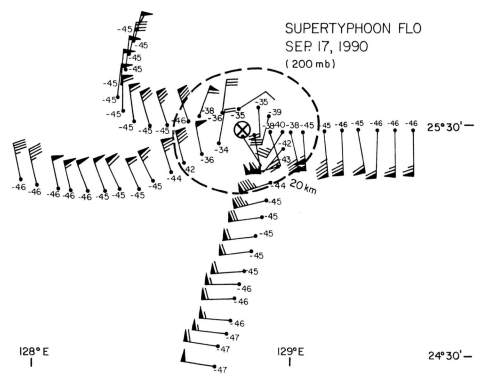

FIG. 11.4. Plan view of inner-core 200-mb winds (in knots) for Supertyphoon Flo at 0700 UTC 16 Sep 1990. Temperature (in °C) is shown next to wind barbs (from analysis by C. Landsea).

COMPARISON OF MEAN TANGENTIAL WINDS

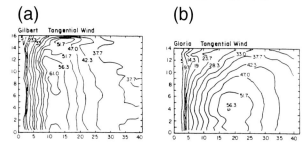

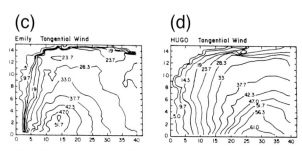

FIG. 11.5. Cross section of tangential winds for hurricanes at different intensities: (a) Hurricane Gilbert at 1000 UTC, 14 Sep 1988 with a central pressure of 890 mb; (b) Hurricane Gloria at 0000 UTC, 25 Sep 1985 with a central pressure of 920 mb; (c) Hurricane Emily at 1800 UTC, 22 Sep 1987 with a central pressure of 958 mb; (d) Hurricane Hugo at 2100 UTC, 17 Sep 1989 with a central pressure of 941 mb. Wind analysis is obtained from velocity track display (VTD) Doppler analysis (Fitzpatrick 1996). Contour intervals are 4.7 m s^{-1}. Diagram courtesy of Frank Marks of the NOAA Hurricane Research Division. Ordinate and abscissa in km.

$$F_\theta = \frac{C_D \rho \overline{V_T}^2}{\rho \Delta z} = \frac{C_D \overline{V_T}^2}{\Delta z}, \qquad (11.1)$$

where Δz is taken over the lowest 100-mb depth. Calculations of F_θ are area integrated over the domain bounded by the radius of twice the RMW:

$$\overline{F_\theta} = 2\pi \int_o^{2 \times RMW} \frac{C_D}{\Delta z} \overline{V_T}^2 r\, dr.$$

We assume that wall cloud updraft air parcels originate just inside the RMW and increase their upward vertical velocity due to condensation and freezing-induced buoyancy. Parcels overshoot their zero buoyancy level near 300 mb, rise farther into negative buoyancy, and decelerate. Most are then advected outward at levels between 250 and 125 mb. A small portion of the updraft

mass moves inward and sinks in the eye. Updraft air is accelerated outward by the weaker pressure gradient that the updrafts rise into and the resulting supergradient winds that are developed.

We will assume that this upward and outward eyewall cloud motion takes place under conservation of absolute angular momentum $DM_a/Dt = 0$, where

$$M_a = rV_T + 1/2r^2 f, \qquad (11.2)$$

and where r = radial distance from the cyclone center and f = Coriolis parameter. We recognize that entrainment into the updraft can potentially alter this conservation assumption. Wall cloud entrainment of eye air brings in weaker momentum air. But entrained wall cloud air that comes from larger radius brings in higher momentum. This inner and outer wall cloud entrainment should lead to a momentum compensation and should preserve a close conservation of M_a within the mean upward and outflowing wall parcels.

Close to the cyclone's center where r is small, rV_T is always $\gg (1/2)r^2 f$. For practical purposes the $(1/2)r^2 f$ term may be ignored, and M_a assumed $\approx rV_T$. Air parcels that move outward under the assumption of constant M_a (or rV_T) reach a radius where their tangential wind drops to half that of the low-level inner-core wind (V_{Tmax}). This occurs at an approximate radius of 2 × (RMW).

The following analysis is concerned with the inner-core hurricane area, which is defined as the area inside of the region 2 × (RMW). Most of our calculations will be for the region inside 60-km radius.

4. Budget analysis

We first calculate the net frictional dissipation in the lowest 100-mb layer within the 2 × (RMW) area, or

$$\text{Net frictional dissipation} = \int_o^{2\pi} \int_{r=o}^{r=2\times(RMW)} F_\theta\, \delta A, \qquad (11.3)$$

where $\delta A = r\, dr\, d\theta$ is an element of area. For simplicity we assume all boundary layer frictional dissipation occurs in the lowest 100 mb. But this is not a strict requirement. Net frictional dissipation would be the same if F_θ were half as strong over a 200-mb-thick layer, or a quarter as strong over 400 mb.

For a cyclone in steady state, this frictional dissipation must be balanced by the difference in the import minus the export of tangential momentum across the 2 × (RMW) radius or,

(Net import of V_T Momentum across the 2 × RMW Radius)

Low Level Import   Upper Level Export

$$= \int_{sfc}^{top\ of\ inflow} \int_0^{2\pi} rV_R V_T\, \delta\theta\, \frac{\delta p}{g} - \int_{bottom\ of\ outflow}^{top\ of\ outflow} \int_0^{2\pi} rV_R V_T\, \delta\theta\, \frac{\delta p}{g}, \qquad (11.4)$$

where r is radius, and where the radial wind (V_R) and tangential wind (V_T) are evaluated at the 2 × (RMW) radius. It is convenient (but not mandatory) to assume identical pressure-weighted inflow and outflow layer thicknesses. Equal mass occurs between equal pressure thicknesses. For instance, it matters not whether radial inflow would be 50% as much over a 200-mb-thick layer as it is over a 100-mb-thick layer. Conservation of mass dictates that the net mass inflow at lower levels be equal to the net mass outflow at the upper level, such that

$$\int_{sfc}^{100\,mb} V_r \frac{\delta p}{g} = 0. \tag{11.5}$$

For simplicity we will assume all inflow and all outflow occurs over a 100-mb-thick layer. This makes it possible to more simply express the net import of tangential momentum across the $2 \times$ (RMW) radius that is needed to balance frictional dissipation as

Net Momentum Import at Low Levels   Net Momentum Export at Upper Levels   Net Frictional Dissipation

$$\int_{sfc}^{top\,of\,inflow} \int_0^{2\pi} \int_{r=0}^{2\times(RMW)} rV_R V_T \,\delta\theta\, \frac{\delta p}{g} - \int_{bottom\,of\,outflow}^{top\,of\,outflow} \int_0^{2\pi} \int_{r=0}^{2\times(RMW)} rV_R V_T \,\delta\theta\, \frac{\delta p}{g} = \int_0^{2\times(RMW)} F_\theta \,\delta A \quad \text{with}$$

Mass inflow at $2 \times$ (RMW)   Mass outflow at $2 \times$ (RMW)

$$\int_{sfc}^{top\,of\,inflow} \int_0^{2\pi} rV_{R(inflow)} \,\delta\theta\, \frac{\delta p}{g} = \int_{bottom\,of\,outflow}^{top\,of\,outflow} \int_0^{2\pi} rV_{R(outflow)} \,\delta\theta\, \frac{\delta p}{g}, \tag{11.6}$$

because the terms on the left of Eq. (11.6) can, assuming mass balance across $2 \times$ (RMW), be taken as the difference in V_T between the inflow and outflow, or can be more simply expressed as

$$\int_0^{2\pi} [(V_R V_T)_{inflow} - (V_R V_T)_{outflow}] r \, d\theta$$

$$= \int_0^{2\times(RMW)} F_\theta \,\delta A, \quad \text{or}$$

$$\int_0^{2\pi} (0.29 V_R V_T) r \, d\theta = \int_0^{2\times(RMW)} F_\theta \,\delta A, \tag{11.7}$$

where (as will be discussed) the V_T of the outflow layer at $2 \times$ (RMW) is approximately 71% of the inflow layer V_T at this radius. This is the direct assumption of the 0.5 power law for x and the conservation of outflow rV_T with radius.

With assumed values of V_{Tmax}, RMW, the radial profile of V_T (value of x in equation $V_T r^x = const.$), and surface frictional assumptions, it is possible to solve for the transverse circulation inside the radius of $2 \times$ (RMW), which is necessary to maintain the tangential wind in steady state. This has been done for the following idealized cases:

1) RMW = 30 km, $x = 0.5$, $F_\theta = (C_D/\Delta Z)\overline{V_T^2}$, where $C_D = 1.5 \times 10^{-3}$, $\Delta Z = 1$ km, for cases of $V_{Tmax} = 80, 50$ and 30 m s^{-1};
2) the V_T at low levels at $2 \times$ (RMW) radius = 35 m s^{-1}, and the radial function of V_T (x values) is 0.3, 0.5, and 0.8;

3) for the standard case of $V_T = 50$ m s^{-1}, $x = 0.5$, but varying values of RMW of 10, 15, and 20 km;
4) for the standard case of $V_{Tmax} = 50$ m s^{-1}, $x = 0.5$, RMW = 30 km, but where the drag coefficient (C_D) is 3×10^{-3} rather than 1.5×10^{-3}, and the case where F_θ is represented in terms of the 2.5 power of V_T rather than the square of this wind; in addition the combination of $C_D = 3 \times 10^{-3}$ and 2.5 wind power is made.

5. Illustrative examples detailing the calculation of vertical motion

Special budget calculations will now be made to determine the characteristics of the hurricane's inner-core circulation. Figure 11.6 is an idealized portrayal of the required in-up-and-out tangential momentum flux through the hurricane's 0 to $2 \times$ (RMW) radius for the maintenance of maximum tangential winds of V_{Tmax} of 30, 50, and 80 m s^{-1} at 30-km radius. This in-up-and-out tangential momentum flux can be broken into two parts,

1) that of the low-level inflow from a radius of $2 \times$ (RMW) to the RMW and
2) that from low levels at the RMW to the upper-level outflow at $2 \times$ (RMW).

During the inward circuit it is assumed [from extensive aircraft support, Riehl (1963), Gray and Shea (1973), and Kossin (1997)] that the tangential wind (V_T) follows a radial distribution given by a $V_T r^{0.5} = $ constant wind relationship. The second part of this radial circulation goes from the RMW at low levels up and outward to

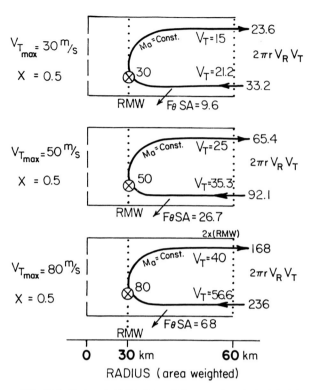

FIG. 11.6. Portrayal of the tangential momentum budget that is required to maintain the tangential wind structure of steady-state cyclones of three different maximum wind (V_{Tmax}) classes. Shown is the in-and-outward flux ($2\pi r V_R V_T$) at the 2 × (RMW) radius, which is required to balance the dissipation of friction ($F_\theta \delta A$). Numbers are expressed in units of 10^3 m^3 s^{-2} kg^{-1}.

the original radius of 2 × (RMW). It is assumed that this second branch of the circulation occurs under a quasi conservation of absolute angular momentum (M_a), or $M_a \approx rV_T$. The radius of 2 × (RMW) has been chosen as a special case because, under the $M_a = rV_T$ assumption, this is the radius for which the outflowing V_T would be just half that of the low level V_{Tmax} at the RMW.

The dissipation by tangential friction inside the 2 × (RMW) radius is given by $\int_A F_\theta \, \delta A$, where $\delta A = \int_o^{2\pi} \int_{r=0}^{r=2(RMW)} r \, \delta r \, \delta\theta$. Momentum changes are expressed in units of 10^3 m^3 s^{-2} kg^{-1}. Figure 11.6 shows the tangential momentum inflow and outflow across the 2 × (RMW) radius for the three different values of V_{Tmax}. Frictional dissipation inside the 2 × (RMW) radius is also shown. Note how rapidly this inner-area frictional dissipation rises with increase of V_{Tmax}, from 9.6 (V_{Tmax} = 30 m s^{-1}) to 26.7 (V_{Tmax} = 50 m s^{-1}) to 68 (V_{Tmax} = 80 m s^{-1}).

If we take the ratio of upper-to-lower-level V_T at the radius of 2 × (RMW) we obtain identical ratios of 15/21.2 = 0.71 for V_{Tmax} = 30 m s^{-1}, 25/35.3 = 0.71 for V_{Tmax} = 50 m s^{-1}, and 40/56.6 = 0.71 for V_{Tmax} = 80 m s^{-1}. This explains the 0.29 factor in Eq. (11.7). It shows (see the three examples of Table 11.1) that there must be an additional tangential momentum flux into and out of the 2 × (RMW) radius, which is nearly 2.5 times larger than what is required to balance boundary layer frictional dissipation (23.6/9.6, 65.4/26.7, 168/68). A large portion of the low-level tangential momentum flux ($+rV_R V_T$) across the 2 × (RMW) radius must be utilized to balance the substantial amount of tangential momentum outflow that occurs aloft. The remaining imbalance of the inward minus outflow tangential momentum flux (about 29%) goes to frictional dissipation. All values are expressed in units of 10^3 m^3 s^{-2} kg^{-1}.

TABLE 11.1. Total inflow of relative tangential angular momentum import ($rV_R V_T$) inside the 2 × RMW radius (column a), the frictional-driven portion of this angular momentum inflow inside 2 × (RMW) (column b), buoyancy driven angular momentum inflow (column c), the ratio of import to frictional dissipation (column d), and the ratio of total inflow to friction inflow (column e). Values in 10^3 m^3 s^{-1}

V_{Tmax}	x	RMW	C_D (10^{-3})	Power law for m s^{-1}	a 100-mb-thick total inflow at 2 × (RMW)	b Frictional-driven inflow	c Buoyant-driven inflow	d Ratio of buoyancy-driven to frictional-driven inflow	e Ratio of total inflow to friction inflow
30	0.5	30	1.5	2	33.2	9.6	23.6	2.44	3.5
50	0.5	30	1.5	2	92.1	26.7	65.4	2.44	3.5
80	0.5	30	1.5	2	236	68	168	2.44	3.5
50	0.3	30	1.5	2	54	21	33	1.57	2.6
50	0.5	30	1.5	2	84	24	60	2.50	3.5
50	0.8	30	1.5	2	240	31	209	6.74	7.7
43	0.5	20	1.5	2	29.0	8.5	20.5	2.44	3.4
49.5	0.5	15	1.5	2	21.6	6.3	15.3	2.44	3.4
61	0.5	10	1.5	2	17.1	5.1	12.0	2.44	3.4
50	0.5	30	3.0	2	184	53	131	2.44	3.5
50	0.5	30	1.5	2.5	600	174	426	2.44	3.5
50	0.5	30	3.0	2.5	1203	349	854	2.44	3.5

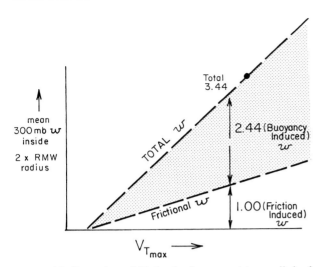

FIG. 11.7. Comparison of frictional to buoyancy-driven wall cloud vertical motion across 300 mb as a function of $V_{T\max}$ in order that steady-state conditions be maintained as cyclone maximum winds increase.

Note that in order to balance boundary layer frictional dissipation inside the 2 × (RMW) radius it is necessary that there be a sizable amount of in and out tangential momentum flux across the 2 × (RMW) radius in all three stratifications. The required transverse circulation is substantially stronger than would be necessary to balance frictional dissipation alone. The majority of the low-level import of tangential momentum goes into balancing a return flow tangential momentum export at upper levels. The ratio of inward tangential momentum flux to frictional dissipation between the top and bottom diagrams ($V_{T\max}$ = 30, 50, and 80 m s^{-1}) goes up with the same proportion (33.2/9.6, 92.1/26.7, 236/68 = 3.5). Irrespective of the maximum wind speed the ratio of the tangential momentum inflow to frictional dissipation inside 2 × (RMW) remains the same (~3.5 to 1). Table 11.1 and Fig. 11.7 show how if steady-state conditions were to be maintained, it would be required that the wall cloud's buoyancy-induced vertical motion to frictionally driven vertical motion across the 300-mb level also remain the same (~2.44 to 1) with increasing $V_{T\max}$. As discussed later, it will be shown that such buoyancy requirements cannot continue to be met as the cyclone becomes very intense and a limit is reached at which steady-state conditions cannot continue to be maintained.

To balance the tangential momentum export at upper levels, it is necessary that wall cloud vertical motion increase in direct proportion to increased $V_{T\max}$. For instance, the vertical averaged tangential wind equation may be written as $\partial V_T/\partial t = -V_R\zeta_a + F_\theta$. At inner radii ($r < 2$ RMW), the relative vorticity (ζ_r) is a close approximation to the absolute vorticity (ζ_a). Beyond the RMW, ζ_r may also be expressed as $\approx(V_T/r)(1 - x)$, where x is the exponent in the $V_T r^x$ = constant wind law. Assuming steady-state conditions, $x = 0.5$, $F_\theta =$

$-(C_D/\Delta z)V_T^2$, and a net tangential wind outflow to inflow (V_T) ratio of 0.71, the steady-state tangential equation may be written as

$$0 = V_R(1 - 0.71)(0.5)\frac{V_T}{r} - \frac{C_D}{\Delta z}V_T^2. \quad (11.8)$$

Solving for V_R (or net radial mass inflow or mass outflow), which is the net required inflow which is proportional to the upward wall cloud mass flux across the 300 mb level, we obtain

$$V_R = KV_T, \quad (11.9)$$

where $K = [C_D r/(\Delta z)(0.29)(0.5)]$ is a constant for specified values of C_D, Δz, x, and ratio of $V_{T\text{outflow}}/V_{T\text{inflow}}$ at 2 × (RMW).

Equation (11.9) demonstrates that for steady-state conditions to be maintained for fixed values of x and the $V_{T\text{outflow}}/V_{T\text{inflow}}$ ratio, the radial mass inflow across the 2 × (RMW) radius (and the upward wall cloud mass flux across 300 mb) must be directly proportional to the low-level tangential wind. A specified V_T, x, RMW, and V_R/V_T ratio assures that there must be a direct proportionality between radial inflow (and wall cloud vertical motion across 300 mb) and the cyclone's maximum wind ($V_{T\max}$). Momentum balance requirements dictate that the strength of the cyclone's inner-core transverse circulation be directly related to the cyclone's maximum tangential wind ($V_{T\max}$). As a tropical cyclone intensifies, so too must its transverse circulation increase by the same percentage amount.

For a cyclone's maximum winds to increase from 35 to 70 m s^{-1} for instance, it is necessary that the 300-mb-level vertical motion inside the 2 × (RMW) also rise by a factor of 70/35 or by 100%. The cyclone with $V_{T\max}$ = 50 m s^{-1} must process approximately 50/30 or 1.67 times more mass upward across the 300-mb level inside its 2 × (RMW) radius as a weaker 30 m s^{-1} $V_{T\max}$ cyclone, etc.

Variations of x. Figure 11.8 portrays the tangential momentum budget inside the radius of 2 × (RMW) for different values of x in the $V_T r^x$ = constant wind profile. Figure 11.9 shows these V_T profiles. For higher values of the x coefficient, the ratio of tangential momentum import to frictional dissipation increases. Higher x values are more typical of the structural characteristics of small intense cyclones at lower latitudes. For $x = 0.8$ the ratio of tangential momentum inflow to tangential frictional dissipation is 7.7 to 1 (see Table 11.1). For radial coefficients of x lower than 0.5 (top diagram of Fig. 11.8 for $x = 0.3$) the ratio of inflow to frictional dissipation drops to a value of 2.6. These cases are more typical of weaker and higher-latitude cyclones.

Variation of RMW. The basic dominance of inflow over frictional dissipation is nearly identical for cyclones with anomalously small RMW (or eye's)

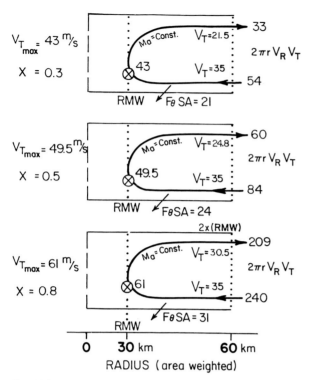

FIG. 11.8. As in Fig. 11.6 but for different values of x in the $V_T r^x$ = constant wind profile.

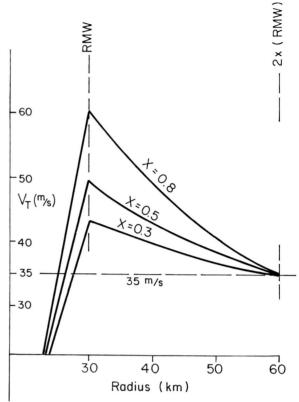

FIG. 11.9. Portrayal of the radial distribution of tangential wind (V_T) for different values of x in the $V_T r^x$ = constant wind relationship.

if one's area of consideration remains inside the 2 × (RMW) radius. Figure 11.10 gives examples of the RMW being reduced from our general case of 30-km radius to 20-, 15-, and 10-km radius. The same dominance of inflow to frictional dissipation of about 3 to 1 is present. Thus, the size of the eye does not alter this analysis provided one only treats the area inside 2 × (RMW).

Assumptions concerning drag coefficients and friction. This analysis has assumed a frictional drag coefficient (C_E) of 1.5 × 10⁻³ and that boundary layer friction is proportional to the square of tangential wind. The precision of these assumptions has no significant effect on this analysis. If the drag coefficient were to be doubled or, if the friction dependence upon wind speed were to be represented by a 2.5 rather than a second power law for wind, the above analysis on the importance of momentum import to frictional dissipation would not change. Examples of these variations are shown in Fig. 11.11. Although these latter calculations indicate an unrealistically large transverse circulation, the dominating influence of inflow tangential momentum flux to frictional dissipation remains unchanged. A lower value of drag coefficient only gives a lower value to the strength of the transverse circulation. The ratio of inflow to friction is not affected by the precise magnitude of the drag coefficient or by the wind power dependence of friction. Thus, the usual accepted value of the drag

coefficient (C_D = 1.5 × 10⁻³) and the second power of the wind law as used here appear to be appropriate. Higher values of the drag coefficient or the wind power give unrealistic results.

Table 11.2 shows the same stratification as for Table 11.1 except for values being expressed in required inward radial wind through a 100-mb-thick layer at the 2 × (RMW) radius in meters per second and in net mass flux. Inflow has been separated into the amount of inflow needed to balance frictional dissipation and the amount of inflow that is required to be drawn into the inner-core region by the wall cloud's condensation and freezing-induced upward mass acceleration due to buoyancy. Buoyancy accounts for the larger part of the inflow. The ratio of buoyancy to frictional-driven inflow is shown in the last column. There can be no question but that the larger part of the hurricane's inflow inside the 2 × (RMW) radius is driven by buoyancy processes associated with the wall cloud's convection.

6. Wall cloud buoyancy as the forcing mechanism of eye subsidence

The sinking motion within the hurricane's eye is a surprising phenomena in the sense that the subsiding air

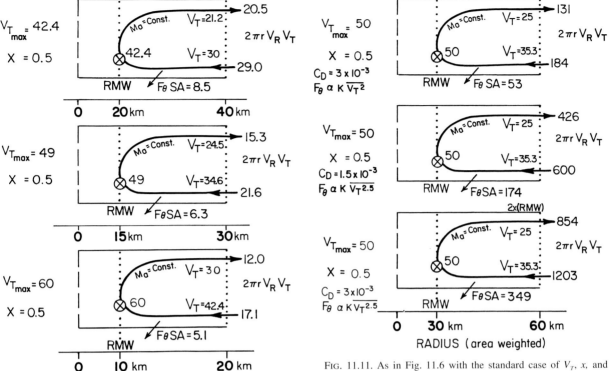

FIG. 11.10. As in Fig. 11.6 but for reduced values of the RMW and 2 × (RMW).

FIG. 11.11. As in Fig. 11.6 with the standard case of V_T, x, and RMW, but with increased drag coefficient (to 3×10^{-3}) and with increase to 2.5 wind power law. Assumed values are indicated on the left side of each diagram.

is less dense than its surroundings. This lower-density warm air is forced downward by mechanical forces associated with the entrainment of air from the eye into the wall cloud. The wall cloud's strong upward accelerating air parcel entrains in large amounts of eye air

as they undergo upward acceleration to their zero buoyancy level at about 300 mb (~10 km). Were it not for the wall cloud's entrainment of eye air, there could be no eye subsidence and no mechanically forced adiabatic compression warming. It is this compressional warming that brings about the very low eye surface pressures.

TABLE 11.2. Mean 100-mb-thick inflow at the 2 × (RMW) radius (column a), and the portion of inflow due to friction (column b), and buoyancy (column c). Column d gives the ratio of buoyancy to frictionally driven inflow. Values in m s^{-1}. Values in parentheses express this required mass inflow (and outflow aloft) in units of 10^9 kg s^{-1}.

V_{max} m s^{-1}	x	RMW	C_D 10^{-3}	Exponential power law for friction	a 100-mb-thick inflow at 2 × (RMW) m s^{-1}	b Frictional-driven component of inflow m s^{-1}	c Buoyancy-driven component of inflow m s^{-1}	d Ratio of buoyancy-to friction-driven inflow at 2 × (RMW)
30	0.5	30	1.5	2	4.16 (1.56)	1.21 (0.45)	2.95	2.4
50	0.5	30	1.5	2	6.92 (2.60)	2.01 (0.76)	4.91	2.4
80	0.5	30	1.5	2	11.1 (4.17)	3.22 (1.21)	7.88	2.4
50	0.3	30	1.5	2	4.10 (1.54)	1.60 (0.60)	2.50	1.5
50	0.5	30	1.5	2	6.92 (2.60)	2.01 (0.76)	4.91	2.4
50	0.8	30	1.5	2	18.24 (6.86)	2.36 (0.89)	15.9	6.7
43	0.5	20	1.5	2	3.85 (1.45)	1.12 (0.42)	2.73	2.4
49.5	0.5	15	1.5	2	5.31 (2.00)	0.97 (0.36)	4.34	4.4
61	0.5	10	1.5	2	3.21 (1.21)	0.39 (0.15)	2.82	7.2
50	0.5	30	3.0	2	13.8 (5.19)	4.00 (1.50)	9.80	2.4
50	0.5	30	1.5	2.5	45.1 (17.0)	13.10 (4.93)	32.0	2.4
50	0.5	30	3.0	2.5	90.2 (33.9)	26.20 (9.85)	64.0	2.4

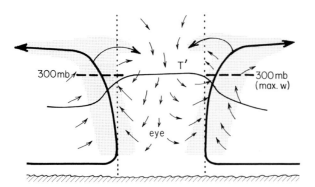

FIG. 11.12. Idealized portrayal of how entrainment into strong sloping eyewall convection drives compensating eyewall subsidence. This subsidence causes hydrostatic warming along the vertical dotted lines, causing larger pressure drops at the low-level RMW than would occur from a vertical eyewall cloud. This picture follows conceptual picture proposed by Malkus (1958) and is verified by inner-core aircraft measurements.

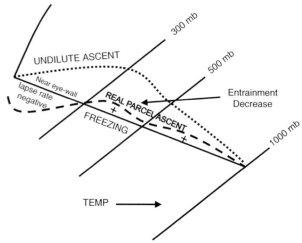

FIG. 11.14. Tephigram portrayal of near eyewall cloud lapse rate with the temperature of real and undilute eyewall cloud parcel ascents. The difference is due to entrainment. Increased buoyancy between 300 and 500 mb is due to freezing of liquid water particles.

Figures 11.12–11.14 give illustrations of such wall cloud parcel entrainment. It is to be expected that the freezing of updraft eyewall cloud particles enhance buoyancy between the 500–300-mb layer by about 4°C, as indicated in Fig. 11.15. Lord et al. (1984) and Ooyama (1990) have indicated that updraft particle freezing makes a significant contribution to eyewall buoyancy. The amount of upward mass acceleration that the wall cloud is able to process through buoyancy is the determining factor in how intense a cyclone can become. By contrast, frictional forcing of vertical motion is established at lower levels and produces no updraft accel-

eration above the friction layer. Frictional forced vertical motion cannot drive updraft acceleration and entrainment and is thus incapable, by itself, of producing eye subsidence. But frictional forcing can initiate deep convection from which buoyancy adds much further updraft enhancement. Wall cloud condensation and freezing-driven updraft acceleration are the primary processes, of eye formation, maintenance, and eye strengthening.

a. Enhanced wall cloud vertical motion from values specified by budget estimates

The previously discussed calculation of net buoyancy-driven upward vertical motion at 300 mb inside the 2 × (RMW) radius underestimates the amount of actual wall cloud vertical motion because the subsidence occurring in the eye was not explicitly accounted for. When the estimated eye subsidence across the 300-mb level is added to the net 300-mb-level ascending motion inside the 2 × (RMW), one finds that the wall cloud

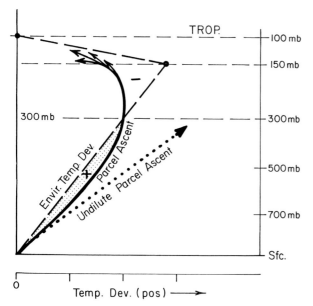

FIG. 11.13. Idealized portrayal of the mechanism by which entrainment into the eyewall cloud updraft causes a general weakening of the undiluted parcel updraft such that maximum updraft velocity and zero buoyancy is reached at 300 mb. Negative buoyancy above 300 mb causes the updraft to weaken and spread horizontally. There is no undilute parcel ascent in the eyewall cloud.

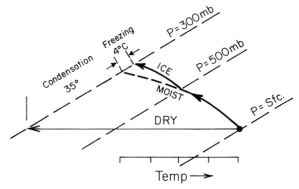

FIG. 11.15. Schematic illustration of how freezing of eyewall cloud updrafts between 500 and 300 mb can enhance updraft buoyancy by about 4°C.

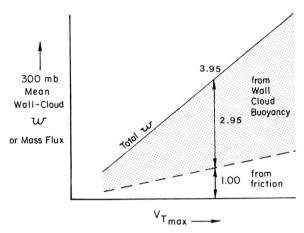

300 mb
Mean
Wall-Cloud

w

or Mass Flux

Total w

3.95

2.95

1.00

from
Wall
Cloud
Buoyancy

from
friction

$V_{T \, max} \longrightarrow$

FIG. 11.16. Relative comparison of required wall cloud mean upward vertical motion or mass flux at 300 mb from frictional forcing and from condensation-induced wall cloud buoyancy for steady-state maintenance as $V_{T \, max}$ increases.

has a larger amount of buoyancy-driven upward vertical motion. Indeed, this additional upward vertical motion is necessary to support the forced eye subsidence. The more intense the wall cloud's buoyancy-driven deep convection, the larger is the wall cloud's entrainment of mass from the eye and the larger is the rate of eye subsidence warming and pressure drop. The smaller the eye and the RMW, the larger is the ratio of wall cloud entrainment to eye area. This can lead to an even stronger eye subsidence and yet lower eye surface pressure. Smaller eyes typically have the lowest central pressures.

For our standard cyclone ($V_{T \, max} = 50$ m s^{-1}, $x = 0.5$, RMW = 30 km) we find that when eye mass subsidence is subtracted from the mean vertical motion calculations inside the 2 × (RMW) radius, the comparison of wall cloud buoyancy induced vertical motion to frictional forced vertical motion raises from 2.44 to 1 (as previously discussed) to 2.95 to 1 (see Fig. 11.16). This emphasizes even more the importance of wall cloud buoyancy induced vertical motion.

One can calculate this extra wall cloud buoyancy and upward vertical motion across the 300-mb level by accounting for the estimated eye subsidence across the 300-mb level. This is done by an estimate of the wall cloud's inner-radius mass entrainment. Wall cloud entrainment is assumed to be a function of the wall cloud's buoyancy-driven updraft velocity at 300 mb minus the buoyancy-driven updraft velocity at cloud base, which is assumed to be zero, or

$$\text{Wall Cloud Entrainment} = \frac{w_{b \text{ at 300 mb}}}{\Delta z},$$

where w_b is the buoyancy-driven wall cloud vertical motion at 300 mb and Δz is the vertical distance from cloud base to 300 mb (~10 km). Wall cloud buoyancy-induced vertical motion at 300 mb is less than the total vertical motion by the amount of frictional forcing.

b. Character of entrainment

Assuming that all vertical motion out to 2 × (RMW) radius goes up within a 10-km-wide wall cloud, we calculate for our standard case ($V_{T \, max} = 50$ m s^{-1}, $x = 0.5$, RMW = 30 km) a mean wall cloud 300-mb vertical motion of 3.18 m s^{-1}. The buoyancy-driven portion of this is 300-mb vertical motion is 2.26 m s^{-1} or 71% of the total for our standard case. The frictionally driven portion of this upward motion is thus 0.92 m s^{-1} or about 29% of the total. To support the vertical mass divergence associated with a 10-km-wide wall cloud upward wind acceleration to 300 mb of 2.26 m s^{-1}, it is necessary that there by a compensating cloud base to 300-mb-level mass convergence into the wall cloud of $\approx 2.26 \times 10^{-4}$ s^{-1}.

The percentage of net entrainment from the inner radial side of the wall cloud to the total wall cloud entrainment is a crucial component in determining the magnitude of the cyclone's eye subsidence and its associated eye warming and pressure drop. It is to be expected that the mass entrainment occurring on the inner side of the wall cloud is significantly less than that on its outer side. There are special factors that restrict the wall cloud's inner radius entrainment. These include the following.

1) The circumference of the cloud's outer radius side is larger than that of the circumference of the wall cloud's inner side. There is more area available for entrainment on the outer side. The upward accelerating wall cloud vertical motion entrains in air from both its inner and outer radial sides. Entrainment is influenced by the area of the eye and the circumference of the wall cloud. The circumference of the outer edge of the wall cloud ($2\pi r_{\text{outside}}$) is larger than the circumference of its inside ($2\pi r_{\text{inside}}$). This dictates that the amount of entrainment occurring on the wall cloud's inside be less than that on the outside by the ratio of $r_{\text{inside}}/r_{\text{outside}}$. If we assume that the eyewall cloud is 10 km wide, then with our standard cyclone case ($V_{T \, max} = 50$ m s^{-1}, $r = 0.5$, RMW = 30 km), the ratio of $r_{\text{inside}}/r_{\text{outside}} = 25/35$. By this reasoning entrainment on the inside of the wall cloud would be expected to be 25/60 or 42% of the total. The entrainment on the outside is 35/60 or 58% of the total wall cloud entrainment.

2) The inertial stability [$(\zeta_a)(2V_T/r + f)$] of the area on the inside of the wall cloud is much higher than the inertial stability of the air on the outer side. The inner side of the wall cloud is inside the RMW and the outer side beyond the RMW. Vorticity is much higher and particle resistance to radial movement is greater on the inner side of the wall cloud. The cloud particles on the inner side of the eyewall cloud are thus more inhibited from moving radially than are the air parcels on the outer edge of the wall cloud. It is to be expected that air particles can be entrained more readily into the wall cloud from their outer

than their inner sides. These inertial stability differences cause the wall cloud's inner edge to resist horizontal motion more than the wall cloud's outer edge. It is to be expected that the rate of entrainment on the inside of the wall cloud would be less than on the outside.

It is not possible to precisely specify the difference between inner versus outer edge wall cloud entrainment. However, we will assume that inertial stability causes the wall cloud's inner-edge entrainment rate per circumference length to be half that of the wall cloud's outer-edge entrainment rate per circumference length. For our standard case ($V_{T\text{max}} = 50$ m s^{-1}, $r = 0.5$, RMW $= 30$ km, $r_{\text{inner wall cloud}} = 25$ km, $r_{\text{outer wall cloud}} = 35$) these assumptions would require that the inner wall cloud mass entrainment rate be $(1/2)[r_{\text{inner}}/(r_{\text{inner}} + r_{\text{outer}})] = (1/2)$ $(25/60)$ or 21% of the total wall cloud entrainment. Seventy-nine percent of the entrainment occurs on the outer edge. Even though small, this percentage inner wall cloud entrainment is a very fundamental feature of the cyclone's inner-core structure. The mass equivalent of 21 percent wall cloud entrainment causes substantial eye subsidence. If 21% of the wall cloud's entrainment is fed by its inner edge, then it is required that the wall cloud have 21% more buoyancy than that calculated for the wall cloud without consideration of eye subsidence.

The ratio of net buoyancy to frictionally forced vertical motion across the 300-mb level is not 2.44 to 1, but rather more like 2.95 to 1. Part of this additional buoyancy-driven mass flux across the 300-mb level is advected inward at levels above 300 mb to support eye subsidence through the 300-mb layer (see Fig. 11.17). But most of the wall cloud's upward mass flux is carried outward.

Eye subsidence that is driven by 21% of the wall cloud's mass entrainment causes a substantial rate of ventilation of mass through the eye. In our standard case we find that the mean divergence from the 0–25-km radius eye is about 3.2×10^{-5} s^{-1}. This gives a 0–25-km radius sized eye a mean ventilation rate of all mass in the eye below 300 mb of about 8.7 h. This agrees with previous estimates by Gray and Shea (1973) but is at odds with recent arguments by Willoughby (1998). Willoughby hypothesizes that the warm eye air of the center is advected with the cyclone as it moves. Perhaps some of the primary warm air at the upper levels of the eye is maintained in the eye. But this analysis indicates that wall cloud entrainment requirements dictate that most of the eye mass be ventilated in 6–10 h or so. For smaller eyes or more intense cyclones the ventilation time would be substantially less. For weaker and larger eyes the ventilation times would be larger. The eye is a dynamic entity. Mass is continuously entering and leaving it at a rapid rate.

It appears that the wall cloud's buoyancy-induced updraft acceleration can cause significant mass out of the eye due to entrainment on the wall cloud's inner side.

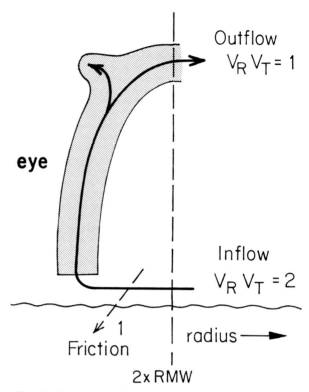

FIG. 11.17. Portrayal of idealized vertical motion and slope of the wall cloud with some return flow to support the subsidence in the eye. Due to frictional dissipation of one unit, conservation of mass, and conservation of relative angular momentum (rV_T), the tangential wind of the air crossing the vertical dashed line must be twice as strong for the inflowing air as for the outflowing air.

This inner-side entrainment is able to mechanically drive the eye's forced subsidence motion despite the eye air being warm and negatively buoyant. The stronger the wall cloud's buoyancy-driven upward acceleration of air parcels, the stronger the eye's subsidence and the lower the eye's central pressure. Cyclones with equal amounts of wall cloud buoyancy but small eyes have from mass flow considerations larger amounts of eye subsidence, lower central pressures, and shorter eye mass ventilation times. The opposite occurs with larger eyes.

7. Buoyancy versus frictionally forced vertical motion in the hurricane's outer-core (60–250-km radius) region

There is no question but that nonfriction or buoyancy-driven vertical motion is much stronger than frictional-driven vertical motion inside the hurricane's inner-core area of $2 \times$ (RMW). For the wall cloud itself this ratio for the typical hurricane is approximately 3 to 1. And for very intense and/or intensifying hurricanes this ratio can be substantially larger. For weakening cyclones it is less. But this assessment is valid only for the hurricane's inner core.

Beyond the inner core [0–2 $\times$ (RMW), or about 60-

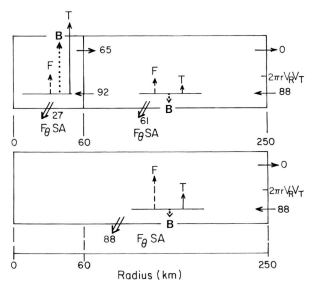

FIG. 11.18. Graphical comparison of the area-weighted relative vertical circulations through the 300-mb level resulting from surface friction (*F*), and buoyancy (*B*) process such as condensation. The total (*T*) vertical motion is a combination of these two processes. Calculations are made for the standard cyclone case (V_{Tmax} = 50 m s^{-1}, x = 0.5, RMW = 30 km. Top diagram) gives a break down between the 0–60- and 60–250-km radius area of these vertical motion values and of the frictional dissipation in each area ($F_\theta \delta A$). Vertical motion is less than frictional forcing at the outer radius. Also shown are the in-and-out fluxes of tangential momentum across the 60- and 250-km radii. Units are 10^3 m^3 s^{-2} kg^{-1}. The bottom diagram portrays this same information but as an average for the whole 0–250-km region. In this latter case, nonfrictional induced vertical motion is near zero.

km radius], the hurricane's upward vertical motion is less than that specified by its frictional forcing. This can be demonstrated from our standard hurricane model (V_{Tmax} = 50 m s^{-1}, RMW = 30 km, $V_T r^{0.5}$ = constant) through the analysis of the hurricane region, which is located inside the radius where the outflowing tangential wind becomes zero (the area approximately inside 250-km radius). And the dominant role of frictional forcing is more the case for the larger region of the hurricane beyond the zero V_T outflow radius, or approximately the region encomposed by the area between 250- and 700-km radius.

Conservation of absolute angular momentum ($M_a = rV_T + 1/2 fr^2$ = constant) considerations dictate that the hurricane's upper-level outflow act to reduce the V_T as it moves to larger radius. It becomes zero (V_T = 0) at a radius of about 250 km [or 8.3 × (RMW)]. At this radius, no net upper-level outflow of tangential momentum (or $rV_R V_T$) occurs. All low-level inflow of tangential momentum at the 8.3 × (RMW) radius must go to balance frictional dissipation inside this radius (see the bottom of Fig. 11.18). Inside the 250-km radius all vertical motion can be thought of as a consequence of frictional forcing. The sum of all the processes involved with buoyancy-driven vertical motion must add up to close to zero. But with the high value of nonfriction or

buoyancy-driven vertical motion inside the 2 × (RMW) radius of 60 km, this implies that condensation's influence on buoyancy-induced vertical motion between 60- and 250-km radius must be negative. Here the upward vertical motion is less than that specified by frictional forcing. This is indicated in Fig. 11.19.

Negative buoyancy in the 60–250-km region is consistent with the large upper-level convergence of tangential momentum into this region from the cyclone's inner core. The large low-level angular momentum advection out of the outer core (60–250-km radius) of 92 units (Fig. 11.18) is significantly compensated for by this upper-level advection from the inner to the outer core. The outer core (60–250 km) cannot maintain itself against frictional dissipation and advect such large amounts of low-level momentum to the inner-core without this upper-level import. Less low-level momentum convergence is required to balance the outer-core's (60–250-km radius) frictional dissipation. The right portion of diagram Fig. 11.19 compares the net vertical motion of the outer core, which is due to frictional processes (*F*), and that due to nonfriction (NF or *B*) processes. The net or total vertical motion is designated by a *T*. Notice that the sum of all nonfriction processes including condensation–freezing induced plus and minus buoyancy is negative.

It is important that budget balances for each of the basic radial units of the hurricane be separately studied. One should not attempt to generalize for the cyclone as a whole or the entire core of the cyclone alone. When one considers the entire circulation of the hurricane inside 250-km radius, buoyancy-induced vertical motion need play no role with regard to the overall region vertical motion. Yet, buoyancy- or nonfriction-induced vertical motion plays a dominate role in the cyclone's inner core [r < 2 × (RMW)]. It would be unwise to conclude that buoyancy-driven net upward vertical motion is not a fundamental feature of the hurricane—even though its importance is a feature only of the inner core. The hurricane could not exist as we observe it if its wall cloud buoyancy-induced upward-accelerating vertical motion were not the dominant driving mechanism of the inner-core region vertical motion.

Nonfrictional-induced negative vertical motion is a prominant feature of the outer core. Here condensation-induced upward vertical motion is being counterbalanced by evaporation- and melting-induced downdrafts. Clear region subsidence from both dynamic forcing and radiational cooling is also occurring in the outer core. The cyclone's net upward vertical motion is still positive in the outer core but is less than that specified by the amount of boundary layer frictionally forced convergence.

8. Outer radius circulation features

Beyond the radius of zero V_T outflow (r > 250 mb) the general symmetrical aspect of the hurricane's wind

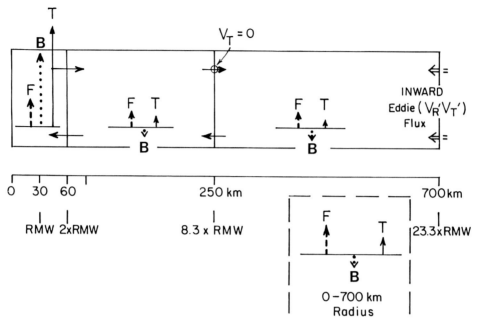

FIG. 11.19. As in Fig. 11.18 but for the upward vertical motion for the larger hurricane area inside 700-km radius where the inward horizontal flux of eddy tangential momentum ($rV_R'V_T'$) is an important component. Total upward vertical motion (T) is less than would be accomplished by frictional forcing (F). Nonfrictional or negative buoyancy (B) processes cause sinking vertical motion. The lower-right diagram is the area-weighted vertical motion for the entire 0–700-km area.

field and its dynamics can no longer be assumed. Horizontal eddy processes ($V_R'V_T'$) play an important role in sustaining the hurricane's outer radius wind field. The importance of inward eddy fluxes of tangential momentum for maintaining the hurricane's outer circulation against frictional dissipation has been demonstrated by Frank (1977), Pfeffer and Challa (1992) and Lee (1984, 1989a, b). Inward eddy flux of tangential momentum at these outer radii make it possible for the hurricane's outer radius transverse circulation and vertical motion to be weaker than what is required to balance frictional dissipation. The contribution of buoyancy-enhanced vertical motion to the maintenance of the hurricane's outer-core ($r > 250$ km) circulation is, in the net, negative. This is portrayed in Figs. 11.18 and 11.19 where the total vertical motion (T) in the 60–250- and 250–700-km radial areas is less than that specified by frictional forcing. The lower right diagram of this figure shows relative vertical motion for the entire 0–700-km area. Note that in an overall sense the hurricane's upward vertical motion is less than that specified by frictional forcing. The inner core of the hurricane, where strong buoyant-induced upward vertical motion is so fundamental, takes up but a small areal portion of the hurricane's circulation. In an integrated sense there is more negative buoyancy (or upward vertical motion being less than required by friction) in the outer region of the hurricane than can be compensated by the small area of strong positive buoyancy of the inner core. Thus, in an overall sense the hurricane is not a buoyancy-driven

system. Only in its inner core can it be classified as buoyancy driven.

9. Summary discussion of inner-core vertical motion

The hurricane's inner-core transverse circulation is much stronger than that specified by frictional influences within the inner core (0–60-km radius) and weaker than that specified by frictional forcing in the outer core (~60–250- and 250–700-km radius). It is important to realize that the convective and frictional forcing processes of the hurricane's inner core and its outer core are quite different. One should not generalize for the cyclone as a whole. Maintenance of the inner-core structure of the tropical cyclone requires substantially more in-up-and-out transverse circulation than that specified by the local forcing of frictional convergence, as specified by the inner-core relative vorticity, while the opposite is true of the cyclone's outer core. Inner-core momentum budget requirements specify that the strength of this transverse circulation increase in direct proportion to the increase in the cyclone's maximum winds provided that the power law on x remains constant. This places a severe requirement for large amounts of wall cloud potential buoyancy. Such wall cloud buoyancy is responsible for about 70%–80% of the wall cloud's 300-mb vertical motion. Strong wall cloud buoyancy-driven upward vertical motion acceleration is also a fundamental requirement for eye subsidence. As the cyclone becomes very intense, such high eye-wall cloud buoyancy requirements have difficulty being met, however. This places a limit on how intense a cyclone can become.

It is important not to infer an overly idealized view of the hurricane's transverse circulation in which inflow is driven by boundary layer friction with inner-core air rising in a neutral moist environment on the mesoscale with little upward vertical acceleration. The nature of the cumulus convective process does not permit this. The mesoscale upward vertical motion in and around the hurricane's eyewall region consists of concentrated updrafts in a near saturated environment. Updrafts are more frequent and stronger than downdrafts. The hurricane region beyond the wall cloud region is drier and has more accelerating downdrafts. Downdrafts of the outer core spread out near the surface to force new cumulus updrafts. Intense squall lines are frequently generated. The up-and-down motion processes of the outer-core region are quite chaotic. The processes of the inner core are less varied but cannot be conceptualized as consisting of near saturated, mesoscale gentle upward motion with little acceleration.

Inside the radius of zero outflow tangential wind (~250-km radius) it is required that the cyclone's net upward vertical motion be a consequence only of frictional forcing. This implies that the net influence of updraft buoyancy would be zero. This is true, but it would be quite inappropriate to conclude from this knowledge (as some have done) that buoyancy does not play a fundamental role in the hurricane's inner core or that in the hurricane's outer core ($\approx$60–250-km radius) upward vertical motion is not substantially weaker than that specified by boundary layer frictional forcing.

Outflow upper-level convergence into the outer core from the wall cloud region causes a dynamic subsidence in the outer core (60–250-km radius). This, combined with the negative buoyancy of downdrafts, produces outer core (60–250 km) sinking motion, which causes the upward vertical motion in the outer core to be less than that specified by frictional forcing. This outer-core (60–250 km) negative buoyancy occurs over an area 16.4 times larger than the inner core (0–60 km) and more than balances out the inner-core's wall cloud high levels of positive buoyancy. The near cancellation of such inside and outside buoyancy does not by any means negate the separate and opposite sign importance of buoyancy to each of the core subregions, however.

Most previous analysis of the hurricane's inner circulation have not emphasized the importance of the inner-core's outflow of positive tangential momentum and the fundamental role that this positive outflow tangential momentum plays in the required enhancement of the cyclone's inner-core transverse circulation and upward vertical motion to magnitudes much beyond what can be provided by frictional forcing alone.

It is of interest to note the lack of low-level angular momentum convergence ($2\pi r V_R V_T$) in the outer core

(60–250-km radius). The absolute amount of inward momentum convergence at the 250-km radius is about the same as that at the 60-km radius. The outer-core's (60–250 km) angular momentum balance against frictional dissipation can be maintained only by upper-level import of tangential momentum from the inner core (0–60 km). The inner-core region robs momentum from the outer core (60–250 km) at low levels but returns about two-thirds of it at upper levels. This requires an upper-level to lower-level downward momentum flux in the outer core (60–250 km). Much of this downward flux is accomplished by dynamically driven subsidence from the extreme inner-core convection and by negatively buoyant downdrafts in the dryer outer-core region.

This analysis shows that it is not at all satisfactory to conceptualize the entire hurricane core region (0–250-km radius) as functioning under quasi-similar physical processes.

Acknowledgments. The author is very grateful to Matthew Eastin and James Kossin for their many calculations involving the NOAA P-3 aircraft data and for many hours of very beneficial discussion. The author appreciates the open access to NOAA/HRD P-3 flight data. This research was supported by a grant from the National Science Foundation.

REFERENCES

Black, M. L., R. W. Burpee, and F. D. Marks, Jr., 1996: Vertical motion characteristics of tropical cyclones determined with airborne doppler velocities. *J. Atmos. Sci.,* **53,** 1887–1909.

Black, R. A., H. B. Bluestein, and M. L. Black, 1994: Unusually strong vertical motion in a Caribbean hurricane. *Mon. Wea. Rev.,* **122,** 2722–2739.

Colón, J. A., and Coauthors, 1961: On the structure of Hurricane Daisy. National Hurricane Research Project Rep. 48, 102 pp.

Eastin, M. D., 1999: Instrument wetting errors in hurricanes and a re-examination of inner-core thermodynamics. Dept. of Atmospheric Sciences Paper 683, Colorado State University, 203 pp.

——, P. G. Black, and W. M. Gray, 2001a: Flight-level thermodynamic instrument wetting errors in hurricanes: Part I: Observations. *Mon. Wea. Rev.,* **130,** 825–841.

——, ——, and ——, 2001b: Flight-level thermodynamic instrument wetting errors in hurricanes: Part II: Implications. *Mon. Wea. Rev.,* **130,** 842–851.

Fitzpatrick, P. J., 1996: Understanding and forecasting tropical cyclone intensity changes. Dept. of Atmos. Sci. Paper 598, Colorado State University, Ft. Collins, CO 346 pp.

Frank, W. M., 1977: The structure and energetics of the tropical cyclone. Part I: Storm structure. *Mon. Wea. Rev.,* **105,** 1119–1135.

Gray, W. M., and D. J. Shea, 1973: The hurricane's inner core region. II. Thermal stability and dynamic characteristics. *J. Atmos. Sci.,* **30,** 1565–1576.

——, and ——, 1976: Data summary of NOAA's hurricanes inner-core radial leg flight penetrations 1957–1967, and 1969. Dept. of Atmospheric Sciences Paper 257, Colorado State University, 245 pp.

Kossin, J., 1997: Maintenance of balance and steady state in the hurricane inner-core. Preprints, *22d Conf. on Hurricanes and Tropical Meteorology,* Fort Collins, CO, Amer. Meteor. Soc., 492–493.

——, and M. D. Eastin, 2001: Two distinct regimes in the kinematic and thermodynamic structure of the hurricane eye and eyewall. *J. Atmos. Sci.,* **58,** 1079–1090.

Lee, C. S., 1984: The bulk effects of cumulus momentum transports in tropical cyclones. *J. Atmos. Sci.,* **41,** 590–603.

——, 1989a: Observational analysis of tropical cyclogenesis in the western North Pacific. Part I: Structural evolution of cloud clusters. *J. Atmos. Sci.,* **46,** 2580–2598.

——, 1989b: Observational analysis of tropical cyclogenesis in the western North Pacific. Part II: Budget analysis. *J. Atmos. Sci.,* **46,** 2599–2616.

Lord, S. J., H. E. Willoughby, and J. M. Piotrowicz, 1984: Role of parameterized ice-phase microphysics in an axisymmetric non-hydrostatic tropical cyclone model. *J. Atmos. Sci.,* **41,** 2836–2848.

Malkus, J., 1958: On the structure and maintenance of the mature hurricane eye. *J. Meteor.,* **15,** 337–349.

Marks, F. D., Jr., 1985: Evolution of the structure of precipitation in Hurricane Allen (1980). *Mon. Wea. Rev.,* **113,** 909–930.

——, and R. A. Houze, 1984: Airborne Doppler radar observations in Hurricane Debby. *Bull. Amer. Meteor. Soc.,* **65,** 569–582.

——, ——, and J. F. Gamache, 1992: Dual-aircraft investigation of the inner core of Hurricane Norbert. Part I: Kinematic structure. *J. Atmos. Sci.,* **49,** 919–942.

Ooyama, K. V., 1990: A thermodnamic foundation for modeling the moist atmosphere. *J. Atmos. Sci.,* **47,** 2580–2593.

Pfeffer, R. L., and M. Challa, 1992: The role of environmental asymmetries in Atlantic hurricane formation. *J. Atmos. Sci.,* **49,** 1051–1059.

Riehl, H., 1963: Some relations between wind and thermal structure of steady-state hurricanes. *J. Atmos. Sci.,* **20,** 276–287.

——, and J. Malkus, 1961: Some aspects of Hurricane Daisy, 1958. *Tellus,* **13,** 181–213.

Shea, D. J., and W. M. Gray, 1972: The structure and dynamics of the hurricane's inner-core region. CSU Dept. of Atmospheric Sciences Pap. 182, 105 pp.

——, and ——, 1973: The hurricane's inner-core region. Part I: Symmetric and asymmetric structure. *J. Atmos. Sci.,* **30,** 1544–1564.

Willoughby, H. E., 1998: Tropical cyclone eye thermodynamics. *Mon. Wea. Rev.,* **126,** 3053–3067.

Chapter 12

Some Aspects of Midlevel Vortex Interaction in Tropical Cyclogenesis

ELIZABETH A. RITCHIE

Department of Meteorology, Naval Postgraduate School, Monterey, California

ABSTRACT

The mechanisms by which mesoscale midlevel vortices that form in the stratiform anvil regions of mesoscale convective systems develop downward in the atmosphere are explored in the context of tropical cyclone genesis. Using simple two- and three-dimensional models, a theory for the processes by which midlevel vortices may interact both with each other, and with their large-scale environment in order to develop a storm-scale vortex, is developed. It is found that absorption of the circulation of one vortex by another results in a vortex of greater horizontal and vertical extent. Embedding the vortices in an enhanced vorticity environment such as might be found in the monsoon trough results in more efficient merger and greater downward development of the circulation associated with the merged vortex.

This theory is used to interpret a real case of the development of Tropical Cyclone (TC) Oliver in the Australian region during the Tropical Ocean Global Atmosphere Coupled Ocean–Atmosphere Response Experiment (TOGA COARE) experiment in 1993. High-resolution flight-level and dropwindsonde data were collected during the interaction and merger phase of two large mesoscale convective systems that were embedded in the monsoon trough. Multiple mesoscale vortices were observed to interact and merge during the development phase of TC Oliver with consequences for the downward development of the vortex, and subsequent eye development.

1. Introduction

The physical processes associated with tropical cyclone development have been a subject of considerable interest. Despite this, the mechanisms involved in the evolution of the immature storm are not well understood. Although this is partly due to a lack of data in the sparsely populated tropical regions, it may also be due to the stochastic nature of the dynamics of the tropical atmosphere (Ooyama 1982). Necessary climatological conditions for tropical cyclone formation include a region of disturbed weather with upper-level divergence, lower-level convergence, and little vertical shear over the disturbance center (Gray 1968, 1975). However, these conditions are satisfied over long timescales in the tropical regions; yet potential disturbances rarely form into tropical cyclones. Presumably, other mechanisms, unresolved by large-scale dynamics, must be necessary to transform a disturbance into a tropical storm. Therefore, whereas the general region where tropical cyclones may potentially develop can be identified from well-known large-scale conditions, the precise location and timing of tropical cyclone development is most likely dictated by mesoscale processes. With requirements for longer-term track and intensity forecasts of tropical cyclones increasing, the ability to predict genesis accurately is essential.

A typical prestorm weather disturbance in the western Pacific is generally characterized by one or more mesoscale convective systems with embedded deep convection loosely organized about a central point (e.g., Fig. 12.1). These mesoscale convective systems provide a favorable environment for the development of associated mesoscale convective vortices that have a scale of a few hundred kilometers (Menard and Fritsch 1989; Chen and Frank 1993). They have been shown to be common in the western Pacific (Miller and Fritsch 1991) with recent studies identifying the associated vortices as potentially important in the formation of tropical cyclones (e.g., Ritchie 1995; Harr and Elsberry 1996; Harr et al. 1996; Ritchie and Holland 1997; Simpson et al. 1997; Bister and Emanuel 1997).

While many studies have attempted to elucidate the processes by which mesoscale convective vortices might develop into a tropical cyclone, no unified theory has been achieved. Here, we examine the problem from a purely dynamic viewpoint. It is assumed that the main atmospheric response to the warming associated with a mesoscale convective system is the development of a midlevel vortex. The theory behind this process will be examined further in the next section. We then examine dynamic processes by which the mesoscale vortex might develop into a tropical cyclone. Although we acknowledge that convection is modulating the system during development, it will be shown that for the processes we are interested in, dynamic processes alone can achieve our goal. However, we consider convective processes to be extremely important for intensification and de-

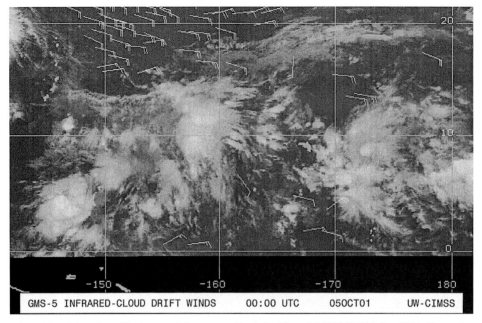

FIG. 12.1. Infrared satellite image of the western North Pacific valid 0000 UTC 5 Oct 2001, illustrating the widespread development of mesoscale convective systems in an active monsoon (image courtesy of Cooperative Institute for Meteorological Satellite Studies).

velopment of the eye after the vortex has developed into the boundary layer.

The paper is organized in the following way. Section 12.2 discusses possible consequences of mesoscale convective systems and associated vortex formation in the Tropics in the context of previous results and simple model experiments. An example of multiple mesoscale convective systems in cyclogenesis is described in section 12.3. The overall findings are summarized and consequences of the results for tropical cyclone development are discussed in section 12.4.

2. Theory and background

a. Mesoscale convective systems

While the thermodynamic conditions for tropical cyclogenesis are often met over the tropical oceans, tropical cyclone formation is still rare (Gray 1975). Certainly there is a need for a region of disturbed weather. In addition, there is a need for a mechanism to lower the Rossby radius of deformation (L_R) from its typical value in the tropical regions (~ 2000 km in the monsoon trough) to that of a tropical cyclone (~ 500 km). In general, L_R is defined (Schubert and Hack 1982; Ooyama 1982) by

$$L_R = \frac{C_g}{I}, \qquad (12.1)$$

where $C_g \propto NH$ is the phase speed of an inertial–gravity wave, I is inertial stability, N is the Brunt–Väisälä frequency, and H is the scale height of the disturbance. In general, if the radius of a local heat source L is much

less than L_R, then the atmospheric response to the heat source is dominated by the formation of gravity waves, which propagate the mass anomaly away leaving the rotational wind largely unaffected by the heating. However, if L is comparable to L_R, then the response is dominated by the inertial mode and the wind field adjusts to the mass field perturbation.

So the question is what are the physical processes occurring in a mesoscale convective system ($L \ll L_R$) that result in a stable, midlevel vortex. Chen and Frank (1993) argue that in the saturated, cloudy atmosphere associated with the large stratiform anvil regions of mesoscale convective systems, L_R is reduced in magnitude locally as a result of the reduction in N. Thus, the warming in the stratiform region results in development of a rotational component to the wind. This results in an increase in the inertial stability locally, and a nonlinear feedback between the warming and the wind response can develop. Such an effect would not be found in a region of similar magnitude filled with purely convective columns since the air is largely unsaturated. In this case the local Rossby radius is not reduced and the warming from latent heat release is largely dispersed by gravity waves [see Chen and Frank (1993) for a detailed discussion]. Numerous studies have now documented that there are midlevel vortices in association with these stratiform regions both in the midlatitudes and the deep Tropics (e.g., Fritsch et al. 1994; Bartels and Maddox 1991; Menard and Fritsch 1989; McKinley 1992; Harr and Elsberry 1996; Harr et al. 1996; Ritchie and Holland 1997; Simpson et al. 1997).

Examination of three years of tropical cyclogenesis

cases in the western North Pacific (Ritchie and Holland 1999) revealed that in the 72-h period prior to designation as a tropical depression 98% of the pregenesis weather disturbances developed at least one mesoscale convective system. It seems reasonable to suppose that midlevel vortices associated with these mesoscale convective systems may commonly be the seedlings from which tropical cyclones develop. However, there is still the question of by what processes these midlevel vortices develop vertically so that the associated circulation extends into the boundary layer and triggers convection.

The depth of the circulation associated with a midlevel potential vorticity (PV) anomaly in the absence of diabatic forcing is given by a modified Rossby–Prandtl–Burger relationship (Hoskins et al. 1985; Shapiro and Montgomery 1993),

$$D = \frac{(f_{loc}\zeta_a)^{1/2}L}{N}, \qquad (12.2)$$

for motions small compared with the Rossby radius of deformation $L \ll L_R$. Here D is the vertical depth of the circulation associated with a PV anomaly of horizontal scale L, ζ_a is absolute vorticity, f_{loc} is the background rotation in the vicinity of the potential vorticity anomaly, and N is the Brunt–Väisälä frequency, a measure of the atmospheric resistance to vertical displacements. Thus, the depth of the circulation associated with a midlevel vortex is proportional to its horizontal scale and the local background rotation, and inversely proportional to the static stability of the atmosphere. The low Coriolis parameter in the Tropics results in vorticity perturbations being more vertically constrained than in higher latitudes. Due to the limited horizontal size of mesoscale convective systems, convectively induced potential vorticity anomalies will have a vertical influence of only a few kilometers in normal atmospheric conditions.[1]

Subsequent convective activity may reinforce or increase the size of the existing midlevel vortex. Fritsch et al. (1994) concluded that the local environment created by a continental mesoscale convective system that they studied was responsible for forcing redevelopment of convection in the associated midlevel vortex and, thus, a continued cycle of regeneration and strengthening of the vortex over a period of days. It has been hypothesized by Bister and Emanuel (1997) that the circulation associated with the midlevel vortex might extend to the boundary layer during a convective event if the column of air within and below the potential vorticity anomaly is completely saturated to prevent strong downdrafts that result in a surface mesohigh. Alternatively, the static stability associated with the midlevel vortex could be decreased between convective events

as the surface mesohigh is eroded by air–sea fluxes allowing the circulation associated with the potential vorticity anomaly to extend into the boundary layer between convective events.

There are also dynamic theories of how the circulation associated with the midlevel vorticity could develop downward to the boundary layer. This could be achieved through increases in both the midlevel vortex and the local background rotation [L and f_{loc} terms in (12.2), respectively]. In this context there are two main hypotheses, not necessarily exclusive, of how the midlevel vortex may develop a surface signature.

One theory proposes that the primary dynamic response to redevelopment of convection on the periphery of the vortex is the development of positive potential vorticity anomalies on the scale of the convection (Montgomery and Enagonio 1998, hereafter ME98). This theory follows from observations that strong convective bursts over a period of a few days frequently precede tropical storm designation (Zehr 1992). Using a dry quasigeostrophic (nondimensional) multilayer model, ME98 demonstrate that the addition of a low-level positive potential vorticity anomaly (and upper-level negative anomaly so that no net vorticity is added) on the periphery of the primary midlevel vortex may result in an interaction and merger of the vortices. Possibly the most exciting result from this work is a demonstration that an idealized "pulsing" simulation where succeeding potential convective anomalies are added to the "seedling" midlevel primary vortex results in a substantial increase in low-level wind speed and temperature anomaly. Although both this theory and the one discussed in this paper argue that vortex merger is a key factor in tropical cyclogenesis, there are two key differences between their work and that described here. ME98 focus on potential vorticity associated with the development of deep convection, which is perhaps a later stage of development than the one we study here. The vortex interaction studies presented here are focused on the development of a storm-scale vortex into the boundary layer so that deep convection can then be focused around it—a very early stage of genesis. The primary motivation here is on potential vorticity that develops in mesoscale convective systems. The real-case example of Tropical Cyclone (TC) Oliver will present observational evidence perhaps to support both of these ideas, because they are complimentary with slightly different foci. But the other important point to make is that although there are real, although perhaps subtle, differences between the ideas presented in ME98 and those presented here and elsewhere, one cannot in the end separate them out in a real developing tropical cyclone because genesis is a continuum of processes in a probablistic setting.

The theory discussed in this paper is developed from observations that multiple mesoscale convective systems often develop within the same prestorm weather disturbance (Ritchie and Holland 1999; Simpson et al.

---

[1] For a standard atmosphere at 15°N, $f \cong 2 \times 10^{-5}$ s^{-1}, $N \cong 0.01$ s^{-1}, $L \sim 500$ km, then $D \sim 1$ km. However, in a monsoon regime such as that in the western North Pacific, $(f_{loc}\zeta_a)^{1/2} \cong 4.5 \times 10^{-5}$ s^{-1} and $D \sim 2.25$ km.

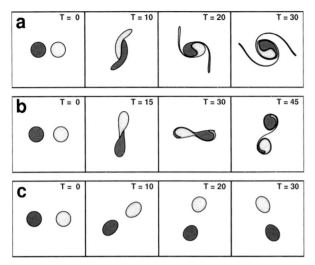

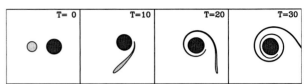

FIG. 12.3. Rapid shearing into a filament of a smaller/weaker vortex patch by a larger/stronger vortex patch. The ratio of the patch radius is 0.7: 1.0, and the ratio of the vorticity is 1.2: 2.0.

FIG. 12.2. (a) Rapid merger of two vortex patches of equal size and vorticity started at a separation distance of 2.7R, where R is the radius of a vortex patch; (b) exchange of fluid between two patches when placed at a separation distance of 3.35R; and (c) interaction of two vortex patches started at a separation distance of 4.0R.

be illustrated next through the use of idealized modeling studies.

b. Two-dimensional vortex interaction

It has been demonstrated by McWilliams (1984) that in an environment without external forcing, turbulent, two-dimensional flow has a tendency to form finite-amplitude, coherent structures. A random distribution of vorticity will concentrate into isolated vortices due to an upscale merger process in which larger vorticity perturbations in the flow sweep up the surrounding vorticity. This provides a potential mechanism for the development and growth of mesoscale vortices and initial development of tropical cyclones without need for convective forcing, if such mergers can be shown to exist.

The simplest cases of vortex interaction are those of binary vortices. Probably the most easily observed example of this process is the binary interaction of tropical cyclones that rotate in a classic Fujiwhara type configuration. Fujiwhara (1931) used a series of laboratory experiments in water to show that vortices of the same rotation placed in near proximity to each other tend to approach in a spiral orbit that has the same sense of rotation as the original vortices (e.g., Fig. 12.2a). The end result is merger of the vortices to produce a new vortex that is both larger and stronger than the original vortices. Melander et al. (1988) defined three regimes for equal vortex interaction that are illustrated in Fig. 12.2. They are: 1) continuous orbit, but no exchange of vortical fluid (e.g., Fig. 12.2c); 2) vortices touch briefly so there is some exchange of vortical fluid before moving apart again (e.g., Fig. 12.2b) and; 3) merger into a new vortex with filament shedding (e.g., Fig. 12.2a). In the atmosphere, regimes 1 and 3 are most likely to be observed. However, it is rare for interacting atmospheric vortices to be equal in strength and size and so in the case of merger, it is more likely that the circulation associated with a weaker vortex will be absorbed into a stronger vortex as in the example in Fig. 12.3 (adapted from Ritchie and Holland 1993).

Waugh (1992) defined the "efficiency" of merger as a quantitative measure of the resultant circulation of the merged vortex. He found that as L_R decreased relative to the size of the merging vortices, there was less loss of vorticity to filament shedding and thus the resulting vortex would contain a greater percentage of the circulation associated with the original vortices. This result

1997). Although an apparently stochastic process, the location of mesoscale convective systems may be modulated by both mesoscale and large-scale processes within the tropical environment. For example, low-level environmental convergence between the monsoonal westerlies and easterly trade flow (Harr et al. 1996; Holland 1995) produces forced uplifting at the eastern end of the monsoon trough, a location that has been identified as a favorable location for genesis (Briegel and Frank 1997; Ritchie and Holland 1999). In addition, forced uplift due to the presence of pre-existing midlevel vortices (Raymond and Jiang 1990; Fritsch et al. 1994) may contribute to the redevelopment of mesoscale convective systems in the vicinity. Ritchie and Holland (1999) showed that in 70% of cases of cyclogenesis in the western North Pacific during 1990–92, mesoscale convective systems formed at multiple times in the pregenesis disturbance in the 72 h prior to tropical depression designation (see their Table 1). As just stated, there is reasonable evidence to support this observation of multiple developments in a proximal location.

Because of these observations of multiple mesoscale convective system developments during the pregenesis period, the hypothesis that the interaction of midlevel vortices both between themselves and also with the large-scale environment in which they are embedded can result in the development of a tropical depression is explored. In the absence of convection, the vertical depth of the circulation associated with a midlevel vortex is hypothesized to be affected by dynamical changes in the horizontal size of the vortex through interaction both with other similar vortices and in a nonlinear fashion with the background flow through the term f_{loc} in (12.2). The nature and result of these interactions will has important implications for tropical cyclone genesis. It implies that while the local inertial stability associated with a weather disturbance is low, vortex merger will be an inefficient process with loss of vorticity to dissipation. However, as the local inertial stability increases (e.g., the vorticity associated with the monsoon trough increases), vortex merger will become more efficient and a nonlinear feedback between the local background rotation and the amount of circulation retained during the merger process will ensue.

c. Three-dimensional vortex merger

When the above two-dimensional ideas are extended to three dimensions, the problem becomes more complicated. From (12.2) it can be seen that increasing the horizontal extent of the potential vorticity anomaly should directly increase the vertical extent of the associated circulation. Thus, merging two midlevel vortices together should result in a vortex that is both larger and deeper than the original vortices. To test this idea, a five-layer baroclinic model was used to initialize two midlevel vortices separated by a distance of approximately 300 km or 2.5 times their radius of maximum winds on an f plane. This is within the critical separation distance of $3.3R$ for merger of two-dimensional vortex patches (Ritchie and Holland 1993). During the simulation, the vortices initially weaken and some vertical adjustment of the vorticity occurs because of the weak secondary circulation that existed in each initial vortex. The rotation and merger of the 500-mb vorticity is very similar to that observed for the contour dynamics vortex patches (Zabusky et al. 1979; Overman and Zabusky 1982) in Fig. 12.2a (Fig. 12.4a). As the vortices rotate and interact with each other the main contribution to the vorticity budget is horizontal advection of vorticity. The 24-h vorticity change indicates horizontal advection of vorticity at 500 mb (Fig. 12.4b). However, the anticyclonic change is stronger than the cyclonic, indicating advection of cyclonic vorticity out of the 500-mb level into adjacent levels as a result of convergence at that level supporting vertical advection of mass (Fig. 12.4b). This evacuation of mass out of the center of the merging system also results in the classic horizontal filamentation structure as the system relaxes back to axisymmetry.

The first stage of merger is complete when the horizontal advection of vorticity term in the vorticity budget diminishes in value. The system then relaxes back to axisymmetry, evacuating mass from the core horizontally by shedding horizontal fluid filaments. Adjustment also continues vertically as convergence into the center is still continuing to support vertical motion and vortex stretching. The horizontal scale of the combined system contracts in size as the vortex tube stretches. The divergence term dominates during this time and there is significant cyclonic vorticity increases at 700 mb in the center of the vortex (Figs. 12.4c,d). However, there is no significant increase of vorticity in the lowest model level, although the resulting vortex is deeper and stronger than either of the original vortices (Fig. 12.5). Thus the hypothesis that merger of vortices will produce a deeper vortex is supported by this idealized simulation.

As mentioned above, another important aspect of the problem is the local background vorticity. Waugh (1992) showed that merger was more efficient—that is, more vorticity was retained in the merged system rather than being shed by filaments during merger—if the background inertial stability was increased. For our study, this term would be represented best in (12.2) by the f_{loc} term. So, in a final idealized simulation here, the background Coriolis is increased by a factor of $3 \times f$ from the surface to 500 mb and then reducing linearly to 1 $\times f$ by 100 mb, in a thought experiment that represents the vertical structure of the vorticity of a basic monsoon trough. This very simplistic representation of large-scale vorticity structure has the advantage of affecting the vortex behavior while remaining invariant throughout the simulation.

Introducing this simple vertical structure into the background rotation results in an increase in the downward development of cyclonic vorticity but no upward increase. Embedding the vortices in this vertical background vorticity structure combines two effects. The vorticity gradient with height in the upper levels produces downward development of each vortex circulation even before merger begins, and the constant, increased rotation with height below this level facilitates efficient merger of the vortices. The result is a significant finite-amplitude circulation at 900 mb (Fig. 12.6). If the vertical structure of the background rotation had been of the opposite sense, the circulation would have developed upward, resulting in cyclonic circulation at 100 mb (Ritchie 1995).

These simple experiments support the hypothesis that midlevel vortices can develop downward as a result of interaction between each other as well as with the local background vorticity. Furthermore, the stronger the local background vorticity, the more efficient the merger and the deeper the mesoscale circulation can develop in a nonlinear feedback process as suggested by Waugh (1992). These ideas will be examined in the next section in the context of a case of genesis in the Australian region.

3. A real case: Tropical Cyclone Oliver

Tropical Cyclone Oliver developed in February 1993 during the Tropical Ocean Global Atmosphere Coupled Ocean–Atmosphere Response Experiment (TOGA COARE) field experiment in close proximity both to the mainland of Australia where the National Aeronautics and Space Administration (NASA) DC-8 and ER-2 aircraft were stationed, and to the Australian Bureau of Meteorology's radar site at Willis Island, just off the coast. The difficulty of predicting the genesis position

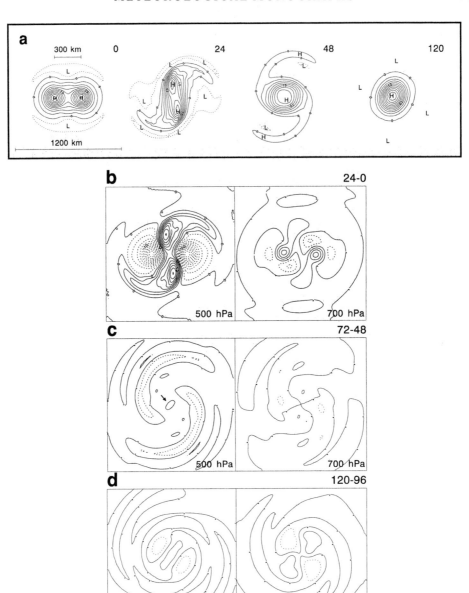

FIG. 12.4. Baroclinic simulation of the interaction between two equal midlevel vortices. Initial separation distance is 300 km (~2.5R): (a) 500-mb vorticity at 0, 24, 48, and 120 h of simulation; (b)–(d) 24-h vorticity difference at 500 and 700 mb (contours = 2.0×10^{-5} s^{-1}).

of a tropical cyclone is highlighted in this case as the location for TC Oliver had very little convective activity in the vicinity only 72 h prior to genesis (not shown) despite a vigorous monsoon trough that extended across the entire north of Australia and well out into the South Pacific. A time series of the average relative vorticity and divergence in a rectangle that encompasses the genesis region reveals the structure of the relatively strong monsoon trough going into the beginning of February (Fig. 12.7). By 2 February an increase in the midtropospheric vorticity is indicated in the time series corresponding to the first organization in the convection

with two weak mesoscale convective systems that developed on the northwest edge of the surface low. The midtropospheric vorticity reached a maximum around 3 and 4 February and then developed toward the surface. The convection redeveloped during 4 February with organization concentrated into two highly structured mesoscale convective systems. The midlevel vorticity increase was accompanied by midlevel convergence probably associated with the stratiform rain regions of the mesoscale convective systems. As the vorticity developed down to the surface, a rapid increase in surface convergence, perhaps indicative of increased frictional

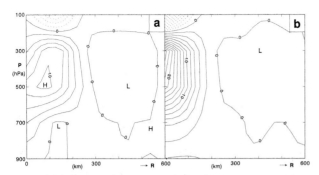

FIG. 12.5. Radial–vertical cross section of azimuthally averaged vorticity centered on the center of mass of the interacting vortices: (a) 0 and (b) 120 h (contours = 2.0×10^{-5} s^{-1}).

forcing, the initiation of air–sea fluxes, and organized convection into the tropical cyclone, was indicated in the time series.

During 3 and 4 February, several small vortices were tracked cyclonically in the C-band radar located on Willis Island, indicated by rotation in the associated precipitation returns. Attempts to retrieve the wind fields using correlation techniques were unsuccessful due to smoothing of the data for archival. The radar imagery indicated that two of the vortices (c, d) were merged into a third (A), while a fourth (B) made a tight inner loop (Fig. 12.8). The radar beam angle indicated that vortex A had an associated circulation between about 400 and 700 mb. Flight-level data from the NASA DC-8 documents the structure of both vortices at 200 mb, and analysis of the dropwindsonde data documents the

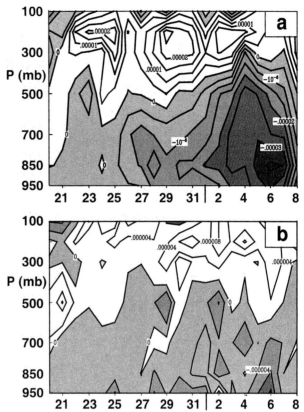

FIG. 12.7. Time–vertical cross sections of parameters averaged over a $5° \times 10°$ latitude box centered on the genesis location for TC Oliver 1993 for the period 21 Jan to 8 Feb 1993: (a) relative vorticity (s^{-1}) and (b) divergence (s^{-1}).

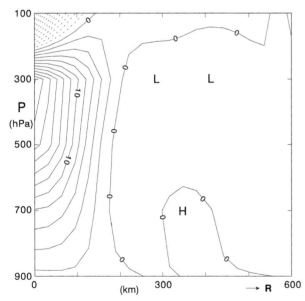

FIG. 12.6. Radial–vertical cross section of azimuthally averaged vorticity centered on the center of mass of the interacting vortices at 120 h of simulation. The vortices have been embedded in an idealized background vorticity environment equal to $3 \times f$ from 900 to 500 mb, then decreasing linearly to $1 \times f$ at 100 mb (contours = 2.0×10^{-5} s^{-1}).

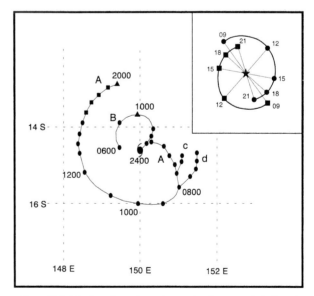

FIG. 12.8. Tracks of vortices observed prior to and during the interaction of the mesoscale convective systems. Positions are indicated at irregular UTC times corresponding to observations from radar (●), aircraft/satellite (▲), and interpolated (■): (inset) centroid-relative tracks of the major mesovortices (A squares, B circles).

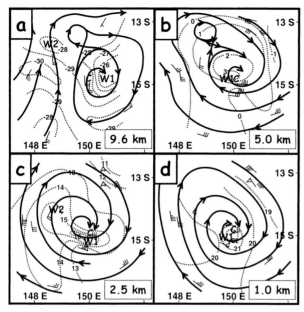

FIG. 12.9. Mesoscale temperature and wind analysis with vector winds (full barb, 5 m s⁻¹), streamlines (solid lines), and temperature (°C, dashed contours) for constant height surfaces of (a) 9.6, (b) 5.0, (c) 2.5, and (d) 1.0 km (adapted from Simpson et al. 1997).

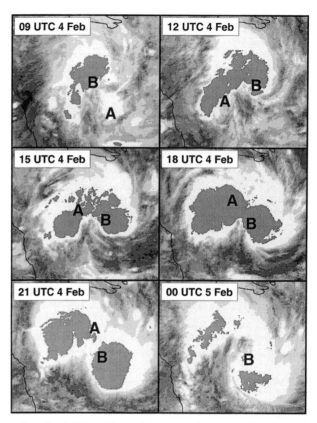

FIG. 12.10. The positions of the two major mesovortices A and B involved in the development of TC Oliver superposed on 3-hourly satellite imagery for the period 0900 UTC 4 Feb to 0000 UTC 5 Feb 1993 (from Ritchie 1995).

vertical structure of both vortices. Vortex B disappears from the data below about 4 km. Thus, both vortices were fairly deep, although the warm core associated with vortex A extended farther down in the atmosphere, as did the associated circulation (Fig. 12.9). It seems likely that the increase in midlevel vorticity in Fig. 12.7 was associated with the development and intensification of these midlevel vortices. In addition, a centroid-relative plot of the positions of vortices A and B reveals the classic Fujiwhara-type mutual rotation and interaction indicating that these vortices were interacting with each other, and not just being advected by the strong low-level monsoonal flow (e.g., Fig. 12.8).

A 3-h loop of infrared satellite imagery from 0900 UTC 4 February to 0000 UTC 5 February shows the development of the two large mesoscale convective systems during the period of increasing midlevel vorticity and subsequent development to the surface in the time series. An overlay of the vortex positions from radar onto the satellite imagery (Fig. 12.10) shows that not only did the mesoscale convective systems develop in conjunction with the vortices as the vortices moved into low-level southerly flow, but the vortices existed prior to the development of the mesoscale convective systems and probably triggered their development. As mentioned in section 12.2a it has been speculated that the retriggering of convection and stratiform cloud within a preexisting vortex acts to reinforce and intensify the vortex through the same processes that resulted in the original development of the vortex.

In a centroid-relative sense, vortices A and B rotated ~200° about each other over a period of 12 h. This rotation was initially quite fast but slowed as the mesoscale convective systems developed. In this case, vortex B was or became stronger than vortex A, and the vortex interaction was in the nature of an unequal merger similar to that depicted in Fig. 12.3. Thus, vortex A and its associated cloud structures sheared into a rainbandlike feature, while vortex B intensified and developed a deeper structure that reached into the boundary layer. Note that during the aircraft flight the strongest recorded dropwindsonde winds were located near 2.5 km (Fig. 12.9c). However, a dropwindsonde near the center of the vortex indicated dry air and weak subsidence rather than the saturated air that is predicted by Bister and Emanuel (1997) during the downward developing stage. This either indicates that the vortex developed downward through unsaturated processes, or that by this time, nascent eye processes were already in progress even though the vortex had barely developed to the surface. It is interesting that ME98 analyzed subsidence and significant warming in the center of their simulated primary low that was a direct result of the axisymmetrization process. Although they achieved a significant amount of subsidence only after several bursts of convection were simulated, it is an exciting conclusion that the dynamic interaction between potential vorticity anomalies may be at least partially responsible for the initiation of subsidence into the nascent eye of a developing tropical cyclone.

4. Conclusions

Despite the longevity of favorable environmental conditions for genesis, and numerous potential weather disturbances, tropical cyclogenesis is a relatively rare event. Presumably, other mechanisms, unresolved by large-scale dynamics, must be necessary to transform a potential weather disturbance into a tropical cyclone. However, until recently, routine observational networks could not capture the mesoscale processes that research studies increasingly support as fundamental to the development of a tropical cyclone.

The thesis here is that the interaction of midlevel vortices that develop in mesoscale convective systems, both with other midlevel vortices, and with the environment, is basic to the development of the tropical cyclone–scale vortex. Although these dynamic processes are modified by convection, the key argument is that downward development of the circulation associated with an initially midlevel vortex may be achieved through dynamic processes alone. Recent studies are reiterating the importance of the development of hot towers of convection in the development of the tropical cyclone eye. Some key questions that remain are as follows: how sensitive is the downward vortex development to the triggering of convection; how does the convection trigger in conjunction with the downwardly developing vortex; how does this feed back on the developing warm core aloft—is it through convective transport of high momentum, high θ_E air aloft, or a dynamic eddy momentum flux forcing due to the axisymmetrization of convectively generated potential vorticity anomalies into the primary vortex; does the wrapping of the decaying mesoscale convective system about the center of the vortex produce weak subsidence into the core, triggering a nascent eye? New remote sensing instrumentation, such as the Advanced Microwave Sounding Unit A (AMSU-A) and AMSU-B, are enabling scientists to investigate these questions.

The case study demonstrates that tropical cyclogenesis is a process of vortex interactions on different scales. Tropical Cyclone Oliver is an example of a simple interaction between two mesoscale convective system (MCSs) and the monsoon environment in which they formed. Model results demonstrate that a large proportion of the vortex dynamics can be captured in a 2D model (Ritchie 1995). The contour dynamics model indicated the overall aspects of the mesoscale vortex interaction and satellite imagery showed that the results were remarkably accurate. Baroclinic simulations demonstrate that midlevel vortex interaction is facilitated by a background rotational environment such as the monsoon trough in which Oliver developed. This environment facilitates the merger process and aids downward

development of the midlevel vorticity. However, vortex interaction is not the only factor in Oliver's development. Diabatic forcing, upper-level divergence, and latent heat release undoubtedly play important roles in the development and maintenance of the MCSs and their associated vortices. However, vortex interaction provides a simple, clear understanding of the basic dynamics of cyclone development as observed in this study.

Clearly, the types of special data sets that are occasionally available do not answer all the questions about tropical cyclone formation. However, one important aspect is clearly shown: that significant mesoscale processes are occurring within the weather disturbance on scales not resolvable by normal observing systems. Without continuous, instantaneous three-dimensional coverage of a potential weather disturbance, it is extremely difficult to ascertain what are the distinguishing mesoscale processes that will develop the system into a tropical cyclone, and whether they are occurring. In addition, there is clearly an interaction between the mesoscale processes occurring within the weather disturbance and the favorable large-scale environment in which the tropical cyclone forms, and it is this interaction that will dictate when and where formation will proceed.

Acknowledgments. This work was conducted during the author's graduate studies under the advisement of Dr. G. J. Holland and in conjunction with Dr. J. Simpson. The author would like to particularly acknowledge Dr. Simpson's continued enthusiastic support and encouragement of her career. The work has greatly benefited from discussions with Dr. P. May and Prof. R. L. Elsberry and from reviews by Profs. M. Montgomery and J. Molinari. The five-layer model was kindly provided by Dr. Y. Wang. Financial support for this research was provided by the Office of Naval Research under Grants N00014089-J1737 and N0001401WR20175.

REFERENCES

Bartels, D. L., and R. A. Maddox, 1991: Midlevel cyclonic vortices generated by mesoscale convective systems. *Mon. Wea. Rev.,* **119,** 104–118.

Bister, M., and K. A. Emanuel, 1997: The genesis of Hurricane Guillermo: TEXMEX analyses and a modeling study. *Mon. Wea. Rev.,* **125,** 2662–2682.

Briegel, L. M., and W. M. Frank, 1997: Large-scale influences on tropical cyclogenesis in the western North Pacific. *Mon. Wea. Rev.,* **125,** 1397–1413.

Chen, S. S., and W. M. Frank, 1993: A numerical study of the genesis of extratropical convective mesovortices. Part I: Evolution and dynamics. *J. Atmos. Sci.,* **50,** 2401–2426.

Fritsch, J. M., J. D. Murphy, and J. S. Kain, 1994: Warm core vortex amplification over land. *J. Atmos. Sci.,* **51,** 1780–1807.

Fujiwhara, S., 1931: Short note on the behaviour of two vortices. *Proc. Phys. Math. Soc. Japan Ser. 3,* **13,** 106–110.

Gray, W. M., 1968: Global view of the origin of tropical disturbances and storms. *Mon. Wea. Rev.,* **96,** 669–700.

——, 1975: Tropical cyclone genesis. Dept. of Atmospheric Sciences Pap. 234, Colorado State University, 121 pp.

Harr P. A., and R. L. Elsberry, 1996: Structure of a mesoscale convective system embedded in Typhoon Robyn during TCM-93. *Mon. Wea. Rev.,* **124,** 634–652.

——, M. S. Kalafsky, and R. L. Elsberry, 1996: Environmental conditions prior to formation of a midget tropical cyclone during TCM-93. *Mon. Wea. Rev.,* **124,** 1693–1710.

Holland, G. J., 1995: Scale interaction in the western Pacific monsoon. *Meteor. Atmos. Phys.,* **56,** 57–79.

Hoskins, B. J., M. E. McIntyre, and A. W. Robertson, 1985: On the use and significance of isentropic potential vorticity maps. *Quart. J. Roy. Meteor. Soc.,* **111,** 877–946.

McKinley, E. J., 1992: An analysis of mesoscale convective systems observed during the 1992 Tropical Cyclone Motion Field Experiment. M.S. thesis, Dept. of Meteorology, Naval Postgraduate School, 101 pp.

McWilliams, J. C., 1984: The emergence of isolated coherent vortices in turbulent flow. *J. Fluid. Mech.,* **146,** 21–43.

Melander, M. V., N. J. Zabusky, and J. C. McWilliams, 1988: Symmetric vortex merger in two dimensions: Causes and conditions. *J. Fluid. Mech.,* **195,** 303–340.

Menard, R. D., and J. M. Fritsch, 1989: A mesoscale convective complex-generated inertially stable warm core vortex. *Mon. Wea. Rev.,* **117,** 1237–1261.

Miller, D., and J. M. Fritsch, 1991: Mesoscale convective complexes in the western Pacific region. *Mon. Wea. Rev.,* **119,** 2978–2992.

Montgomery, M. T., and J. Enagonio, 1998: Tropical cyclogenesis via convectively forced vortex Rossby waves in a three-dimensional quasigeostrophic model. *J. Atmos. Sci.,* **55,** 3176–3207.

Ooyama, K., 1982: Conceptual evolution of the theory and modelling of the tropical cyclone. *J. Meteor. Soc. Japan,* **60,** 369–379.

Overman, E. A., II, and N. J. Zabusky, 1982: Evolution and merger of isolated vortex structures. *Phys. Fluids,* **25,** 1297–1305.

Raymond, D. J., and H. Jiang, 1990: A theory for long-lived mesoscale convective systems. *J. Atmos. Sci.,* **47,** 3067–3077.

Ritchie, E. A., 1995: Mesoscale aspects of tropical cyclone formation. Ph.D. Dissertation, Monash University, Melbourne, Australia, 167 pp.

——, and G. J. Holland, 1993: On the interaction of tropical cyclone-scale vortices. II: Discrete vortex patches. *Quart. J. Roy. Meteor. Soc.,* **119,** 1363–1379.

——, and ——, 1997: Scale interactions during the formation of Typhoon Irving. *Mon. Wea. Rev.,* **125,** 1377–1396.

——, and ——, 1999: Large-scale patterns associated with tropical cyclogenesis in the western Pacific. *Mon. Wea. Rev,* **127,** 2027–2043.

Schubert, W. H., and J. J. Hack, 1982: Inertial stability and tropical cyclone development. *J. Atmos. Sci.,* **39,** 1687–1697.

Shapiro, L. J., and M. T. Montgomery, 1993: A three-dimensional balance theory for rapidly rotating vortices. *J. Atmos. Sci.,* **50,** 3322–3335.

Simpson, J., E. A. Ritchie, G. J. Holland, J. Halverson, and S. Stewart, 1997: Mesoscale interactions in tropical cyclone genesis. *Mon. Wea. Rev.,* **125,** 2643–2661.

Waugh, D. W., 1992: The efficiency of symmetric vortex merger. *Phys. Fluids,* **4A,** 1745–1758.

Zabusky, N. J., M. H. Hughes, and K. V. Roberts, 1979: Contour dynamics for the Euler equations in two dimensions. *J. Comput. Phys.,* **30,** 96–106.

Zehr, R. M., 1992: Tropical cyclogenesis in the western North Pacific. NOAA Tech. Rep. NESDIS 61, 181 pp. [Available from U.S. Department of Commerce, NOAA/NESDIS, 5200 Auth Rd., Washington, D.C. 20233.]

Chapter 13

My View of the Early History of TRMM and Dr. Joanne Simpson's Key Role in Winning Mission Approval

JOHN S. THEON

Institute for Global Environmental Strategies, Arlington, Virginia

ABSTRACT

By 1984, more than a decade had passed since the National Aeronautics and Space Administration (NASA) weather and climate program had won approval for a new research mission. There was concern that it would be difficult to justify the budget of the program, so ideas were requested for a new research mission aimed at advancing our understanding of the weather and/or climate. More than a dozen proposals were submitted, including one by North, Wilheit, and Thiele for a mission to observe rainfall directly from space. They called it the Tropical Rainfall Measuring Mission (TRMM).

Studies were conducted to demonstrate that the proposal was feasible by deploying airborne versions of the proposed precipitation radar, microwave radiometer, and visible-infrared radiometer over carefully documented ground-based observations of rainfall. Sampling studies were undertaken to assure that one satellite could adequately sample precipitation events, and advanced mission studies were undertaken to define the mission as well as its cost.

When it became obvious that the cost of the mission would severely limit chances of winning approval, it was decided to invite an international partner to share the cost. With the support of Dr. Bert Edelson, the NASA associate administrator, and through the cooperation of Dr. Nobuyoshi Fugono of Japan, it was possible to study the mission as a joint enterprise. Although the one-year joint mission study concluded that the mission was feasible, obtaining the funding in both countries was anything but simple. When Dr. North decided to leave NASA, Dr. Simpson was suggested as his successor as project scientist. Dr. Simpson's energy and determination were key to winning approval of TRMM by the U.S. Congress. Dr. Simpson had, as President of the American Meteorological Society, briefed Congressman Green of New York on the enormous potential scientific benefits of TRMM. The fiscal year 1991 NASA budget was amended, mandating a new start for TRMM. Once NASA had approval for the mission, Japan agreed to share the costs, and the rest is history. TRMM was launched in 1997 and continues to acquire unprecedented rainfall data on a global scale.

As I reflect on the events that shaped my role in bringing the Tropical Rainfall Measuring Mission (TRMM) to reality, I thought about how I first became involved in meteorology. Although I had earned a degree in aeronautical engineering and had begun working in that profession, the U.S. Air Force decided to train me as a meteorologist. It was during that training at The Pennsylvania State University that my mentor, Professor Alfred Blackadar, interested me in studying the effects of precipitation and latent heating on the atmosphere (Theon 1962). Thus my interest in the importance of this subject was established, and years later it would influence my decision (perhaps subconsciously) to select and support TRMM.

The beginnings of TRMM can, in some ways, be traced to 1972 when National Aeronautics and Space Administration (NASA) scientists saw the first images acquired by the Electronically Scanning Microwave Radiometer (ESMR), the first imaging microwave radiometer of its type in space. I was fascinated by the instrument's ability to "see through" cloud cover, and

its response to what appeared to be the outlines of rainfall within clouds when the clouds occurred over a "cold" ocean background. These early precipitation observations from space were described by Theon (1973a, b).

Dr. Tom Wilheit, who was the ESMR principal investigator, explained the physics of the observations theoretically, and as more and more of these observations became available, Wilheit et al. (1976) quantified the observations experimentally for the first time by comparing ESMR observations of rainfall off the east coast of Florida with rainfall estimates obtained from the Melbourne, Florida, weather radar. This research effort marked the beginning of the direct, quantitative, global satellite observations of rainfall from space that are available today.

Rao et al. (1976) generated a global oceanic rainfall atlas by compiling the data from the ESMR instruments on *Nimbus 5* and *Nimbus 6*. Although it was known that this atlas was flawed, it was more complete than anything available at that time, and we believed that it was "better to light a candle than to curse the darkness."

In 1978, I moved to NASA Headquarters and in 1982 I became the leader responsible for overseeing all of NASA's weather and climate research programs. In addition to analyzing satellite-acquired meteorological data, these programs developed much of NASA's new meteorological remote sensing technology. It bothered me that very few, if any, of these advanced capabilities were being utilized or even planned for deployment in space. I had observed how my boss, Dr. Shelby Tilford, forged the upper atmosphere community's needs to define the Upper Atmosphere Research Satellite (UARS) mission, and later, how Dr. Stanley Wilson did the same for the oceanography community with the TOPEX/Poseidon mission. However, when I sought to initiate a research satellite mission for the meteorology community, I was told, "Your programs have the National Oceanic and Atmospheric Administration (NOAA) operational satellites to provide data for NASA's research and analysis programs, so you don't need a new mission." That response was not very satisfying because NOAA was using old remote sensing technology, and NASA had discontinued its Nimbus research satellite series. Thus, we had no space flight platforms on which to demonstrate new sensors being developed in the research programs, and no advanced observations to improve our understanding of atmospheric behavior.

The operational weather satellites were using technologies that were 10–15 years old, and the forecast models had surpassed observational capabilities with the result that initializing weather forecast models with operational satellite data made no positive impact on forecast skill. There had been no new weather and climate research instruments flown in space since *Nimbus 7* was launched almost a decade earlier. I became concerned that without some new research satellite mission, the continuation of our weather and climate research programs would be difficult to justify. I believed that we were in a going-out-of-business mode, and that something had to be done to rejuvenate NASA's weather and climate research programs.

Therefore, during the summer of 1984, I sent a letter to each of the NASA field centers that were involved in weather and climate research, calling for the submission of new ideas for a space flight research mission. I knew that this mission would be difficult to sell to NASA management because Dr. Tilford was focused on gathering scientific community support for what would eventually become the Earth Observing System (EOS). I gave the centers guidelines that the new proposals should cost less than $100 million. By October, I had received more than a dozen proposals on a broad range of weather and climate topics. I convened a peer review panel to examine the proposals and selected the TRMM proposal by Drs. Gerald North, Tom Wilheit, and Otto Thiele of Goddard Space Flight Center for further study. My selection was based on the fact that measuring rain from space had never been done well previously, because of the importance of precipitation to weather and climate, and, perhaps subconsciously, because of my long-held interest in the measurement of rainfall and in the effects of latent heating on the atmosphere. I believed that TRMM observations of rainfall on a global scale had the potential to enable us to make a quantum leap forward in understanding and predicting atmospheric behavior. I took a big risk because I acted without the formal approvals normally required for such a significant undertaking. Since Dr. Tilford gave his subordinates wide latitude in managing their programs, I felt that it was easier to ask forgiveness for possibly exceeding my authority than to obtain permission in advance for initiating the mission study.

I was able to reprogram $500 000 for the first year of studies of TRMM. As originally proposed, TRMM included a mechanically scanned, two-frequency radar, a refurbished *ESMR-5* microwave radiometer, a modified Advanced Very High-Resolution Radiometer (AVHRR), and an off-the-shelf spacecraft bus. The $100 million guideline was the first to go, and none of these proposed components survived the myriad budget reviews, scientific reviews, and technical reviews. Fortunately for TRMM, Mr. Tom Keating was assigned to conduct the preliminary feasibility studies (pre-phase-A studies) by the advanced mission studies group at Goddard. Mr. Keating is distinguished not only by his ongoing important contributions to TRMM, but for being the only person other than the principal investigators (North, Wilheit, and Thiele) to serve on the team from its inception through launch.

As the engineering studies progressed, I initiated a parallel effort to prove to skeptical NASA Headquarters management that we could actually measure rainfall remotely from space (here I use the term "we" to refer to the TRMM team). Following the precedent established by other programs, we decided that making the measurement from an airborne platform would demonstrate its feasibility. That meant developing an airborne radar and combining it with the airborne ESMR that had been flown on the NASA *CV-990/DC-8* aircraft to validate the spaceborne *Nimbus-5* ESMR measurements. Such flights would have to be made over well-instrumented surface-based sites to validate the airborne measurements.

We attempted to promote TRMM both internally in NASA and externally to the scientific community, and Dr. North undertook the work of study scientist for TRMM. We were very fortunate to have Dr. North, who was (and is) highly respected by his peers, do a superb job of educating and convincing the research community about the virtues of TRMM as he traveled far and wide giving TRMM briefings. Subsequently, Dr. North demonstrated that TRMM sampling would be adequate to determine monthly average rainfall amounts for climate purposes, based on his studies of the Global Atmospheric Research Program Atlantic Tropical Experiment (GATE) radar data. Dr. Wilheit, who was (and is) very well versed in the physics of the remote sensing of rain, concerned himself with developing the required hardware and the algorithms that would convert the raw observations into useful rainfall parameters. Mr. Thiele, who served as the study manager in the earliest days of TRMM, used his vast knowledge of radars, rain gauges, and ground truth experience to develop a plan and implement a program for understanding the measurement of rain and a system for validating the TRMM data.

In January 1985, I traveled to India as part of a delegation from several U.S. government agencies to meet with our counterparts there in support of a presidential initiative on scientific cooperation in weather and climate research. Before I departed, Dr. David Atlas mentioned that he had read a paper by Dr. Nobuyoshi Fugono of the then Radio Research Laboratory (RRL) in Japan reporting RRL's success in remotely observing rainfall using a combined radar–radiometer instrument. Dr. Fugono's interest in rainfall was the result of the fact that rain interfered with Japan's satellite communications. What was noise to RRL was signal to us. After discussing it with me, Dr. Atlas contacted Dr. Fugono to ask whether I could meet with him on my way home from India. While I was in India, I briefed the Indian scientists on TRMM and asked if they had any interest in cooperating on the mission. They politely declined.

On my way home from India, I stopped in Tokyo to meet Dr. Fugono for the first time. He was most gracious and hospitable. Little did I realize then that we would become kindred spirits with the common goal of making TRMM a success. Dr. Fugono agreed to cooperate with us by lending us RRL's radar–radiometer for flight aboard NASA aircraft. Further, he agreed to send Dr. Kenji Nakamura to NASA Goddard as the first in what was to become a continuous succession of top notch visiting Japanese scientists and engineers who worked with the Goddard TRMM team. In return, we agreed to send NASA scientists to RRL (later renamed the Communications Research Laboratory, CRL) as part of an ongoing exchange program.

By the late summer of 1985, it became painfully evident that the mission cost had grown to exceed anything we had the faintest hope of obtaining in the NASA budget. Dr. Tilford was fully engaged in working on EOS as a new start mission, so when I raised the subject of TRMM, he was less than enthusiastic. Dr. Tilford has been much maligned for his failure to support TRMM in the early days. However, nothing could be further from the truth. I wish to set the record straight: almost from its inception, he was aware of what I was doing about TRMM, and although he could not openly endorse it, he allowed me to continue funding it and even quietly encouraged my efforts. He did not openly endorse TRMM because, as he had told me on several occasions, to start any major new program would require taking funds out of the ongoing program. Major new programs require new start funding. One word from Dr. Tilford would have killed the TRMM study program instantly,

including the engineering studies, the field campaigns, the cooperation with RRL, and the growing number of scientists who had submitted successful proposals to study TRMM-related problems. He even allowed me to establish a separate funding line in the budget that legitimized the TRMM research program. This permitted me to expand the TRMM research budget each year such that by fiscal year 1994, the annual TRMM budget had grown to more than $6 million. Thus TRMM never would have happened without Dr. Tilford's approval and support.

When Dr. Fugono visited Washington, DC, in September 1985, we were still struggling with the projected cost of the TRMM mission. At Dr. Wilheit's suggestion, I asked Dr. Fugono if he thought Japan might become a partner in developing TRMM by sharing the costs. Dr. Fugono was taken by surprise, and he immediately declined politely. When we met the next day, however, he had changed his mind. He said that he would pursue the idea when he returned to Japan.

E-mail was just becoming available to both of us at that time, and it turned out to be an indispensable asset. Without e-mail, communicating with colleagues in Japan would have been tedious, time consuming, and costly. E-mail enabled us to communicate on a daily basis with relative ease despite the 14-hour time difference between Washington and Tokyo, greatly facilitating our cooperation.

Having witnessed firsthand how time consuming it is to obtain the required internal consensus on any issue before a Japanese government agency will commit to action, I have come to appreciate what a mover and a shaker Dr. Fugono is. Without him, TRMM never would have happened. He even had a good effect in Washington. When he visited NASA Headquarters in September 1986, he briefed then NASA Associate Administrator Dr. Bert Edelson on TRMM. Dr. Fugono asked Dr. Edelson to write a letter to Professor Saito, the chairman of the Space Activities Commission of Japan, a panel that advises the Prime Minister of Japan on space issues. To his credit, Dr. Edelson did so without hesitation. The letter had a powerful effect, and TRMM was included on the agenda of the Senior Standing Liaison Group (SSLG), which coordinated all United States–Japan cooperative space activities. TRMM then had recognition status at the highest levels in the space programs of the two countries. At the next SSLG meeting, Dr. Edelson proposed a joint one-year feasibility study of TRMM, which was accepted by the Japanese. A year later, the study report concluded that a joint TRMM mission was feasible and recommended that studies of a joint flight mission studies be continued.

In August 1986, Dr. North decided to accept a position with Texas A&M University. At first, his departure seemed to me to be a death blow to TRMM. How could we replace such a valuable, highly respected, articulate advocate for TRMM? This development worried me considerably. Dr. Marvin Geller, then chief of Goddard's Laboratory for Atmospheres, had become an ardent supporter of TRMM. He was an enormous asset, not only for his considerable scientific skills, but also for his political skills and the manpower he was able to bring to bear on TRMM. Dr. Geller suggested that Dr. Simpson would be an excellent choice to succeed Dr. North as TRMM study scientist.

I had known Dr. Joanne Simpson for many years not only because of her considerable scientific contributions, but because early in my NASA career, when she served as the chair of the American Meteorological Society (AMS) Scientific and Technical Activities Commission (STAC), she had appointed me to serve on the AMS Committee on the Upper Atmosphere. I first met her in person when she joined the Goddard Laboratory for Atmospheres in 1979, after I had accepted a permanent appointment at NASA Headquarters. We became better acquainted after the responsibility for the Severe Storms Program (under which Dr. Simpson's research was funded) was assigned to my branch in 1982. I had great respect for Dr. Simpson, and although I thought that she would be a good choice for the job, I knew that she was deeply involved in her own research and that she might not wish to accept the additional responsibilities that TRMM entailed. Although I hesitated to ask, I thought the offer might be more palatable if I asked her to serve for only one year. Little did I realize that I had just made one of the best and luckiest decisions of my entire professional career. Fortunately for me and a lot of the people who work on TRMM, she not only accepted, but played a major role in making TRMM a success.

Once Dr. Simpson became the TRMM study scientist, positive progress became palpable, and the inertia that had met our earlier efforts began to give way to the determined and aggressive campaign that she waged to win approval of the mission. She convened a TRMM science team consisting of a group of distinguished scientists chaired by Professor Eugene Rasmussen. They worked together for several years to produce a science plan that justified the mission and documented its scientific rationale in superb fashion. Dr. Simpson used her extensive network of prominent scientists to have TRMM plans reviewed by the National Academy of Sciences Board on Atmospheric Sciences and Climate, which gave it a glowing endorsement. Her sterling reputation and integrity were keys to convincing many reputable scientists, including Professor Verner Suomi, that TRMM should be a NASA priority. Professor Suomi made his enthusiastic support for the mission known to the highest levels of NASA. Dr. Simpson asked the community to write to their congressional representatives about the importance of TRMM, and the community responded in kind. I know because many of these letters were referred to NASA Headquarters for a reply, along with the question, "What are you doing about this?" I was the individual who was assigned the task of preparing responses. The responses were usually written for the associate administrator's signature (by this time, Dr. Lennart Fisk), so he became aware of the growing support for the mission outside of NASA. Although this clamor for TRMM was somewhat embarrassing to NASA in a positive way, all I could get from my superiors was, "Wait until next year."

By 1989, the TRMM community was growing impatient. Then TRMM indirectly received a huge boost when Dr. Simpson was elected president of the AMS. In addition to her superb scientific reputation, she then had the prestige of that office to add to her credentials. One of her duties as president of the AMS, along with the presidents of other national scientific societies, was to meet with the members of the Congressional Subcommittee on Science and Technology. This was significant because one of this subcommittee's responsibilities is oversight of the NASA budget. Knowing Dr. Simpson as I do, I am certain that she did her homework to prepare herself for that breakfast meeting with the subcommittee. She told me later that she had learned that a member of the subcommittee, Representative William Green, formerly a wealthy businessman from upstate New York, was interested in weather and had coauthored a book on the subject. The best of circumstances seated Dr. Simpson next to Congressman Green at the breakfast, so she proceeded to tell him all about TRMM in her charming and intelligent manner. She told him what a great advance it would be, as well as the low priority it was given in the NASA budget. Being a man of action, Congressman Green attached a $50 million rider to the NASA budget appropriation for fiscal year 1991, mandating that the funds be used to initiate the TRMM flight program. I had the feeling that some of my superiors (not Dr. Tilford) would have liked to strangle me. That event, so skillfully orchestrated by Dr. Simpson, was the final U.S. action that made TRMM a reality.

Having won the struggle at home, it was essential that our Japanese colleagues become official partners in the mission. Dr. Fugono arranged for me to brief a number of key officials in the Science and Technology Agency (STA) of Japan [which oversees the National Space Development Agency (NASDA), Japan's equivalent of NASA], NASDA, and CRL. In addition, Dr. Fugono had enlisted the support of some very influential allies in the Japanese academic community, such as Professor Taroh Matsuno of Tokyo University and Professor Toshibumi Sakata of Tokai University. Dr. Fugono, Professor Sakata, and I had organized two symposia in Tokyo to promote TRMM to both the Japanese scientific community and the Japanese government. Thanks to the superb arrangements by Dr. Fugono and Professor Sakata, there was impressive participation by Japanese researchers, and I saw to it that U.S. scientists were well represented. These two symposia, the proceedings of which were subsequently published as hardcover books (Theon and Fugono 1988; Theon et al. 1991) contributed significantly to winning support for TRMM in Japan.

There is an anecdote that I wish to record that demonstrates the political savvy of Dr. Geller. After NASA had begun to comply with the congressional mandate to move forward with the TRMM flight project, and Dr. Fugono had successfully won approval for the mission in Japan, Dr. Geller and I journeyed to Tokyo to meet with Dr. Fugono and Mr. Kawasaki, the Director of Advanced Research for STA. Mr. Kawasaki was to sign the memorandum of understanding (MOU) officially confirming the joint partnership and specifying the roles of NASA and NASDA in developing, launching, and operating TRMM. The MOU had already been signed for NASA by Dr. Fisk, the NASA Associate Administrator for Space Science and Applications. As soon as Mr. Kawasaki signed the document, Dr. Geller pulled a beautiful, bronze-faced, walnut plaque out of his briefcase to present to Mr. Kawasaki. A photocopy of the MOU had been etched on the bronze face of the plaque. To the surprise of everyone except Dr. Geller, the plaque MOU already bore the signatures of both Dr. Fisk and Mr. Kawasaki. While he had a good laugh over this, Mr. Kawasaki must have wondered about what he was getting into with these guys from NASA.

After TRMM received a new start for fiscal year 1991, the Project Office was established and progress accelerated. Dr. Tilford assigned the TRMM project to Goddard where it was decided that the spacecraft would be built in house both to keep Goddard's engineers proficient and to keep costs manageable. Mr. Thomas La Vigna was appointed to manage the TRMM Project in the fall of 1990, an appointment that turned out to be another stroke of good luck, for he proved to be an outstanding manager. He overcame the many challenges and difficulties inherent in an international cooperative program and delivered the TRMM spacecraft on time and within budget.

The Tropical Rainfall Measuring Mission is the result of the contributions by many people, each making unique, essential, and sustained efforts. A few of the key leaders and their teams deserve special mention: thanks to Mr. La Vigna and his project team and their NASDA and CRL counterparts, the superb science leadership of Dr. Simpson and the science team, the excellent management of the scientific research by Dr. Ramesh Kakar, the budgetary priority given to TRMM by Drs. Jack Kaye and Ghassem Asrar, and the hard work of many dedicated and competent scientists who process and analyze the data, TRMM data are making many outstanding and unprecedented scientific results possible.

A final thought in closing: Dr. Joanne Simpson is one of the finest people with whom I have ever had the pleasure of working during my entire professional career, now exceeding 45 years. She has proven time and again that her word is her bond. When she agrees to do something, it can be considered done, and it will be done well and on time. She is clearly a world class scientist, having made enormous and continuous contributions to our knowledge and understanding of the atmosphere, and to our profession, for more than a half century. I am truly honored to count her as a cherished friend and colleague. I wish her many more happy, healthy, and productive years at NASA Goddard Space Flight Center (or at whatever endeavor she undertakes).

REFERENCES

Rao, M. S. V., W. Abbott, and J. Theon, 1976: *Satellite-Derived Global Oceanic Rainfall Atlas.* NASA SP-410, 186 pp.

Theon, J. S., 1962: On the inclusion of latent heat liberation in a graphical model of the atmosphere. M. S. thesis, Dept. of Meteorology, Pennsylvania State University, 37 pp.

——, 1973a: Microwave Radiometer on Nimbus 5. *Atmos. Technol.,* **2.**

——, 1973b: A multispectral view of the Gulf of Mexico from Nimbus 5. *Bull. Amer. Meteor. Soc.,* **54,** 934–936.

——, and N. Fugono, Eds.,1988: *Tropical Rainfall Measurements.* A. Deepak, 528 pp.

——, T. Matsuno, T. Sakata, and N. Fugono, Eds.,1991: *The Global Role of Tropical Rainfall,* A. Deepak, 486 pp.

Wilheit, T., J. Theon, W. Shenk, L. Allison, and E. Rodgers, 1976: Meteorological interpretations of images from Nimbus 5 electrically scanned microwave radiometer. *J. Appl. Meteor.,* **15,** 166–172.

Chapter 14

Dr. Joanne Simpson and the Beginning of the TRMM Project

TOM KEATING

Silver Spring, Maryland

ABSTRACT

The beginnings of the Tropical Rainfall Measuring Mission (TRMM) project are outlined, in particular the role of Joanne Simpson as TRMM study scientist and then TRMM project scientist. Her important work in developing the TRMM science rationale and interacting with our Japanese partners, along with her skill in finding solutions to project problems, is described.

In September of 1985, Dr. John Theon of National Aeronautics and Space Administration (NASA) Headquarters initiated a phase-A study of the feasibility of measuring precipitation from space. The mission was titled the Tropical Rainfall Measuring Mission (TRMM). The Advanced Mission Analysis Office (AMAO) of Goddard Space Flight Center (GSFC) was assigned the phase-A study and I became the study manager. The study progressed in a low-key manner until August of 1986 when Dr. Joanne Simpson of GSFC became the project scientist, and the United States–Japan Standing Senior Liaison Group (SSLG) formed a United States–Japan "expert panel" to study the feasibility of a joint endeavor. My personal recollection is that Dr. Simpson brought a certain "scientific power" that was a combination of not only scientific knowledge, but also scientific standing in the precipitation measuring scientific community. This latter attribute was invaluable in mustering the scientific support necessary to move any scientific mission from the drawing board to a satellite in orbit.

The initial TRMM design had the following performance characteristics and mission parameters:

- mission life, 3 years;
- orbital altitude, 300 km;
- inclination, 28.5°;
- launch vehicle, Space Transportation System (STS);
- launch date, 1994;
- platform, space station or free flyer.

The instrument complement was

- dual-frequency precipitation radar,
- Special Sensor Microwave Imager (SSM/I),
- Advanced Very High-Resolution Radiometer (AVHRR), and
- Electronic Scanning Microwave Radiometer (ESMR).

The first meeting of the expert panel was held in Japan in January 1987. The U.S. scientific leader was Dr. Simpson. It was obvious to me that Dr. Simpson was held in high esteem by her Japanese counterparts. This observation meant a lot, especially when it is considered in the context of the Japanese culture.

In March of 1988 the TRMM phase-A study was in trouble. The mass of the spacecraft was much too large for a phase-A design and the cost had far exceeded NASA Headquarters' willingness to seriously consider funding such a mission. It was at this time that the Dr. Simpson–led scientific community discovered the means of retaining the principal scientific objectives and allowing considerable "descoping" of the designed spacecraft.

The major changes that resulted in a design that was highly feasible and of acceptable cost were as follows:

- precipitation radar, change from dual frequency to single frequency;
- data rate, change from 370 to 171 kbps;
- altitude, change from 300 to 350 km;
- inclination, change from 28.5° to 35°;
- Tracking and Data Relay Satellite System (TDRSS) link, change from Ku and S-band to S-band only;
- propulsion system, change from a Cosmic Background Explorer (COBE) to an Magnetospheric Multiscale Mission (MMS);
- launch vehicle, established the National Space Development Agency of Japan (NASDA) *H-II* as a given.

In March and May the United States–Japan joint feasibility study was signed off by the United States and Japan, respectively. In July 1988 the phase-A study was completed and accepted by NASA Headquarters.

In the time period of July 1988–91 the TRMM project was formed and the scientific community working with

181

NASA Headquarters defined the TRMM instrument complement as follows:

- precipitation radar (single frequency);
- TRMM Microwave Imager (TMI; this was a modified SSM/I with an additional 10.6-GHz channel);
- Visible and Infrared Scanner (VIRS; this was similar to the AVHRR);
- Clouds and the Earth's Radiant Energy System (CERES; this was added to maintain earth radiation measurements);
- Lightning Imaging Sensor (LIS; this was added as a complement to the precipitation measurements).

Throughout this period, Dr. Simpson reviewed and evaluated the impact of changes to the scientific mission.

She marshaled the scientific community in both support of the changes and in speaking in one voice to provide coherent and certain guidance to the TRMM project as we worked to develop a design that was ready for phase-C/D approval.

On 27 November 1997 (EST) the TRMM spacecraft was successfully placed in orbit by the Japanese *H-II* launch vehicle and is now close to successfully completing its three-year orbit life requirement. Many personnel made vital contributions to the success of the mission and the success of the scientific products; certainly, high on the pantheon of vital personnel is Dr. Joanne Simpson. From 1986 to the present she has "stayed the course." Working with her was an inspiration because she was always energetic, enthusiastic, and most of all appreciative of the work of others.

Chapter 15

Working with Dr. Joanne Simpson while Managing the TRMM Project

THOMAS A. LAVIGNA*

Bowie, Maryland

ABSTRACT

The development and challenges of the Tropical Rainfall Measuring Mission (TRMM) project are defined, and the role of Joanne Simpson as the project scientist is described as the project progressed from early development to a highly successful operational satellite system for rainfall measurement. Close interaction between the project scientist and the project staff of TRMM was one key to the success of the mission.

1. Introduction

I am pleased to honor Dr. Joanne Simpson for her 50 years of pioneering work in the fields of earth science and meteorology. She has made major contributions to the advancement of these areas. Joanne worked very hard to get the Tropical Rainfall Measuring Mission (TRMM) started and her persistent and diligent efforts paid off in 1991 when Congress approved funding for TRMM.

I worked with Joanne from April 1991, when TRMM officially started hardware development, through April 1998, some six months after launch. In those seven years, the project encountered many challenges, many of them due to the international partnership with Japan. Joanne worked very closely with the project and, as such, contributed significantly to its success. This paper will address the specifics of her actions in helping to resolve key project issues. A description of the TRMM characteristics and challenges are included to give an understanding of the project complexity and to highlight Joanne's contributions.

2. TRMM characteristics

The centerpiece of TRMM is a large satellite including five scientific instruments as illustrated by the artist's concept in Fig. 15.1. A brief description of the TRMM characteristics follows.

The baseline mission operations is three years but, based on propellant usage for the expected solar activity, the observatory life is estimated to be more than six years.

As a joint mission with National Space Development Agency of Japan (NASDA), they provided the Precipi-

tation Radar (PR), launch vehicle and services, and science participation. We at the National Aeronautics and Space Administration (NASA) provided project management; the observatory; TRMM Microwave Imager (TMI), Visible and Infrared Scanner (VIRS), Clouds and the Earth's Radiant Energy System (CERES), Lightning Imaging Sensor (LIS) instruments; science data processing; mission operations; and science participation. The instrument complement can be broken down into the rain package consisting of TMI, VIRS, and PR, and the additional instruments CERES and LIS.

The observatory was launched on 27 November 1997, Thanksgiving Day, from Tanegashima Island, Japan, by an H-2 launch vehicle. The launch was delayed from August due to late changes in NASDA's launch manifest.

TRMM, measuring $17' \times 16' \times 45'$ when deployed in orbit, was the largest observatory designed, developed, assembled, and tested "in house" at Goddard Space Flight Center where the project performed as "prime contractor" and "system integrator." As an in-house project, TRMM required "hands-on" management with daily decisions to resolve problems and keep it "on-track." Over 400 people worked on TRMM at the peak.

Another major element designed and developed at Goddard was the TRMM Science Data Processing System. The mission operations planning and implementation was also done in house.

Figure 15.2 is a photo of the TRMM observatory in its final "all-up" launch configuration just before shipment to the launch site. It gives the viewer a good feel for its size with a person shown for comparison. The observatory is $17'$ tall and $13'$ in diameter in this launch configuration with all appendages undeployed.

---

* NASA GSFC TRMM Project Manager (ret.)

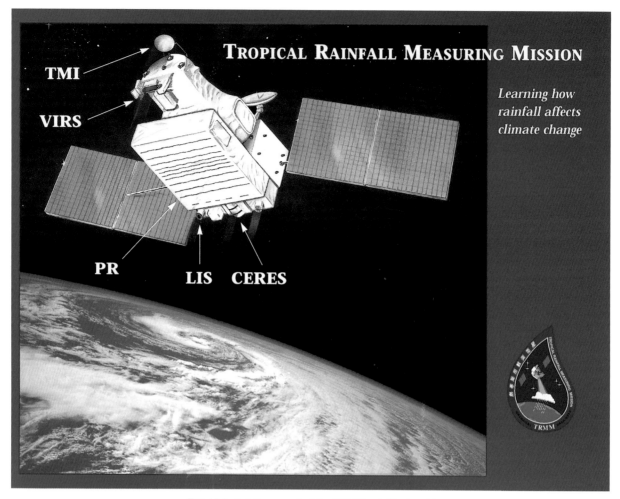

FIG. 15.1. Artist's concept of the TRMM satellite in orbit.

3. Significant challenges

A brief description of the major challenges of TRMM follows.

From the start, the project had a very tight budget with very little contingency. The schedule was also aggressive for a program of this large size and complexity, and it was driven by the limited launch opportunities of the H-2 launch vehicles, which were only available twice per year with six months between them.

Working with Japan required an intense effort not only in the hardware and software areas, but also in establishing and implementing a cooperative science program. The United States and Japan had separate TRMM science teams, and how they used the data and released the results needed to be defined and agreed upon. Joanne led the activity to establish the working relationships between the teams and the process for reporting results. Her work set in place the framework for sharing the data and release of the findings. Joanne did a great job of organizing the U.S. science team and providing direction for development of the science plan. She provided leadership to a diverse group of over 35 scientists from the United States including NASA, the National Oceanic and Atmospheric Administration (NOAA), international universities and other organizations. Through her efforts, top scientists in the precipitation field were chosen to participate in TRMM.

Designing the observatory to withstand the high levels of atomic oxygen at the 350-km orbit and development of the Precipitation Radar were significant technical challenges that were overcome.

Another major challenge was to develop a Science Data Processing System that could handle the large data volumes and support both research and regular generation of data products. This was a major undertaking that NASA Headquarters believed could not be done for the limited budget allocated by the project. Joanne led a key activity, which refined the science requirements and established acceptable boundaries. She also worked very intensely in defending the budget and gaining support

FIG. 15.2. Photo of TRMM observatory in launch configuration.

from the NASA Headquarters scientists. The data system led by Erich Stocker was completed within budget and it was ready for the launch and has done an outstanding job of processing the science data and producing various rainfall products. Without Joanne's work we would have had major difficulty building the Science Data Processing System and, especially, keeping it within budget.

Joanne made sure that the science was not compromised and that the mission provided the expected data from the observatory. I appointed her a member of the TRMM Project Configuration Control Board and she reviewed proposed changes bringing the science point of view to the board's deliberations.

A good case in point was the need for a backup system

for the earth sensor. It is critical to mission success since it keeps the observatory pointing to earth. The initial solution by the Attitude Control System (ACS) engineers was to add a star tracker at a cost of $2 million. That would take a big chunk out of our budget contingency, so I asked the ACS engineers to look for other alternatives. After some study, an approach was proposed that used the Fine Sun Sensor with revised attitude control software. However, the initial look indicated that the pointing accuracy would have to be relaxed. Joanne and her team evaluated the pointing requirements based on the science, and after some detailed analyses, provided a reduction in requirements for the backup mode. Using this information, the ACS engineers developed the software at a cost of less than $400 000, five times less than the star tracker approach. System tests in orbit with the backup mode showed that it meets the initial requirements and is ready for use if needed.

The close working relationship between Joanne and the project made this software solution possible saving substantial dollars. In fact, Joanne was a key member of my project management team and she attended my weekly meetings and reported on the progress and status of the science development. Her reports were very informative and the project people looked forward to hearing from her.

4. Schedule aspects

The project schedule was aggressively managed to ensure meeting the August launch date. Special project tracking and control software was utilized to keep on track and uncover problems as early as possible. This worked very well and, with effective management, the TRMM program was kept on plan for the August launch. However the launch was delayed from August to November because of launch manifest issues with the Japanese H-2 launch vehicle. NASDA wanted to delay the TRMM launch for six months from August 1997 to February 1998. This delay would result in a substantial cost to the project and also compromise the science. Joanne led an evaluation of the impact of the delay to the TRMM science. She worked closely with key U.S. scientists and Japan's project scientist and determined that a launch delay to February would cause loss of important science data from "El Niño" which was ex-

pected in January 1998. In addition, there was potential for loss of mission lifetime below the 3-yr requirement due to the increased solar activity for a later launch. Joanne's study activity and leadership established a strong position by the science community that a delay to February 1998 was not acceptable. This result, together with the strong project objection based on increased cost and loss of mission life, led NASDA to negotiate with the Fisherman's Union for an earlier launch, which ultimately resulted in the November 1997 launch.

5. Summary

The project has overcome major challenges on the road from development to launch and operations. TRMM has been an extremely successful program, completing the development and implementation within cost and schedule and is meeting mission objectives. The observatory continues to operate very well and has exceeded the 3-yr mission life requirement. The science data have given the TRMM scientists a better understanding of rainfall and its influence on climate.

Dr. Joanne Simpson has contributed significantly to TRMM's outstanding success. From inception of the TRMM program to the present, Joanne has provided very effective leadership to the science team. Under her direction, a well-focused science plan was developed and implemented and has resulted in exciting new findings. She was instrumental in getting NASA Headquarters to accept a dedicated Science Data Processing System for TRMM. In working behind the scenes, she also helped maintain the budget for the data processing system when attempts were made by headquarters to cut the budget. Joanne has not only been an effective science leader but also has set a fine example for how the project scientist should work with the project. It has been a distinct pleasure and rewarding experience working with her.

Acknowledgments. The author wishes to acknowledge the TRMM Project Team, civil servants, and support contractors for their dedication and hard work, which has contributed significantly to TRMM's outstanding success. I also wish to thank my wife, Ann, for her constant support and encouragement, which got me through the tough times.

Chapter 16

A Short History of the TRMM Precipitation Radar

KEN'ICHI OKAMOTO

Osaka Prefecture University, Osaka, Japan

ABSTRACT

The Tropical Rainfall Measuring Mission (TRMM) satellite carried aboard the world's first spaceborne precipitation radar (PR). This paper describes a short history of the TRMM PR. It describes the Communications Research Laboratory's (CRL's) airborne dual-frequency rain radar/radiometer system, some results of the airborne experiments, and considerations of system design and system parameters of the PR. It also describes data processing and analysis algorithms for the PR, and examples of PR rain measurements.

1. Introduction

The potential of spaceborne weather radar has been recognized since the era of the early visible and infrared (IR) meteorological satellites. Although the visible and IR sensors give some information on the precipitation beneath the clouds, the lack of direct measurements of precipitation by these sensors strongly suggested the need for a satellite-borne precipitation radar. Measurements from ground-based meteorological radars have clearly demonstrated that the instrument represents the best means of mapping the three-dimensional structure of precipitation. It is quite natural then to ask whether it is possible to combine the radar's ability to observe precipitation with the advantages offered by a satellite platform. Despite a recognition of these advantages, there were also strong concerns regarding cost, reliability, and effectiveness of spaceborne weather radar. It was only with the launch of the Tropical Rainfall Measuring Mission (TRMM) that a spaceborne precipitation radar became a reality and concerns about the reliability and effectiveness of such an instrument were put to rest.

Several types of satellite-borne precipitation radars were proposed in the early design studies. Keigler and Krawitz (1960) recognized that some of the major technological problems associated with a satellite-borne weather radar could be solved by the use of a pencil-beam antenna. Katzenstein and Sullivan (1960) proposed a fan-beam Doppler radar. Eckerman (1975) proposed a push-broom type of multiple-beam precipitation radar for the space shuttle. The push-broom antenna has the merit in that the large increase in the number of samples per resolution volume that this design offers can improve the attainable measurement accuracy. However, in order to implement the design, a complex antenna feed with a large antenna aperture is required.

Okamoto et al. (1979) proposed a dual-frequency spaceborne precipitation radar with a scanning pencil-beam antenna to observe the same resolution volume simultaneously by two frequencies (10 and 34.45 GHz) that were allocated at that time for satellite-borne precipitation radar. The study concluded that rain rates could be retrieved with fairly good accuracy from the rain attenuation at the higher frequency. It was also shown that a combination of the echo time delay and low sidelobe levels (about 30 dB below the maximum gain) could be used to separate echoes from precipitation layers in the main lobe of the antenna from ground clutter entering along the sidelobes.

Based on this system design study, the Communications Research Laboratory (CRL) in 1980 developed an airborne dual-frequency rain radar/radiometer system as a first step toward the design of a spaceborne rain radar (Okamoto et al. 1982). After a series of airborne experiments conducted in Japan a joint experiment to study "the feasibility of measuring rain from space" was initiated in 1985 between CRL and National Aeronautics and Space Administration/Goddard Space Flight Center (NASA/GSFC). This cooperative experiment was critical in bringing about what later was to be called the joint U.S.–Japan TRMM.

After completion of a feasibility study in 1987, development of the TRMM precipitation radar (PR) began. The TRMM, carrying aboard the world's first spaceborne weather radar, was launched on board the *H-II Launch Vehicle No. 6* from the Tanegashima Space Center in Japan on 28 November 1997.

It is worth mentioning that, in parallel with the TRMM PR development, several advanced airborne rain radars were also developed and used to study various atmospheric phenomena and to simulate a spaceborne rain radar. They are the Electra Doppler radar (EL-

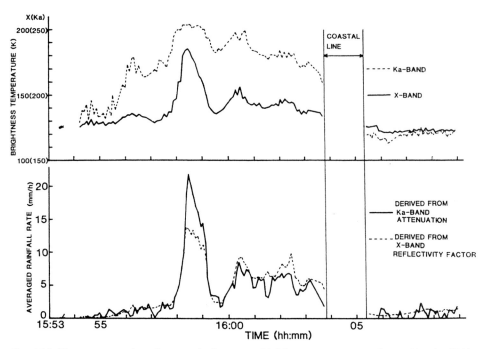

FIG. 16.1. Time sequence plots of averaged rain rates and brightness temperatures observed by the CRL's
airborne rain radar/radiometer system.

DORA)/ASTRAIA developed by the National Center for Atmospheric Research (NCAR) and the Centre for the study of Earth and Planets Environments (CETP) in France (Hildebrand et al. 1994; Roux et al. 1993), the ER-2 Doppler radar (EDOP) developed by NASA/ GSFC (Heymsfield et al. 1993), the airborne rain-mapping radar (ARMAR) developed by NASA/Jet Propulsion Laboratory (JPL) (Li et al. 1993; Durden et al. 1994) and the CRL Airborne Multiparameter Precipitation Radar (CAMPR) developed by CRL (Kumagai et al. 1996).

The paper is organized as follows. A description of the results obtained from the CRL's airborne radar/radiometer is followed by discussions of the design considerations and system parameters of the PR. The paper concludes with description of the PR data processing algorithms and examples of rain measurements by the PR.

2. Airborne dual-frequency rain radar/radiometer system and some results of airborne experiments

An airborne rain radar/radiometer instrument that measures rain from a down-looking platform can be used to simulate measurements from a satellite platform. The measurements from such an instrument are important in understanding the critical issues in measuring rain from space and in developing algorithms to analyze the data. CRL developed an airborne rain radar/radiometer system in 1980 consisting of a dual-frequency radar and radiometers that operate at X-band and Ka-

band. A major objective of the design was to provide dual-wavelength radar and radiometer measurements over the same region of space nearly simultaneously. With this instrument, over 100 hours of airborne observations were obtained in Japan from 1980 to 1982. The experiments demonstrated that simultaneous observations of rain by radar and microwave radiometer were quite useful and that this design concept could be adopted in the mission definition of TRMM.

Figure 16.1 shows an example of rain observation results by the airborne rain radar/radiometer system in Japan (Okamoto et al. 1982). The lower two curves in the figure show time sequences of averaged rain rates derived from the 10-GHz radar reflectivity factor (dotted curve) and from the surface attenuation data at 34.5 GHz (solid curve). The averaged rain rates derived from 10- and 34.5-GHz data agree well in the weak rain area. However, large differences of the derived rain rates are observed in the heavy rain area. This may be partly due to the insufficient attenuation correction at 10 GHz. The correlation coefficient between averaged rain rates derived from 10 and 34.5 GHz is 0.907. The upper two curves show a time sequence of brightness temperatures from the 9.9- (solid curve) and 34.2-GHz (dotted curve) channels. For the data shown in the figure, the correlation coefficient between brightness temperatures at 9.9 GHz and averaged rain rates derived from 34.5-GHz attenuation data is 0.993. On the other hand, the correlation coefficient between brightness temperature at 34.2 GHz and the averaged rain rate derived from 34.5-GHz attenuation data is 0.667. These results suggest that the 9.9-GHz radiometer is generally better than the

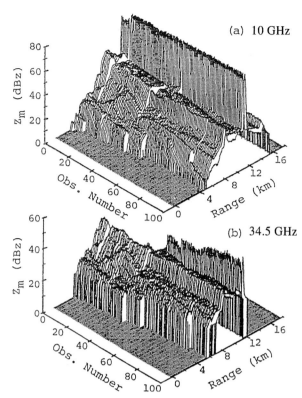

(a) 10 GHz

(b) 34.5 GHz

FIG. 16.2. Example of 3D plots of (a) 10- and (b) 34.5-GHz Z_m profiles on 1 Nov 1998.

TABLE 16.1. Major mission requirements for TRMM.

Requirement	Characteristics
Frequency	13.8 GHz
Sensitivity	S/N per pulse > 0 dB for 0.5 mm h^{-1} at rain top
Dynamic range	Both surface return and 0.5 mm h^{-1} rain should be measured
Horizontal resolution	<5 km at nadir (3 dB one way)
Vertical resolution	<250 m at nadir
Number of independent samples	>64 per resolution volume
Swath width	>200 km, contiguous coverage
Observable range	From surface to a height >15 km 5-km mirror image for nadir

port the algorithm development and science studies for TRMM. The 10-GHz radar was modified to enable measurements at two orthogonal linear polarizations (HH and HV). The linear depolarization ratio (LDR = HV/ HH) was obtained by switching the polarization of the receiving antenna pulse by pulse.

3. System design of TRMM precipitation radar

At the fourth Standing Senior Liaison Group (SSLG) meeting that was held in Washington, D.C., in 1986, the delegations from the United States and Japan approved a cooperative study of the TRMM project. A one-year feasibility study of the TRMM project began in January 1987 under the auspices of the SSLG. The final feasibility study report was completed at the end of March 1988. In this feasibility study, CRL took charge of the conceptual design of the precipitation radar (Okamoto et al. 1988; Nakamura et al. 1990). The mission requirements for the TRMM PR, basically specified by NASA and used as guidelines in the design, are summarized in Table 16.1. CRL studied both single-frequency (13.8 GHz) and dual-frequency (13.8 and 24.15 GHz) radar systems. As the size, weight, and power requirements of the dual-frequency radar proved to be too large, a single-frequency radar (13.8 GHz) was finally adopted. The frequency of the TRMM PR, 13.8 GHz, was allocated by the International Telecommunication Union (ITU) for spaceborne radar observations of earth. The frequency was selected by considering the requirement for a sufficient dynamic range of measurable rain rates, and the requirement for an antenna beamwidth narrow enough to achieve the desired spatial resolution.

The PR has provided various benefits to the overall TRMM program. With a range resolution of 250 m and a horizontal resolution of 4–5 km, the PR is the only instrument of TRMM that captures the three-dimensional structure of rain. This is important in that the vertical distribution of rain can help determine the vertical profile of latent heating. The vertical structure information is also useful in identifying rain type and in improving radiometer-based rain-rate retrieval algo-

34.2-GHz radiometer for estimating path-averaged rain rate. On the other hand, the 34.2-GHz radiometer is more sensitive to light rain rates.

A series of joint airborne experiments between CRL and NASA/GSFC were conducted from 1985 to 1994. In the experiments, CRL's airborne rain radar/radiometer system was installed on NASA aircraft. Figure 16.2 shows an example of a 3D plot of the radar reflectivity factors without rain attenuation correction, Z_m, at 10 and 34.5 GHz measured on 1 October 1988, over the Atlantic Ocean from the T-39 aircraft (Kozu 1991; Kozu et al. 1991). The observation number represents the distance along the flight course where a set of 100 observations corresponds to about 8 km. The range axis represents the distance measured from the aircraft along the nadir direction. Direct rain echoes are observed in the region from 4 to 11 km. The sharp spike near a range of 11 km below the aircraft represents the surface return. In most cases, the 10-GHz surface return is almost constant, indicating little rain attenuation at 10 GHz. In the corresponding 34.5-GHz plot, much stronger rain attenuation is evident from the large variation in the surface return.

A typhoon experiment was conducted in the western Pacific in September 1990 using CRL's airborne instrument aboard the NASA/DC-8 aircraft (Kumagai et al. 1993). One of the objectives of this experiment was to make radar/radiometer measurements over rain to sup-

rithms. Unlike the radiometers, the PR estimates of rain rate are independent of the background microwave emission of the land or ocean.

In the conceptual design of the PR, three major items were considered in trade-off studies. They were 1) conventional nonpulse compression versus pulse compression, 2) an active array radar with many solid state power amplifiers (SSPAs) versus a passive array radar with a traveling wave tube amplifier (TWTA), and 3) planar array antenna versus a cylindrical parabolic antenna. After a detailed study of the options, a conventional pulse (nonpulse compression) with a planar, active array antenna was selected for the TRMM PR.

Pulse compression has the advantage that the required number of independent samples at each field of view can be easily achieved. The most serious problem with this strategy is that the accompanying range sidelobes can cause surface clutter interference with the rain echo. In order to ensure that rain can be detected close to the surface, it is necessary to reduce the range sidelobe levels to at least 60 dB down from the peak power. Because this was difficult to accomplish at the early phase of the TRMM development, a conventional short pulse radar, which is free from the range sidelobe problem, was selected for the TRMM PR. However, JPL's airborne rain radar ARMAR achieved such low-range sidelobe performance by applying digital pulse compression technique and was used regularly during major international experiments like TOGA COARE and Convection and Moisture Experiment (CAMEX-3) (Im et al. 1992; Tanner et al. 1994).

A passive array radar is simple in structure and relatively light in weight. However, it is not easy to develop a high power pulsed TWTA and its power supply for space use. Moreover, a malfunction of the TWTA can result in a total failure of the radar. Although an active array radar is intricate in structure and relatively large and heavy, it offers the significant advantage in that the loss of one or two transmitters or receivers will cause only a modest degradation in radar performance. To provide a more robust, fail-safe system, the active array radar was selected. The planar array antenna was selected because its size is smaller than that of a cylindrical parabolic antenna with the same beamwidth.

Beginning in 1988, CRL developed the key devices of the Bread Board Model of the TRMM PR. These key devices include the slotted waveguide antenna elements, the five-bit PIN diode phase shifters, the solid state power amplifiers (SSPA) and the low-noise amplifiers (LNA). These components were integrated into the eight-element Bread Board Model (Okamoto et al. 1992).

The Engineering Model and Proto-Flight Model of TRMM PR were developed by National Space Development Agency of Japan (NASDA) in cooperation with CRL (Oikawa et al. 1997; Kummerow et al. 1998; Kozu et al. 2001). The PR is a 128-element active phased array system, operating at 13.8 GHz. The transmitter/

TABLE 16.2. Major parameters of TRMM PR.

Item	Specification
Radar type	Active phased array
Frequency	13.796, 13.802 GHz
	(two-channel frequency agility)
Polarization	HH (Horizontal)
Antenna	
Type	128-element WG planar array
Beamwidth	$0.71° \times 0.71°$
Aperture	2.1×2.1 m
Scan angle	$\pm 17°$ (cross-track scan)
Gain	47.7 dB
Sidelobe level	< -30 dB
Transmitter	
Type	SSPA (128 units)
Peak power	>578 W
Pulse width	$1.67 \ \mu s \times 2$ ch. (total 3.34 μs)
PRF	2778 Hz
Receiver	
Type	LNA (128 units)
Noise figure	<5.5 dB
Bandwidth	0.6 MHz $\times$ 2 ch
Dynamic range	70 dB
Others	
Total feed loss	<2 dB
Number of independent samples	64
Data rate	93.5 kbps
Mass	465 kg
Power consumption	250 W
Swath width	215 km
Observable range	Surface to 15-km altitude
Horizontal resolution	4.3 km (nadir)
Vertical resolution	250 m (nadir)

receiver (T/R) consists of 128 SSPAs, LNAs, and five-bit PIN diode phase shifters. Each T/R element is connected to a 2-m slotted waveguide element, from which the 2×2 m planar array is constructed. To achieve low antenna sidelobe levels (i.e., to suppress sidelobe-coupled surface clutter), the SSPA output powers and LNA gains are weighted to achieve a Taylor distribution (SL $= -35$ dB, $\bar{n} = 6$) with a 1-dB quantization. Major parameters of the PR are listed in Table 16.2.

The mechanical structure of the PR, consisting of the platform and the antenna, is approximately $2.3 \times 2.3 \times 0.7$ m, weighing 465 kg. The PR uses a frequency-agility technique to obtain 64 independent samples with a single PRF of 2778 Hz, in which a pair of 1.6-μs pulses having two different frequencies 6 MHz apart are transmitted. During the normal observation mode, the PR antenna scans in the cross-track direction over $\pm 17°$ (215-km swath). The cross-track scan, which takes 0.6 s, consists of 49 angle bins with an angle-bin interval of $0.71°$. At each angle bin 32 pairs of 1.6-μs pulses are transmitted and received. The observation concept of the PR is illustrated in Fig. 16.3 along with those of the TRMM Microwave Imager (TMI) and the TRMM Visible and Infrared Scanner (VIRS).

The radar returns from the PR consists of the rain,

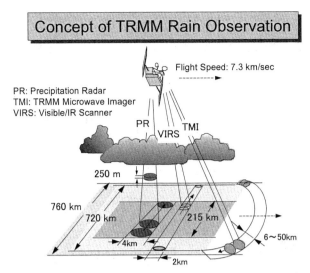

FIG. 16.3. Concept of rain observation with PR, TMI, and VIRS.

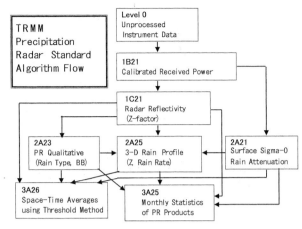

FIG. 16.4. TRMM PR standard algorithm flow.

surface, and mirror image components. Because of the importance of the surface echo for estimating the total path attenuation and for providing the range to the surface along the radar beam, the surface echo is also measured in addition to the rain echo. The mirror image, which is the rain echo received through a double reflection of the signal from the surface, is measured at nadir incidence and provides an alternative means of estimating path attenuation and rain rate. The radar echo sampling is performed over range gates between the sea surface and an altitude of 15 km for each observation angle. For nadir incidence, the mirror image is also collected up to an altitude of 5 km. To improve the retrieval of those portions of the profile with high spatial frequency such as the surface return and bright band, oversampled data (125-m interval) are collected at the surface for scan angle within $\pm 9.94°$ and for near-nadir rain echoes ($\pm 3.55°$) up to a height of 7.5 km.

The Active Radar Calibrator (ARC) was developed for calibration of TRMM PR (Kumagai et al. 1995). The ARC functions in one of three modes, either as a radar receiver, beacon-signal transmitter, or delayed transponder. Using the ARC, external calibration of the PR yielded measurements that were within 1 dB of the calibration results calculated from the PR system parameters. In the initial checkout phase, on-orbit functions and performance specifications of the TRMM PR were evaluated. It was confirmed that the PR is functioning normally and was within specifications (Kawanishi et al. 1998; Kozu et al. 2001). CRL developed a new airborne rain radar, CAMPR, for validation of TRMM PR through simultaneous observation of rain from CAMPR and TRMM PR (Kumagai et al. 1996). CAMPR operates at the same frequency as the TRMM PR and also has polarimetric and Doppler capabilities.

4. Data processing and analysis algorithms for TRMM precipitation radar

Because of limitations on the size and mass of the antenna, the PR must use a frequency much higher than a typical ground-based radar to attain the desired spatial resolution. The signal at higher frequencies, however, is attenuated by the rain and partially melted hydrometeors. Correction of this attenuation is one of the major problems that must be solved in the rain-rate retrieval algorithms. The primary goal of the TRMM PR is to estimate rain rate at the resolution of the instantaneous field of view (IFOV). However, it is also used to judge the existence of rain, to classify rain type, and to calculate the surface scattering coefficients. The PR data are also used to estimate monthly averaged rain rate over an area either by the straightforward use of the high-resolution data or by an application of one or more of the statistical methods that have been proposed for rain estimation over large time- and space scales (Chiu 1988).

Since 1991, the international TRMM PR team has been organized in accordance with the TRMM research announcements issued by NASA and NASDA. The standard data processing and analysis algorithms of TRMM PR have been developed by an international TRMM PR team and were in place to begin processing and analysis of the TRMM data immediately after the initial instrument checkout (Okamoto et al. 1998; Kummerow and Okamoto 1999). The standard algorithms are classified into levels 1, 2, and 3. Levels-1 and -2 products are quantities measured at the highest resolution of the instrument, determined by the antenna beamwidth of $0.71°$ and range resolution of 0.25 km. Level-3 data provide monthly statistics of rain parameters in $5° \times 5°$ grid boxes, and, for selected products, over $0.5° \times 0.5°$ grid boxes. Figure 16.4 shows the TRMM PR standard algorithm flow. Table 16.3 is organized according to the levels of data processing and lists each of the TRMM PR standard algorithms along with the contact persons, product names, and a short

TABLE 16.3. TRMM PR standard algorithms.

Number/name	Contact person	Products	Algorithm description
1B-21: PR calibration rain/no rain	NASDA/EOC H. Kumagai J. Awaka	Total received power, noise level, rain/no rain flag, storm height, clutter contamination flag.	Converison of the count value of radar echoes and noise level into engineering value; decision of rain/no rain; determination of effective storm height from minimum detectable power value.
1C-21: PR reflectivities	NASDA/EOC	Profiled reflectivity factors Zm in case of rain.	Conversion of the power and noise value to reflectivity factors (Z factors) in case of rain without rain attenuation correction.
2A-21 Surface scattering coefficients σ^0	R. Meneghini	Rain attenuation value of σ^0 (in case of rain), and its reliability; database of σ^0 (ocean/land, in case of no rain).	Estimation of path attenuation and its reliability using the surface as a reference target; spatial and temporal statistics of surface σ^0 and classification of σ^0 into land/ocean, rain/no rain.
2A-23: PR qualitative	J. Awaka	Bright band (presence or no), bright-band height, rain type classification, warm rain.	Whether a bright band exists in rain echoes and determination of bright-band height when it exists. The rain type is classified into stratiform, convective, and others. Isolated rain, the height of which is below the 0° height, is classified as warm rain.
2A-25: PR profile	T. Iguchi	Range profiles of rain rate, average rain rate between predefined heights (2, 4 km)	The rain-rate estimate is given at each resolution cell; this algorithm employs hybrid method of the surface reference method and Hitschfeld–Bordan method.
3A-25: Space–time average of radar products	R. Meneghini	Space–time averages of accumulations of 1C-21, 2A-21, 2A-23, 2A-25. Monthly averaged rain rate over 5° × 5° boxes.	To calculate various statistics over a month from the 1C-21 and level-2 PR output products. Four types of statistics are calculated: (a) probabilities of occurrence (count value), (b) means and standard deviations, (c) histogram, and (d) correlation coefficients.
3A-26: Space–time average using a statistical method	R. Meneghini	Monthly averaged rain rate over 5° × 5° boxes using a statistical method.	Rainfall accumulations and rain rate averages over 5° × 5° × 1-month boxes using a statistical method (multiple-threshold method).

description. The international PR team maintains, validates, revises, and improves these standard PR algorithms.

The primary function of algorithm 1B-21 is to convert the count values of radar echoes (signal + noise) and noise levels into their engineering values. By comparing the received power with a threshold level, a decision is made as to whether rain is present in the IFOV. Algorithm 1B-21 also estimates the effective storm height from the range at which a signal is first detected. Algorithm 1C-21 gives the radar reflectivity factor Z without correction for rain attenuation.

Algorithm 2A-21 estimates the path attenuation by rain and its reliability using the surface as a reference target (Meneghini et al. 2000). The algorithm also computes the spatial and temporal statistics of the surface scattering coefficient σ^0 over ocean and land in the absence of rain. Algorithm 2A-23 tests for the existence of a bright band and, when present, determines its height

(Awaka et al. 1998). This information is used in algorithm 2A-25 and in the combined PR–TMI algorithm. The rain type is classified into stratiform, convective, and "other" categories. If the maximum height at which rain is detected is below the freezing level, it is classified as warm rain. Algorithm 2A-25 estimates the vertical profiles of the attenuation-corrected radar reflectivity factor and rain rate for each radar beam (Iguchi et al. 1998; 2000). A combination of the Hitschfeld–Bordan and the surface reference method is used to correct the attenuation effect in the radar returns. The algorithm also computes the average rain rate between two predefined height levels.

Algorithm 3A-25 computes the space–time averages from the instantaneous, high-resolution data products of algorithms 1C-21, 2A-21, 2A-23, and 2A-25. The most important output product for climate studies is the monthly average rain rate over 5° × 5° grid boxes at altitudes of 2 and 4 km. Algorithm 3A-26 estimates

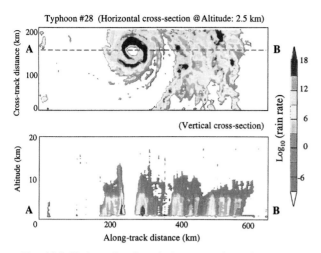

FIG. 16.5. Horizontal and vertical cross sections of rain rate in Typhoon No. 28 named (Paka) measured by the PR (processed by Dr. T. Iguchi, CRL).

monthly rain-rate averages over the 5° × 5° grid boxes using the multiple-threshold method.

Algorithm 2B-31 is the radar (PR) and radiometer (TMI)-combined algorithm. Although it is not part of the TRMM standard PR algorithm, it uses the PR data together with the TMI data. It is designed to improve single-instrument retrieved algorithms by merging data from two sensors into a single retrieval that embodies the strengths of each sensor (Kummerow and Okamoto

1999; Kummerow et al. 2000). The algorithm itself was described in Haddad et al. (1997). It estimates high-resolution radar rain, while constraining the solution to be consistent with the total attenuation derived from passively measured 10-GHz brightness temperature.

5. Examples of TRMM PR data

On 8 December 1997, after the satellite had reached its operational orbit at an altitude of 350 km, the initial system tests were begun. The first storm detected during the initial checkout was "Pam", the 17th tropical cyclone of 1997 that raged in the South Pacific. Although the TRMM PR has observed many more tropical storms since then, the observations of Pam represented the first test of the measurement capabilities of the PR.

Figure 16.5 shows the horizontal and vertical cross sections of rain rate in Typhoon Number 28 named (Paka) measured by the PR on 19 December 1997 around 0400 UTC. The eye of the typhoon was located at approximately (36°N, 136°E). The upper panel shows the TRMM PR horizontal cross section of rain at an altitude of 2.5 km. The lower panel shows the vertical cross section of rain along the line A-B in the upper panel. As this example demonstrates, the PR can observe the three-dimensional structure of rain, while conventional meteorological satellites are sensitive only to hydrometeors near the cloud top.

Figure 16.6 shows the global rainfall distribution for

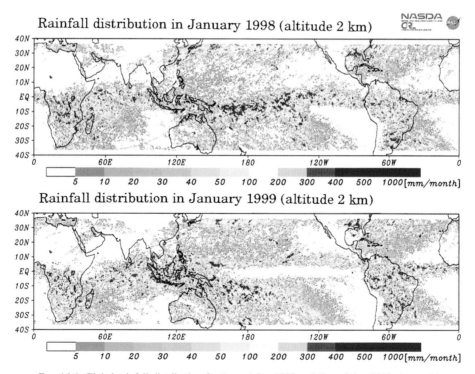

FIG. 16.6. Global rainfall distribution for (upper) Jan 1998 and (lower) Jan 1999 observed by TRMM PR (processed by NASDA/EORC).

January 1998 (upper panel) and January 1999 (lower panel), observed by the PR. Differences in the rainfall distribution associated with the presence and absence of El Niño are clearly seen in these figures. In January 1998 (upper panel), while El Niño was active, heavy rainfall areas in the Pacific shifted from the western to the central Pacific, in contrast to the situation in January 1999, when the El Niño event had ended. These data are examples of the type of observations critical to studies of global energy and water cycle.

6. Conclusions

The TRMM PR was successfully launched in November 1997 and continues to operate quite well. On-orbit functions and system performance were evaluated in the initial checkout period, confirming that the PR was functioning normally and within specifications. Since then, the radar has remained stable and well calibrated. External calibration measurements of the PR continue to be made using the ARC. The results have demonstrated that the PR external calibration agrees with that calculated from the PR system parameters to within ±1 dB. Observations made by the PR indicate the usefulness of spaceborne radar in meteorological, climate, and atmospheric science studies.

The experience gained in the design of the TRMM PR and in the analysis of the data will be valuable in the development of the next generation dual-frequency radar. The dual-frequency radar is one of the primary instruments on the proposed Global Precipitation Mission (GPM).

Acknowledgments. The author would like to express sincere thanks to NASDA/EORC for providing the processed data of TRMM PR. The author also thanks Dr. Robert Meneghini for his careful check of the paper's English.

REFERENCES

Awaka, J., T. Iguchi, and K. Okamoto, 1998: Early results on rain type classification by the Tropical Rainfall Measuring Mission (TRMM) precipitation radar. *Proc. Eighth URSI Commission F Triennial Open Symp.,* Aveiro, Portugal, International Union of Radio Science Commission F, 143–146.

Chiu, L. S., 1988: Rain estimation from satellites: Area rainfall–rain area relation. Preprints, *Third Conf. on Satellite Meteorology and Oceanography,* Anaheim, CA, Amer. Meteor. Soc., 363–368.

Durden, S. L., E. Im, F. K. Li, W. Ricketts, A. Tanner, and W. Wilson, 1994: ARMAR: An airborne rain-mapping radar. *J. Atmos. Oceanic Technol.,* **11,** 727–737.

Eckerman, J., 1975: Meteorological radar facility for the space shuttle. IEEE National Telecommunications Conference, New Orleans, *IEEE pub. 75 CH1015-7 CSCB,* 37-6-37-17, 12 pp.

Haddad, Z. S., E. A. Smith, C. D. Kummerow, T. Iguchi, M. R. Farrar, S. L. Durden, M. Alves, and W. S. Olson, 1997: The TRMM "day-1" radar/radiometer combined rain-profiling algorithm. *J. Meteor. Soc. Japan,* **75,** 799–809.

Heymsfield, G. M., W. Boncyk, S. Bidwell, D. Vandemark, S. Ameen,

S. Nicolson, and L. Miller, 1993: Status of the NASA/EDOP airborne radar system. Preprints, *26th Int. Conf. on Radar Meteorology,* Norman, OK, Amer. Meteor. Soc., 374–375.

Hildebrand, P., C. A. Walther, C. L. Frush, J. Testud, and F. Baudin, 1994: The ELDORA/ASTRAIA airborne Doppler weather radar: Goals, design and first field tests. *Proc. IEEE,* **82,** 1873–1890.

Iguchi, T., T. Kozu, R. Meneghini, J. Awaka, and K. Okamoto, 1998: Preliminary results of rain profiling with the TRMM precipitation radar. *Proc. Eighth URSI Commission F Triennial Open Symp.,* Aveiro, Portugal, International Union of Radio Science Commission F, 147–150.

——, ——, ——, ——, and ——, 2000: Rain-profiling algorithm for the TRMM precipitation radar. *J. Appl. Meteor.,* **39,** 2038–2052.

Im, E., S. L. Durden, F. K. Li, A. Tanner, and W. Wilson, 1992: Pulse compression technique for spaceborne precipitation radars. Preprints, *11th Conf. on Clouds and Precipitation,* Montreal, Quebec, Canada, Amer. Meteor. Soc., 1079–1082.

Katzenstein, H., and H. Sullivan, 1960: A new principle for satellite-borne meteorological radar. Preprints, *Eighth Weather Radar Conference,* San Francisco, CA Amer. Meteor. Soc., 505–515.

Kawanishi, T., H. Kuroiwa, Y. Ishido, T. Umehara, T. Kozu, and K. Okamoto, 1998: On-orbit test and calibration results of TRMM precipitation radar. *Proc. SPIE Conf. on Microwave Remote Sensing of the Atmosphere and Environment,* Beijing, China, International Society for Optical Engineering, 94–101.

Keigler, J. E., and L. Krawitz, 1960: Weather radar observation from an earth satellite. *J. Geophys. Res.,* **65,** 2973–2808.

Kozu, T., 1991: Estimation of raindrop size distribution from space-borne radar measurement. Ph.D. dissertation, Kyoto University, 196 pp.

——, K. Nakamura, R. Meneghini, and W. Boncyk, 1991: Dual-parameter radar rainfall measurement from space: A test result from an aircraft experiment. *IEEE Trans. Geosci. Remote Sens.,* **29,** 690–703.

——, and Coauthors, 2001: Development of precipitation radar on-board the tropical rainfall measuring mission (TRMM) satellite. *IEEE Trans. Geosci. Remote Sens.,* **39,** 102–116.

Kumagai, H., R. Meneghini, and T. Kozu, 1993: Preliminary results from multiparameter airborne rain radar measurement in the western pacific. *J. Appl. Meteor.,* **32,** 431–440.

——, T. Kozu, M. Satake, H. Hanado, and K. Okamoto, 1995: Development of an active radar calibrator for the TRMM precipitation radar. *IEEE Trans. Geosci. Remote Sens.,* **33,** 1316–1318.

——, and Coauthors, 1996: CRL Airborne Multiparameter Precipitation Radar (CAMPR): System description and preliminary results. *IEICE Trans. Commun.,* **E-79B,** 770–778.

Kummerow, C., and K. Okamoto, 1999: Space-borne remote sensing of precipitation from TRMM. *Review of Radio Science, 1996–1999,* W. Ross Stones, Ed., Oxford Science, 487–502.

——, W. Barnes, T. Kozu, J. Shiue, and J. Simpson, 1998: The Tropical Rainfall Measuring Mission (TRMM) sensor package. *J. Atmos. Oceanic Technol.,* **15,** 809–816.

——, and Coauthors, 2000: The status of the Tropical Rainfall Measuring Mission (TRMM) after two years in orbit. *J. Appl. Meteor.,* **39,** 1965–1982.

Li, F. K., S. L. Durden, E. Im, A. Tanner, W. Ricketts, W. Wilson, R. Meneghini, T. Iguchi, and K. Nakamura, 1993: Airborne rain mapping radar and preliminary observations during TOGA/COARE. *Proc. IEEE Int. Geoscience and Remote Sensing Symp.,* Tokyo, Japan, IEEE, 832–834.

Meneghini, R., T. Iguchi, T. Kozu, L. Liao, K. Okamoto, J. A. Jones, and J. Kwiatkowski, 2000: Use of the surface reference technique for path attenuation estimates from the TRMM precipitation radar. *J. Appl. Meteor.,* **39,** 2053–2070.

Nakamura, K., K. Okamoto, T. Ihara, J. Awaka, T. Kozu, and T. Manabe, 1990: Conceptual design of rain radar for the Tropical Rainfall Measuring Mission. *Int. J. Satell. Comm.,* **8,** 257–268.

Oikawa, K., T. Kawanishi, H. Kuroiwa, M. Kojima, and T. Kozu, 1997: Development results of TRMM precipitation radar. *Proc.*

IEEE Int. Geoscience and Remote Sensing Symp., Singapore, IEEE, 1630–1632.

Okamoto, K., S. Miyazaki, and T. Ishida, 1979: Remote sensing of precipitation by a satellite-borne microwave remote sensor. *Acta Astronaut.,* **6,** 1043–1060.

——, S. Yoshikado, H. Masuko, T. Ojima, N. Fugono, K. Nakamura, J. Awaka, and H. Inomata, 1982: Airborne microwave rain-scatterometer/radiometer. *Int. J. Remote Sens.,* **3,** 277–294.

——, J. Awaka, K. Nakamura, T. Ihara, T. Manabe, and T. Kozu, 1988: A feasibility study of rain radar for the Tropical Rainfall Measuring Mission. *J. Comm. Res. Lab.,* **35,** 109–208.

——, T. Ihara, and H. Kumagai, 1992: Development of Bread Board Model of TRMM precipitation radar. *Proc. 18th Int. Symp. on Space Technology Science,* Kagoshima, Japan, ISAS and NASDA; 1919–1924.

——, T. Iguchi, T. Kozu, H. Kumagai, J. Awaka, and R. Meneghini, 1998: Early results from the precipitation radar on the Tropical Rainfall Measuring Mission. *Proc. URSI Commission F Open Symp. on Climate Parameters in Radiowave Propagation Prediction,* Ottawa, ON, Canada, International Union of Radio Science Commission F, 45–52.

Roux, F., and Coauthors, 1993: Preliminary results from airborne dual-beam Doppler radars during TOGA-COARE. Preprints, *26th Int. Conf. on Radar Meteorology,* Norman, OK, Amer. Meteor. Soc., 713–715.

Tanner, A., S. L. Durden, R. Dennings, E. Im, F. K. Li, W. Ricketts, and W. Wilson, 1994: Pulse compression with very low sidelobes in an airborne rain mapping radar. *IEEE Trans. Geosci. Remote Sens.,* **32,** 211–213.

Chapter 17

The TRMM Measurement Concept

THOMAS WILHEIT

Department of Atmospheric Sciences, Texas A&M University, College Station, Texas

ABSTRACT

The relationship between the microwave radiometer and the precipitation radar on the Tropical Rainfall Measuring Mission TRMM satellite is inherently complementary. Neither sensor by itself would be adequate to achieve the TRMM objectives but the match between the strengths and weaknesses of each sensor results in an extremely powerful payload. Here these strengths and weaknesses are discussed and a specific example is examined.

Prima facie, flying a precipitation radar in space is an obviously good idea. However, it took until the launch of TRMM, four decades after *Sputnik,* for it to become reality. Why?

Here we will only consider radar measurements from low orbiting spacecraft. The power and antenna size problems inherent in geosynchronous measurement remove such a capability from the realm of the reasonable for now. From a low orbit, only a small portion of the earth can be viewed at any given time. The bulk of the rain that occurs will not be observed. Thus no rain measurement capability on a single low-orbiting satellite is useful for such applications as flash flood warnings. Although recent developments in assimilation techniques have enabled weather forecast models to make use of limited rainfall data, the primary justification used for spaceborne rainfall measurement has been climate applications. Here we can accumulate observations over an area for an extended period of time and, through sampling, arrive at an estimate of the rainfall total for the space–time box. Using Global Atmosphere Research Program (GARP) Atlantic Tropical Experiment (GATE)[1] data, it was shown that the uncertainty in such an estimate with a perfect rain measurement capability with a swath of the order of 1000 kilometers would be of the order of 10% for a cell as large as $5° \times 5° \times 1$ month (North 1988). Chang et al. (1993) showed from Special Sensor Microwave Imager (SSM/I) data that this was too optimistic. This sampling error is the direct result of trying to estimate a rain total from a limited subset of observations of the rain. Thus, achieving useful rainfall measurements from space is a significant challenge.

Now consider a precipitation radar. For viewing directly at nadir, the geometry is reasonably simple. The horizontal resolution is determined by the diffraction of the antenna aperture. A larger aperture will provide a smaller pixel on the surface. Horizontal resolutions of the order of 5 km are achievable with reasonable combinations of wavelength, antenna size, and orbital altitude. The vertical resolution is provided by the range gating of the radar and values of the order of 100 m are straightforward to achieve.

If we confine ourselves to viewing at the nadir, the sampling problem discussed above is greatly exacerbated. To observe any reasonable fraction of the rain, the radar must scan away from nadir. Figure 17.1 illustrates the problem. When viewing away from nadir, the antenna beamwidth has a vertical component. If the horizontal resolution for nadir viewing is 5 km, then scanning enough to provide a swath width equal to half of the orbital altitude will degrade the vertical resolution to worse than 1 km at the edges of the swath. Since the backscatter from the surface is much larger than that from rain, this will obscure the bottom kilometer or so of the rain profile with ground clutter. Realistic antenna problems, such as sidelobes, will exacerbate this. The situation can be improved by making a larger antenna to get better horizontal resolution, but the cost of the instrument and required spacecraft resources quickly gets out of hand. The Tropical Rainfall Measuring Mission (TRMM) radar has a swath width of 200 km, of limited usefulness at the edges. A significantly larger radar would not be economically realistic and the 200 km swath is simply too limited for acceptable accuracy in monthly rainfall products. As long as the radar was considered to be *the* rain measurement capability, it could never happen. The radar that could be afforded was useless and vice versa.

---

[1] Joanne Simpson's contributions to TRMM began long before TRMM had a name.

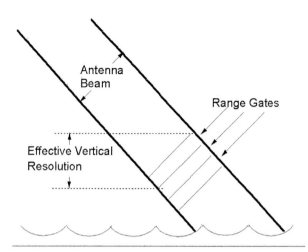

FIG. 17.1. Geometry of the view of a spaceborne precipitation radar away from nadir.

The use of passive microwave measurements to infer rainfall has been going on for some time, beginning with the 1972 launch of the electronic scanning microwave radiometer (ESMR) on the *Nimbus-5* satellite (Wilheit et al. 1977). Radiometers, since they lack a transmitter and the very precise timing circuits, are simpler and less expensive than radars. Since they measure a vertical integral of the rain, the scan angle problem is not an issue. Swath widths of the order of twice the orbital altitude are straightforward to attain. The sampling error estimate made by North et al. (1988) and quoted above is based on a typical radiometer swath.

The passive measurement technique has its own set of problems. The primary issues in the passive measurements have been discussed by Wilheit (1986). Over oceans, rainfall retrievals can be made based on a radiative transfer model. While philosophically satisfying, the model, nevertheless, has many assumptions. Over land, there is no useful signal from the liquid hydrometeors, and we are reduced to inferring rainfall from the scattering of radiation by the frozen hydrometeors. Unfortunately, there is no unique relationship between the frozen hydrometeors aloft and the rainfall at the surface.

When one considers a multitude of issues, it becomes clear that the strengths and weaknesses of the radar and the radiometer are complementary. The radar has a narrow swath: the radiometer wide. The radar, when measuring backscatter, has a large sensitivity to drop size distribution (dsd) while the radiometer is virtually independent of dsd. When a radar is used in an attenuation measurement, it achieves this same independence to dsd. Conversely, the radiometer measurement is somewhat corrupted by nonprecipitating clouds, but the radar is not.

The radiometer provides no profile information but that is what a radar does best. The radiometer provides a poor measurement over land while the background

surface is of little importance to the radar. For a given aperture size and frequency, the radar can attain somewhat greater spatial resolution than the radiometer since it uses the antenna twice, in transmitting and receiving; the antenna gain appears squared in the radar equation. Inhomogeniety within the field of view is the primary source of uncertainty in the radiometer measurements. Because the power-law Z–R relationships are not strongly nonlinear, inhomogeniety is less of an issue for the radar (backscatter) measurement. When the radar is used for an attenuation measurement, it has essentially the same inhomogeniety problem as the radiometer.

The TRMM measurement concept is based on this complementarity of strengths and weaknesses between the radar and the radiometer. In this concept the radiometer, having a relatively wide swath, provides the sampling while the radar provides additional physical insight so that the radiometer observations can be better understood and more credible. Note that the insights provided by the radar can be used outside of the narrow radar swath and, for that matter, on data taken before the radar was launched, on other satellites concurrent with TRMM, and even after TRMM has run out of fuel and reentered the atmosphere.

Here we will explore one example of the use of the radar on TRMM to improve our understanding of the modeling for the radiometer. A key parameter of the radiative transfer models used for interpreting the radiometer is the freezing level. Below the freezing level, the hydrometeors (for the most part) are liquid and will absorb and emit microwave radiation efficiently. On the other hand, above the freezing level the bulk of the hydrometeors are frozen and their absorption and emission of microwave radiation is virtually nonexistent. They can scatter microwave radiation at the shorter wavelengths, but for the best measurements (over the oceans) we choose wavelengths for which this scattering is very weak ($\lambda > \sim 1$ cm). The absorption per unit length is a nearly linear function of the rain rate. The radiometer measurement can be interpreted in terms of the total path absorption. Thus to get a rain rate we need a path length. That path length is from the freezing level to the surface and back again along the view direction of the radiometer. An error in the path length translates inversely into an error in rain rate.

From whence cometh data on the freezing level to use in the rainfall retrieval. The best approach seems to be to use the radiances themselves. The model assumes a temperature lapse rate and relative humidity profile, which serves to connect the water vapor distribution to the freezing level. Water vapor, like hydrometeors, absorbs and emits microwave radiation. There is a relatively weak water vapor absorption feature centered at 22.235 GHz (1.35 cm) but there is some absorption at all of the TRMM Microwave Imager (TMI) frequencies. The sensitivity to the vertical distribution of the water vapor is modest, so the influence of the water vapor can be quantified in terms of the precipitable water (vertical

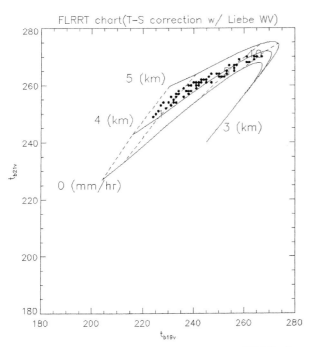

FIG. 17.2. Freezing level, rain rate, temperature (FLRRT) chart over oceanic rain from the TMI. The solid lines in the space of 19.35- and 21.3-GHz brightness temperatures are isopleths of freezing level; the dashed lines, rain rate. The dots are observations from the TMI (after Bellows 1999).

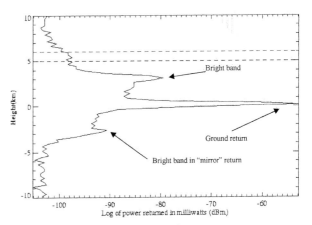

FIG. 17.3. Typical radar return from the TRMM PR over oceanic rain (after Bellows 1999).

integral of the water vapor density). This relationship has long been exploited to retrieve the precipitable water from the observations of the SSM/I. The model assumptions generate a unique mapping between the precipitable water and the freezing level. Thus, if we can infer the precipitable water from the radiances, we likewise know the freezing level. It must be emphasized that this connection is in the model; it must be determined whether this connection applies within the actual atmosphere.

The process of inferring the freezing level from the radiances is illustrated in Fig. 17.2. The lines represent theoretical calculations from the Wilheit et al (1977) model in a space of two TMI radiances, 19.35 and 21.3 GHz vertically polarized. The solid lines are isolines of constant freezing level (3, 4, and 5 km); the dashed lines, rain rate (0, 1, 5, and 10 mm h^{-1}). The dots are actual TMI observations of a raining area. Note that nearly all of the points fall between the 4- and 5-km freezing level contours. If nonraining points had been included, they would have fallen near and to the left of the 0 mm h^{-1} contour and, in many cases, would extend to lower freezing levels. When there is no rain, there is no reason to expect that the thermodynamic assumptions of the model are close to reality so there is no useful relationship between the radiances and the freezing level in the absence of rain. This approach is only useful for the retrieval of freezing level in rain.

Bellows (1999) has used the Precipitation Radar (PR)

on TRMM to validate the freezing level retrievals from the TMI. Figure 17.3 shows a typical radar return in rain. Here we have chosen the scan position directly at nadir. For this beam position, additional range gates are included so that the mirror image return can be viewed. Here we can clearly see a bright band at about 4-km altitude and the very strong return from the surface. We can also see a bright band in the mirror return. Here the energy transmitted from the radar has passed through the rain storm, bounced off the surface of the ocean, been backscattered from the hydrometeors, bounced off the ocean once again, and finally passed through the atmosphere to return to the radar to be detected. The radar operates at a frequency of 13.9 GHz, which is strongly attenuated in heavy rain. This could cause a decrease in the backscatter as the beam penetrates into the storm, which would look superficially like a bright band. However, the attenuation would cause the mirror return to decrease monotonically with height in the absence of a bright band. Indeed, if the attenuation were that strong it is unlikely that any mirror image return would be seen. Thus, seeing the bright band in the mirror image confirms that it is truly a bright band and not an artifact caused by attenuation.

The bright band is caused by the melting of frozen hydrometeors as they fall through the 0°C isotherm. Bellows (1999) chose the point of the maximum increase in the derivative of the backscatter as defining the beginning of this melting process, that is, the freezing level. Bellows (1999) found a wide variety of rain cases in the TRMM data for which he could define a freezing level. Figure 17.4 shows a comparison of the freezing levels derived from the PR with those derived from the TMI data. Although there is significant scatter, the trend is well represented and there is no obvious bias.

The water vapor absorption coefficient model of Liebe (1985) is generally considered to be the best description. It replaces an earlier model published by Waters (1976). In comparing precipitable water retrievals

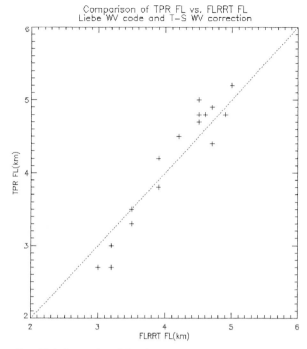

FIG. 17.4. Scatterplot of freezing level derived from the PR and from the TMI (after Bellows 1999).

TABLE 17.1.

Model	Freezing level bias
Waters	0.647
Waters + T-S*	−0.1235
Liebe	0.1647
Liebe + T-S	0.0411

* T-S represents the Thomas-Stahle correction.

sumptions can be considered a minor contributor to the net error of the TMI retrievals.

Here, we have shown one way in which the PR can be used as a probe of the physics underlying the radiometric rainfall retrievals. Many others can be thought of. This is the essential relationship between the spaceborne precipitation radars and radiometers. The radar provides the physics and the radiometer, having a wide swath, provides the statistics.

For future improvements in our rain measurement capability, the strengths of each part should be enhanced. The radar should measure more parameters such as additional frequencies or cross polarization to give a more complete view of the physics of the rainfall, and the radiometer's sampling should be improved by wider swaths and/or multiple satellites.

Acknowledgments. The financial support of the NASA TRMM (NAG5-4746) and AQUA (NAS5-32593) projects is gratefully acknowledged. The results reported here draw heavily on the labors of my present and past graduate students, particularly Jody Thomas-Stahle and Chris Bellows.

from TMI with radiosondes, Thomas-Stahle (2001) found a small correction (ca. 5%) was required in the model to get agreement between the retrievals and the radiosonde-derived precipitable water. Earlier Manning (1997) had found that a similar, but somewhat larger, correction was required to match precipitable water retrievals from SSM/I with radiosondes. It is not clear to what extent this is an error in the water vapor spectroscopy or a bias in radiosondes. At any rate, taking the difference between the two water vapor models (Liebe and Waters) with and without the Thomas-Stahle correction should represent a very generous estimate of the uncertainty in the water vapor absorption coefficient model.

Bellows used all four possible combinations in radiative transfer models to generate the TMI freezing level retrievals and compared them with the PR estimates. The results are given in Table 17.1. The mean differences varied from −0.12 to +0.16 km, a range of only 280 m. The smallest bias was only 0.04 km and occurred for the combination of the newest (and presumably best) water vapor model with the empirical correction. However, with biases this small, it would be difficult to assert that this validates the Liebe model versus the Waters model or the need for the Thomas-Stahle empirical correction. The full range of possibilities covers a range of less than 10% of the freezing level. Thus, uncertainty in the freezing level retrieval and the associated spectroscopic and thermodynamic as-

REFERENCES

Bellows, C. T., 1999: Consistency testing of models used to infer rainfall from TRMM passive microwave observations. M.S. thesis Dept. of Atmospheric Sciences, Texas A&M University, 53 pp.

Chang, A. T. C., L. S. Chiu, and T. T. Wilheit, 1993: Random errors of oceanic monthly rainfall derived from SSM/I using probability distribution functions. *Mon. Wea. Rev., 121,* 2351–2354.

Liebe, H. J., 1985: An updated model for millimeter wave propagation in moist air. *Radio Sci., 2,* 1069–1089.

Manning, N. W. W., 1997: Remote sensing of total integrated water vapor, wind speed and cloud liquid water over the ocean using the special sensor microwave/imager (SSM/I). M.S. thesis, Dept. of Meteorology, Texas A&M University, 100 pp.

North, G. R., 1988: Survey of sampling problems for TRMM. *Tropical Rainfall Measurements,* J. S. Theon and N. Fugono, Eds., A. Deepak, 337–347.

Thomas-Stahle, J. L., 2001: Remote sensing of marine atmospheric parameters using the TRMM Microwave Imager. M. S. thesis, Dept. of Atmospheric Sciences, Texas A&M University, 52 pp.

Waters, J. W., 1976, Absorption and emission by atmospheric gasses: *Meth. Exp. Phys., 12B,* 142–176.

Wilheit, T. T., 1986: Some comments on the passive microwave measurement of rain. *Bull. Amer. Meteor. Soc., 67,* 1226–1232.

——, A. T. C. Chang, M. S. V. Rao, E. B. Rodgers, and J. S. Theon, 1977: A satellite technique for quantitatively mapping rainfall over the oceans. *J. Appl. Meteor., 16,* 1137–1145.

Chapter 18

GATE and TRMM

GERALD R. NORTH

Texas A&M University, College Station, Texas

ABSTRACT

It is natural that a book chapter honoring Joanne Simpson draw the connection between the two most important tropical meteorological observing programs in the history of meteorology: the Global Atmospheric Research Program Atlantic Tropical Experiment (GATE) and the Tropical Rainfal Measuring Mission (TRMM). Both programs were dominated by the influences of Joanne Simpson. When TRMM data are all in, these two grand experiments will have given us more information about the behavior of tropical convection and precipitation over the tropical oceans than all other tropical field campaigns combined. But some may not know how GATE data played a key role in demonstrating the feasibility of a mission like TRMM. This chapter will present a review of a number of studies that connect GATE precipitation data with TRMM, especially in the early planning stages.

1. Introduction

The Global Atmospheric Research Program (GARP) had as one of its components the GARP Atlantic Tropical Experiment (GATE). The major aims were to spell out the details of convective energetics in tropical plumes. Houze and Betts (1981) provided an excellent review article on convection findings based upon GATE. GATE was a fully international program with many nations participating. Among the features of the observing system was an array of stationary ships with precipitation radars aboard. These were positioned in a hexagonal configuration at roughly 8°N in the tropical Atlantic. The ships and accompanying aircraft collected many kinds of data over several intensive observing periods lasting a few weeks each in the summer of 1974. The radars collected space–time snapshots of the precipitation fields at the surface and at higher elevations. The radar data were converted to rain rates and continuously calibrated with point gauges aboard the ships. The eventual product that resulted was a multivariate time series taken at 15-min intervals and averaged over 4- × 4-km boxes or tiles (Arkell and Hudlow 1977; Hudlow and Patterson 1979; Hudlow 1979). The spatial array was 70 × 70 tiles providing an areal coverage of 280 × 280 km.

Some years after the completion of GATE there was a broad-based workshop on the measurements of precipitation from satellites held at Goddard Space Flight Center and hosted by David Atlas and Otto Thiele (1981). One study in the volume of papers presented at this conference stands out in my mind as I later began to prepare presentations for TRMM. This was a little-noticed paper by Charles Laughlin, a communications engineer who retired from the National Aeronautics and Space Administration (NASA) shortly after the workshop. In his watershed paper Laughlin worked up some important statistical summaries of the GATE precipitation data. Among his most important computations was a graph of the autocorrelation function of area-averaged precipitation as a function of the area being considered. Larger areas had longer autocorrelation times and the largest area he considered (280 × 280 km) had an autocorrelation time of about 6 h. Laughlin went on to conclude that the diurnal cycle was small and that a satellite returning to the same averaging box every 12 h could have tolerable sampling error characteristics (of the order of 10%).

In formulating the first presentations for TRMM it was recognized that the most challenging issues before us were 1) statistical sampling errors due to the sparse visiting pattern of a single satellite; 2) inevitable "beam filling" errors due to the finite size of an individual microwave radiometer field of view (FOV) compared to the smaller important scales of tropical precipitation; and 3) ground truth for such a satellite observing system because of the difficulties in deploying the necessary ground observing system. The purpose of this chapter is to show how studies with the same GATE precipitation dataset have helped us to sort out some of the difficulties with the TRMM observing system and to provide some insights into how these might be overcome.

2. TRMM observing system

The original TRMM observing system, as planned in the 1980s, is not so different from the final configuration except for the addition of a lightning instrument and an earth radiation budget instrument (Wilheit 1986; Simpson et al. 1988). In most of the studies to be summarized in this chapter, the attention will be mostly focused on the 19.6-GHz channel of the microwave radiometer. This instrument is to have an FOV of the order of 20 km across and the scanning is to be across track with a scan width of about 700 km (72 FOVs). There are other channels available on the radiometer and there are other instruments on board, but in these studies we assume that the primary measurement, over oceans at least, will come from the 19.6-GHz channel. This is our prototype measurement.

3. Sampling error studies

Laughlin constructed a univariate time series model for GATE area averages. He then estimated the sampling errors for a satellite visiting at different time intervals from a few hours to a few days with 1-week, 2-week, and 1-month averages. For one-month averages and a satellite visiting the whole averaging box, the errors were expected to be about 10%. Following Laughlin, the first GATE-related study to appear on TRMM sampling errors was that of McConnell and North (1987). In this study an imaginary satellite was flown over the GATE region at 12-h intervals and the data recovered were used to estimate the 3-week-long average of rain rate (GATE I was slightly less than 3 weeks and GATE II was a few weeks later in 1974 and of duration about 18 days). An ensemble of satellite overpasses was configured so that the first satellite overflew at 0800 and 2000, etc. The next member began at 0900 and 2100, then 1000 and 2200, etc. In this way an ensemble of nine members was constructed. The month-long averages of light, medium, heavy, and extremely heavy rain categories were estimated from the ensemble. The estimates from the different ensemble members were of course correlated, but visual inspection of the estimates in bar graph form was rather revealing. The bar graphs showed that the sampling errors, as indicated by the spread across the ensemble members, were less than 10% for the lighter rain categories and only somewhat larger for the other categories. Of course, there was no proof that the same kind of statistics would hold for another month, but the objection that the system might miss significant rain events by such a sampling scheme was considerably quieted. Later we found that individual rain storms in GATE lasted 12 to 18 h and most of this lifetime might occur in a single averaging box. The graphics from the McConnell–North study were very useful in the early stages of support building for TRMM.

Soon after the McConnell–North study came the study based on modeling the probability density functions for rain rates (Kedem et al. 1990, hereafter KCN; see also Kedem and Chiu 1987). The idea is based upon the mixed probability density distribution—in this case, a lognormal distribution when raining, and a delta func-

tion at zero rain rate—the latter containing usually more than 90% of the probability. GATE data were used to verify the mixed lognormal distribution; it was already known that many parameters (areas, storm column heights, etc.) in tropical convection follow the lognormal (Lopez 1977). This rain-rate distribution was found to hold for a range of different averaging areas. A good approach to estimating the area-average rain rate was to estimate the parameters of the mixed lognormal distribution. The main advantage of this system is that if the microwave retrieval algorithm "saturates," that is, it cannot really resolve high rain rates, then one can estimate the parameters of the whole distribution from a knowledge of only the lower rain-rate portion of the distribution. In this KCN study it was also found that sampling errors for a TRMM-like configuration of the order of 10% were not an unreasonable expectation. An important side product of the KCN study was that Long Chiu discovered in the GATE data a linear relationship between area-average rain and the area covered by rain (Chiu 1988). Finally, the KCN study was the basis for one of the important retrieval strategies based on probability distribution estimation (Wilheit et al. 1991).

One problem with the Laughlin, McConnell–North, and KCN studies is that they were based on the GATE area, which was only 280 × 280 km. The first data requirements for TRMM were for a 500- × 500-km averaging box for month-long averages as opposed to only 3 weeks (Thiele 1987). Hence, some extrapolating to the larger box and longer averaging time had to be conducted. Moreover, the visits in these primitive studies were always "flush" that is, the satellite covered the entire averaging box on each visit. In reality, the swaths will only partially cover the box on an individual visit. There was a need to construct sampling models that could take these effects into account.

One approach along these lines was to construct a stochastic rain field model with two spatial dimensions and temporal evolution. The model could be tuned to fit GATE data. This program was carried out by Thomas L. Bell. The idea of the model was inspired by turbulence theory in that a Gaussian random field of vertical winds was produced as though from stationary, spatially homogeneous turbulence but with the space–time spectrum to be determined. For areas where the vertical wind exceeded a certain threshold value, it was deemed "raining." The exceedence variate was distributed like the tail of a Gaussian distribution. Bell then converted this random "exceedence" field to lognormal. Finally, the spectrum of the original underlying field was adjusted in both space and time to bring the rain field statistics (space–time autocorrelations, etc.) into agreement with those of the GATE data. This field was very similar in appearance to GATE rainfall data. It was patchy and had the correct spatial correlation structure by design. It still did not have some features, such as bands or waves. Nevertheless, it was a very good model for the purpose intended. The model is described in Bell (1987) and sampling error results are reported in Bell et al. (1990). One important conclusion is that based upon this model the error distributions tended to be Gaussian even though the underlying fields were far from it.

The beauty of this model was in its flexibility. One could produce a long control run of the model and save the output. Hence, more than a few weeks of data could be easily generated. This permitted many studies for different orbit altitudes, inclinations, etc. One could even make partial coverage visits based on real orbit calculations. A major problem was that the simulations were very computer intensive (for the mid-1980s). A much simpler model was that of Shin and North (1988). This model used realistic overflights generated from orbital calculations, but it was a univariate model of the area-averaged rain rate. The fraction of coverage for each visit was calculated and weights were assigned to each visit accordingly to estimate the month-long average over the box. The results were in reasonable agreement with those from the Bell model. Again the sampling errors for orbits in the parameter range envisioned by TRMM were on the order of 10%.

North and Nakamoto (1989; hereafter referred to as NN) introduced a spectral approach to evaluating sampling errors for a general class of sampling designs. A formula for the mean square sampling errors was constructed based upon an integral over all space wavenumbers and frequencies. The integrand consisted of two factors, one a function only of the sampling design and the second depending only on the space–time spectral density of the field being sampled. The formula was interesting because it showed that the only field property needed to evaluate the mean square error was the space–time spectrum. This is especially interesting since it is only a second-order statistic and even though the field may be highly non-Gaussian (such as rain rates) the mean square error only depends on the second-moment statistics of the field. North and Nakamoto presented error estimates for satellite overpasses and for some other configurations such as regular arrays of point gauges. Part of the paper of further interest is a rain-rate model based on a damped diffusion model forced by space–time white noise. The two parameters of this model were tuned to GATE data. While the model solutions were not patchy like real rain, it could be made to have roughly the correct second-moment statistics.

The North–Nakamoto approach led to some space–time spectral analyses of GATE data. Examples include those of Nakamoto et al. (1990) and Polyak and North (1995). There were also a few studies of hydrological interest not directly related to GATE but stimulated by the GATE-based models (Graves et. al. 1993; Valdes et al. 1994). Valdes et al. (1990) used several spectral estimates based upon different rain field models to obtain sampling errors.

How can TRMM data be combined with other data? North et al. (1991) showed, using the NN formalism, that one could easily evaluate a design composed of a combination of satellite overpasses and ground-based gauges. A very good approximation consists of simply averaging the two together using the inverse variances as weights. There is a cross term because the two measurements are correlated, but this term was shown to be very small.

The culmination of this series of papers was one by North et al. (1993) in which the errors were evaluated for configurations of several satellites combining TRMM with sun-synchronous orbiters. The NN rain-rate model was used along with a stochastic model of the fractional visits. Once again, the sampling errors were in the tolerable range (5%–10%). Many of the early results were summarized by North (1988).

An issue always lurking in the TRMM estimation problem has been the diurnal cycle. Bell and Reid (1993) showed that to estimate the amplitude of the diurnal cycle might take many months of data. More recently, Shin and North (2000) investigated the contribution to sampling errors if the field being sampled was diurnally cyclostationary. This allows for the possibility that the variance might have a diurnal cycle along with the mean. This effect does increase sampling errors but only by a few percentage points.

A major problem for all the analyses based upon GATE data is that GATE may not be representative. Shin and North (1991) used Electronic Scanning Microwave Radiometer–5 (ESMR-5) data to show that the autocorrelation lengths of rain rates from one FOV to another along the track were quite homogeneous across the tropical oceans. This property might not be too surprising since the sizes of tropical convective cells are not highly variable from one place to another over the tropical oceans.

On the other hand, GATE was located in the intertropical convergence zone (ITCZ) and this is not typical of most of the Tropics. The rain rates in the ITCZ are much greater and all of the studies listed above were based upon those kinds of statistics. Tropical rain tends to be characterized over a month by a small number of heavy events. In the ITCZ there are many of these events in a month. Away from the equator the number of events in a month will be smaller. This will tend to make sampling errors larger.

Bell and Kundu (1996, 2000) have developed a sampling error model much more appropriate for these kinds of situations. The percentage sampling errors (compared to the local mean) in their model vary as $R^{-1/2}$, where R is the monthly averaged rain rate at the particular location, season, etc. Bell and Kundu (2000) compare in detail a number of the studies referred to above as well as many based on other field experiments. The inverse square root scaling seems to be a satisfactory way of accounting for the evaluation of sampling errors outside the heavy rain areas associated with the ITCZ where GATE data were taken.

4. Beam filling

If only a single microwave frequency is used to estimate rain rate, there is an inevitable problem because of the nonlinearity of the formula relating microwave temperature and rain rate. The satellite-borne radiometer sees an area average of microwave brightness temperature over the FOV (typically 20 km across). But the FOV may contain patchy rain patterns much smaller than the FOV. The result of this is that the distribution of rain rates inside the FOV leads to a measurement of brightness temperature that when inverted does not give the average of the rain rate in the FOV. A bit of analysis shows that there will be a bias and a large random error in this kind of single-channel retrieval. The use of additional channels can sometimes reduce the beam filling error but as long as the retrieval is nonlinear, no amount of channels can completely eliminate it.

GATE data make an excellent test bed to estimate the size of this kind of error. The first recognition (to my knowledge) of the beam filling error in these kinds of retrievals is in an unpublished report by Eric Smith and Stan Kidder (1978, unpublished manuscript) based upon ESMR-5 data compared to GATE. The most revealing study was that of Chiu et al. (1990). In this paper the retrieval errors were estimated from GATE data with different sized FOVs. A simple saturating exponential model of the T–R relationship was adopted:

$$T(R) = A + Be^{-CR}.$$

Using the rain-rate fields from GATE one could insert ensembles of fields inside the FOV to see how the inversion to rain rate worked. As expected the bias was about 40% and the random error accompanying the retrieval was similarly of the same order or larger. In the estimate of monthly averages the latter was not considered a problem, but in smaller aggregates it could be significant.

Another extremely interesting study of the beam filling effect was by David Short in his Ph.D. dissertation (Short 1988) and in the paper by Short and North (1990). In this study, GATE data were compared to ESMR-5 brightness temperatures, overpass by overpass. Because of small positioning errors, the two pictures had to be shifted until the best fit occurred. Once this was done, the expected brightness temperatures could be computed from the GATE data and compared to the satellite data. The satellite data could also be inverted to give the GATE data plus error. It was found that the errors were entirely consistent with the earlier findings by Chiu et al. (1990). In other words, ESMR-5 (19.6 GHz) errors in retrieving rain rates could be completely explained in terms of simple beam filling error.

These studies showed that the prototype measurement based upon the 19.6-GHz channel with FOV about 20 km was a perfectly good means of retrieving rain rates from space if the bias could be removed perhaps by climatology. The addition of other channels could be used to ascertain the freezing level, etc., but the fundamental principle advanced much earlier by Wilheit (1986) was intact.

Another source of error related to beam filling is the random error from FOV to FOV. This kind of random error is usually assumed to be statistically independent (e.g., Bell and Kundu 2000). Polyak and North (1996) examined GATE data for neighboring 20- × 20-km boxes. The beam filling error is mostly accounted for by the variance of rain rate within the FOV (Chiu et al. 1990; Short 1988). If these variances are correlated from one FOV to another, the random part of the beam filling error will be correlated. Polyak and North (1996) found a correlation coefficient of the variances (when raining) between neighboring FOVs of between 0.35 and 0.50. This reduces the number of independent samples by about a factor of 3. When a month of data are aggregated into 5° × 5° boxes, this can add a random error of the order of a 1–2 percentage points to the sampling error.

5. Ground truth

The third major concern raised in the early presentations promoting TRMM was ground truth (some prefer *ground validation*). In fact, probably more resources have been devoted to this than to any other aspect of TRMM other than the actual satellite's construction. Many problems beset the ground truth problem. First is that the tropical oceans are remote and especially difficult to work in for field campaigns. Even though several such campaigns have been undertaken since the launch of TRMM, they are necessarily of short duration and any comparisons of ground and space measurements will be troubled with sampling errors.

All ground-based systems have their own set of biases and random errors. In fact, some have conjectured that the satellite may be a more accurate measure of rain rates than any ground-based observing system. The main tool for ground truth is the surface-based precipitation radar. However, the radar retrieval of rain is usually based upon a Z–R relationship, which is strongly dependent on rain type and additionally complicated by empirical parameters. Nevertheless, the radar is a very useful tool because of its spatial resolution, which is better than or equal to that of the satellite's. One can examine spatial patterns of each observed field and from those draw conclusions, even if the calibration of the radar is not certain.

Perhaps the closest to "truth" is the point rain gauge. Unfortunately, these are usually located on small islands and there is a clear bias associated with small landmasses located in large oceanic regions. Such a gauge will not be representative of the open ocean. Ships offer another alternative, but these are also beset with biases owing to location amidst the proximate geometrical structure of the vessel. One gauge configuration seems promising: the point gauge located on oceanic buoys. There is such a set of gauges located on the Tropical

Ocean and Global Atmosphere Tropical Atmosphere Ocean (TOGA TAO) array along the equatorial East Pacific (Michael McPhadden 1999, personal communication). Several of these gauges have collected more than 12 months of rain-rate data and this may be enough to make a definitive ground truth test of TRMM.

GATE data have afforded us some ideas about how the test might be conducted. First, the satellite and the gauge measure very different quantities: the gauge measures precipitation rate continuously in time at a point on the earth's surface, the satellite measures an area average over an individual FOV. There will be a random sampling error even if both systems are perfect. Furthermore, the location of the gauge will be positioned at a random location within the visiting FOV.

Several theoretical studies have been undertaken to investigate the types of problems that might be encountered in such an approach. The main question to be addressed is how many visit pairs are required to assess a bias of 10% of the natural variability (Ha and North 1994, 1999; North et al. 1994). The basic formalism for the comparison was worked out by North et al. (1994), and stochastic model examples were presented based on GATE data. It appeared that if both instruments were perfect but offset from one another by a bias, the number of pairs might require as much as a year of data. Ha and North (1994) investigated the gain to be made by having several gauges close together and even short microwave attenuation links cutting across the FOV. There was found to be some significant gain from having two gauges, but little more from the attenuation or more than two gauges. Ha and North (1999) conducted further studies along these lines with more realistic models that included the "patchiness" feature of real rain. Finally, a study has been conducted with overflights of actual GATE data (Ha and North 2000). The conclusion is that 8–10 months of data pairs would be required. Actually, we know that the retrieval of an individual FOV rain rate will have a random beam filling error of approximately 50% (e.g., Chiu et al. 1990); therefore the 8–10-month requirement will probably be substantially increased.

6. Conclusions

GATE data gave us a good idea about how tropical convection over the oceans works. But it was in a small limited region, perhaps not representative of the tropical oceans as a whole. The emergence of ENSO as a dominant factor in seasonal forecasting led to an urgent need to extend this kind of measurement to the Pacific and throughout the Tropics. Furthermore, the need to provide databases to compare with climate model simulations has become an important priority. So the Tropical Rainfall Measuring Mission has been approved and successfully launched with now nearly three years of excellent data having been archived. Some of the main problems associated with the TRMM observing system

remain to this day, but the early studies with GATE data (although now somewhat dated by more recent, but less comprehensive, field campaigns) were an essential ingredient in the successful understanding of the TRMM observing system.

Acknowledgments. The author wishes to thank the NASA TRMM Office for its continuous grant support since 1984. I further acknowledge my indebtedness to all of the coauthors listed in the references. Finally, I wish to give my special thanks to Joanne Simpson for her courage, doggedness, scientific leadership, friendship, and encouragement over these years.

REFERENCES

Arkell, R., and M. D. Hudlow, 1977: *GATE International Meteorological Radar Atlas.* U.S. Department of Commerce, NOAA.

Atlas, D., and O. W. Thiele, Eds.,1981: *Precipitation Measurements from Space: Workshop Report.* NASA/GSFC Rep.

——, D. Rosenfeld, and D. A. Short, 1990: The estimation of convective rainfall by area integrals. Part I: The theoretical and empirical basis. *J. Geophys. Res.,* **95,** 2153–2160.

Bell, T. L., 1987: A space–time stochastic model of rainfall for satellite remote-sensing studies. *J. Geophys. Res.,* **92,** 9631–9643.

——, and N. Reid, 1993: Detection of the diurnal cycle of tropical rainfall from satellite observations. *J. Appl. Meteor.,* **32,** 311–322.

——, and P. K. Kundu, 1996: A study of the sampling error in satellite rainfall estimates using optimal averaging of data and a stochastic model. *J. Climate,* **9,** 1251–1268.

——, and ——, 2000: Dependence of satellite sampling error on monthly averaged rain rates: Comparison of simple models and recent studies. *J. Climate,* **13,** 449–462.

——, A. Abdullah, R. L. Martin, and G. R. North, 1990: Sampling errors for satellite-derived tropical rainfall: Monte Carlo study using a space–time stochastic model. *J. Geophys. Res.,* **95,** 2195–2205.

Chiu, L. S., 1988: Estimating rain rates from rain area. *Tropical Precipitation Measurements,* J. Theon and N. Fugono, Eds., A. Deepak, 361–367.

——, G. R. North, and D. A. Short, 1989: Errors in satellite rainfall estimation due to non-uniform field of view of space-borne microwave sensors. *Microwave Remote Sensing of the Earth System,* A. Chedin, Ed., A. Deepak, 95–109.

——, ——, ——, and A. McConnell, 1990: Rain estimation from satellites: Effect of finite field of view. *J. Geophys. Res.,* **95,** 2177–2185.

Doneaud, A. A., S. I. Niscov, D. L. Priegnitz, and P. L. Smith, 1984: The area–time integral as an indicator for convective rain volumes. *J. Climate Appl. Meteor.,* **23,** 555–561.

Graves, C. E., J. B. Valdes, S. S. P. Shen, and G. R. North, 1993: Evaluation of sampling errors of precipitation from spaceborne and ground sensors. *J. Appl. Meteorol.,* **32,** 374–385.

Ha, E., and G. R. North, 1994: Use of multiple gauges and microwave attenuation of precipitation for satellite verification. *J. Atmos. Oceanic Technol.,* **11,** 629–636.

——, and ——, 1995: Model studies of the beam-filling error for rain rate retrieval with microwave radiometers. *J. Atmos. Oceanic Technol.,* **12,** 268–281.

——, and ——, 1999: Error analysis for some ground validation designs for satellite observations of precipitation. *J. Atmos. Oceanic Technol.,* **16,** 1949–1957.

——, ——, C. Yoo, and K.-J. Ha, 2002: Evaluation of some ground truth designs for satellite estimates of rain rate. *J. Atmos. Oceanic Technol.,* **19,** 65–73.

Houze, R. A., and A. K. Betts, 1981: Convection in GATE. *Rev. Geophys.,* **19,** 541–576.

Hudlow, M. D., 1979: Mean rainfall rates for the three phases of GATE. *J. Appl. Meteorol.,* **18,** 958–962.

——, and V. L. Patterson, 1979: *GATE Radar Rainfall Atlas.* NOAA, 158 pp.

Kedem, B., and L. S. Chiu, 1987: On the lognormality of rain rates. *Proc. Nat. Acad. Sci. USA,* **84,** 901–905.

——, ——, and G. R. North, 1990: Estimation of rain rate: Application to satellite observations. *J. Geophys. Res.,* **95,** 1965–1972.

Laughlin, C. R., 1981: On the effect of temporal sampling on the observations of mean rainfall. *Precipitation Measurements From Space, Workshop Report,* D. Atlas and O. Thiele, Eds., NASA/GSFC, D59–D66.

Lopez, R. E., 1977: The lognormal distribution and cumulus cloud populations. *Mon. Wea. Rev.,* **105,** 865–872.

McConnell, A., and G. R. North, 1987: Sampling errors in satellite estimates of tropical rain. *J. Geophys. Res.,* **92,** (D8), 9567–9570.

McPhadden, M. J., 1995: The Tropical Atmosphere Ocean (TAO) Array is completed. *Bull. Amer. Meteor. Soc.,* **76,** 739–741.

Meneghini, R., 1998: Application of a threshold method to airborne–spaceborne attenuating-wavelength radars for the estimation of space–Time ram-rate statistics. *J. Appl. Meteor,* **37,** 924–938.

——, and J. A. Jones, 1993: An approach to estimate the areal rain-rate distribution from spaceborne radar by the use of multiple thresholds. *J. Appl. Meteor.,* **32,** 386–398.

Nakamoto, S., J. B. Valdes, and G. R. North, 1990: Frequency-wave-number spectrum for GATE Phase I rainfields. *J. Appl. Meteor.,* **29,** 842–850.

North, G. R., 1988: Survey of sampling problems for TRMM. *Tropical Rainfall Measurements,* J. S. Theon and N. Fugono, Eds., A. Deepak, 337–348.

——, 1992: Characteristics of tropical precipitation important for its estimation by satellites. *Global Role of Tropical Rainfall,* J. Theon et al., Eds., A. Deepak, 161–181.

——, and S. Nakamoto, 1989: Formalism for comparing rain estimation designs. *J. Atmos. Oceanic Technol.,* **6,** 985–992.

——, S. S. P. Shen, and R. Upson, 1991: Combining gages with satellite measurements for optimal estimates of area-time averaged rain rates. *Water Resour. Res.,* **10,** 2785–2790.

——, ——, and ——, 1993: Sampling errors in rainfall estimates by multiple satellites. *J. Appl. Meteor.,* **32,** 399–410.

——, J. B. Valdes, E. Ha, and S. S. P. Shen, 1994: The ground-truth problem for satellite estimates of rain rate. *J. Atmos. Oceanic Technol.,* **11,** 1035–1041.

Polyak, I., and G. R. North, 1995: The second moment climatology of the GATE rain rate data. *Bull. Amer. Meteor. Soc.,* **76,** 535–550.

——, and ——, 1996: Spatial correlation of beam-filling error in microwave rain-rate retrievals. *J. Atmos. Oceanic Technol.,* **13,** 1101–1106.

Shin, K.-S., and G. R. North, 1988: Sampling error study for rainfall estimate by satellite using a stochastic model. *J. Appl. Meteorol.,* **27,** 1218–1231.

——, and ——, 1991: On the homogeneity of spatial correlation statistics of tropical rainfall. *J. Geophys. Res.,* **96,** 9273–9283.

——, and ——, 2000: Errors incurred in sampling a cyclostationary field. *J. Atmos. Oceanic Technol.,* **17,** 656–664.

Short, D. A., 1988: Remote sensing of oceanic rain rates by passive microwave sensors: A statistical-physical approach. Ph.D. dissertation, Texas A&M University, 85 pp.

——, and G. R. North, 1990: The beam filling error in the Nimbus 5 Electronically Scanning Microwave Radiometer observations of global atlantic tropical experiment rainfall. *J. Geophys. Res.,* **95** (D3), 2187–2193.

Simpson, J., R. F. Adler, and G. R. North, 1988: A proposed Tropical Rainfall Measuring Mission (TRMM) satellite. *Bull. Amer. Meteor. Soc.,* **69,** 278–295.

Thiele, O. W., Ed.,1987: On requirements for a satellite mission to measure tropical rainfall. NASA RP-1183.

Valdes, J. B., S. Nakamoto, S. S. P. Shen, and G. R. North, 1990: Estimation of multidimensional precipitation parameters by areal estimates of oceanic rainfall. *J. Geophys. Res.,* **95,** 2101–2111.

——, E. Ha, C. Yoo, and G. R. North, 1994: Stochastic characterization of space–time precipitation: Implications for remote sensing. *Adv. Wat. Resour. Res.,* **17,** 47–59.

Wilheit, T. T., 1986: Some comments on passive microwave measurement of rain. *Bull. Amer. Meteor. Soc.,* **67,** 1226–1232.

——, 1988: Error analysis for the Tropical Rainfall Measuring Mission (TRMM). *Tropical Rainfall Measurements,* J. S. Theon and N. Fugono, Eds., A. Deepak, 377–385.

——, T. A. Chang, and L. S. Chiu, 1991: Retrieval of monthly rainfall indices from microwave radiometric measurements using probability distribution functions. *J. Atmos. Oceanic Technol.,* **8,** 118–136.

Chapter 19

Performance Evaluation of Level-2 TRMM Rain Profile Algorithms by Intercomparison and Hypothesis Testing

ERIC A. SMITH

NASA Goddard Space Flight Center Greenbelt, Maryland

THROY D. HOLLIS

Department of Meteorology, Florida State University, Talahassee, Florida

ABSTRACT

Currently, satellite algorithms are the methodology showing most promise for obtaining more accurate global precipitation estimates. However, a general problem with satellite methods is that they do not measure precipitation directly, but through inversion of radiation–rain relationships. Because of this, procedures are needed to verify algorithm-generated results. The most common method of verifying satellite rain estimates is by direct comparison with ground truth data derived from measurements obtained by rain gauge networks, ground-based weather radar, or a combination of the two. However, these types of comparisons generally shed no light on the physical causes of the differences. Moreover, ground validation measurements often have uncertainty magnitudes on the order of or greater than the satellite algorithms, motivating the search for alternate approaches. The purpose of this research is to explore a new type of approach for evaluating and validating the level-2 Tropical Rainfall Measuring Mission (TRMM) facility rain profile algorithms. This is done by an algorithm-to-algorithm intercomparison analysis in the context of physical hypothesis testing.

TRMM was launched with the main purpose of measuring precipitation and the release of latent heat in the deep Tropics. Its rain instrument package includes the TRMM Microwave Imager (TMI), the Precipitation Radar (PR), and the Visible and Infrared Scanner (VIRS). These three instruments allow for the use of combined-instrument algorithms, theoretically compensating for some of the weaknesses of the single-instrument algorithms and resulting in more accurate estimates of rainfall. The focus of this research is on the performance of four level-2 TRMM facility algorithms producing rain profiles using the TMI and PR measurements with both single-instrument and combined-instrument methods.

Beginning with the four algorithms' strengths and weaknesses garnered from the physics used to develop the algorithms, seven hypotheses were formed detailing expected performance characteristics of the algorithms. Procedures were developed to test these hypotheses and then applied to 48 storms from all ocean basins within the tropical and subtropical zones over which TRMM coverage is available ($\sim35°$N–$35°$S). The testing resulted in five hypotheses verified, one partially verified, and one inconclusive. These findings suggest that the four level-2 TRMM facility profile algorithms are performing in a manner consistent with the underlying physical limitations in the measurements (or, alternatively, the strengths of the physical assumptions), providing an independent measure of the level-2 algorithms' validity.

1. Introduction

Precipitation is an important weather process because the world's population depends on a regulated amount of freshwater. Not enough rainfall leads to drought and occasionally famine, while too much causes flooding. But these are only the most obvious impacts of precipitation. The not-so-obvious impacts stem from the production of latent heat, with three-fourths of the heat energy of the world's atmosphere being generated from the release of latent heating by the condensation–precipitation process (Simpson et al. 1996). Thus, latent heating is the main diabatic factor in the thermodynamic control of the earth's climate and general circulation.

As important as precipitation is to man's welfare, there are still major problems associated with its measurement. Precipitation is difficult to measure accurately because it does not occur uniformly at area-wide scales. When a rain cell tracks directly over a rain gauge measuring station, that measurement is generally not representative of the surrounding area because of the nonergodic variability in time–space associated with most convective precipitation systems, particularly in the Tropics (Seed and Austin 1990). The same can be said about ground-based radar, albeit to a lesser degree. However, added to the difficulty of radar is that it is an intrinsically remote sensing measurement, highly susceptible to transmitter–receiver calibration errors, based on beam-broadened horizontal signal paths through largely vertically structured rain clouds, often overshooting the main rain layer and generally limited to land areas and coasts (see Smith et al. 1998). Also, since it is unrealistic to deploy rain gauges, radars, or other ground-measuring devices to cover the entire earth's surface, especially over oceans, which make up three-fourths of the global surface, large areas have historically gone without systematic in situ measurements. The only viable means to overcome this limitation is through satellite remote sensing.

Besides the purely scientific search for knowledge, there are also real-world benefits in improving the measurement of precipitation. One of those is improved weather forecast models. Today's models use different methods for estimating precipitation and the accompanying release of latent heat. Because of this, discrepancies occur between large-scale prediction models forced with standardized inputs (Gadgil and Sajani 1998). However, current methodologies for obtaining vertical latent heating structures from rain profile retrieval algorithms have proved viable (Smith et al. 1992; Tao et al. 1993a; Yang and Smith 1999a,b, 2000). Moreover, accurate estimates of precipitation and latent heating used for model initialization have been shown to lead to more accurate forecasts (Hou et al. 1999, 2000). In fact, any new understanding of the vertical structure properties of liquid and frozen precipitation processes and their influence on diabatic heating, some of which are examined in the context of this study, tends to improve predictability. Finally, improvement in global precipitation measuring leads to better physical understanding of important short-term climate phenomena such as the El Niño–Southern Oscillation (ENSO; Simpson et al. 1988).

Free-flying satellites are the technology showing most promise for obtaining more accurate global precipitation measurements. Satellites enable observing a greater area of the earth's surface for a sustained period of time more so than any other type of measuring platform. Of course, the general problem in using satellite instruments to measure precipitation is that they cannot measure rainfall or rain rate directly, instead requiring the use of some type of inversion scheme involving radiative properties of the atmosphere, and the emission, transport, and attenuation processes of the measured radiative quantities. Much research has gone into precipitation retrieval methods employing optical and infrared imagery since the late 1960s (e.g., Barrett and Martin 1981), and more recently with passive microwave imagery (e.g., Wilheit et al. 1994; Smith et al. 1998).

Because satellites do not measure rainfall directly, procedures are needed to verify retrieved estimates (see Dodge and Goodman 1994). The most common method of verifying satellite-retrieved rain estimates is to directly compare to ground validation measurements (ground truth) derived from rain gauge networks, ground weather radar, or a combination of the two. However, there are long-standing errors associated with measuring rainfall with rain gauge networks (Rodriguez-

Iturbe and Mejia 1974; Damant et al. 1983; Bellon and Austin 1986; Seed and Austin 1990; and Morrisey 1991) as well as with ground-based radar systems (Harrold et al. 1973; Damant et al. 1983; Seed and Austin 1990; Smith et al. 1998). In essence, in situ methods do not measure "true" rainfall in the strict interpretation of ground truth and thus calibration-level "ground truthing" via conventional scatter diagram analysis is nearly impossible. As a result, alternate methods for verifying or calibrating satellite algorithms are welcomed.

The purpose of this research is to explore an alternate approach for satellite rainfall validation. This is done by algorithm-to-algorithm intercomparison in the context of physical hypothesis testing. The algorithms in question are four level-2 Tropical Rainfall Measuring Mission (TRMM) facility rain profile algorithms.

On 28 November 1997, the TRMM satellite was launched from the Tanegashima launch range in Japan. The TRMM program is a joint venture between the U.S. National Aeronautics and Space Administration (NASA) and Japan's National Space Development Agency (NASDA; see Simpson et al. 1988, 1996). The TRMM satellite is the first earth-orbiting mission dedicated to studying precipitation and latent heating. As its name implies, the TRMM mission has been designed to measure precipitation over tropical latitudes, actually the latitude zone extending from ~35°S to 35°N (TRMM carries scanning instruments in a 35° inclined prograde orbit). The mission is also designed to obtain the horizontal, vertical, and temporal distribution of latent heating over the Tropics. The tropical regions are important because more than two-thirds of global precipitation occurs there, while diabatic heating associated with precipitation processes strongly controls the large-scale general circulation, tropical waves, and seasonally/intraseasonally modulated weather disturbances.

In the Tropics, there is better correlation between high and cold cloud tops and rain rates (Arkin and Meisner 1987) than at higher latitudes where baroclinically driven circulations and complex ice microphysics predominate; plus there has been greater progress in the microphysical modeling of convective cloud systems (Tao et al. 1993a,b). TRMM's non-sun-synchronous inclined orbit covers the Tropics (as well as subtropics) and permits detection and sensing of daily rainfall variations since any given area is sampled approximately 25–28 times per month at different solar times. Robust sampling is essential in keeping the sampling error low (Shin and North 1988; Bell et al. 1990). Another feature of the TRMM orbit is its relatively low 350-km altitude, which enables increased resolution for the diffraction-limited TRMM Microwave Imager (TMI), one of the three main rain-measuring instruments on the satellite.

The TRMM satellite's instrument package is made up of three primary instruments for rainfall measurement and two additional instruments for lightning detection and cloud-radiation budget measurement (Kummerow et al. 1998). The primary precipitation instruments are the TMI, the Precipitation Radar (PR), and the Visible and Infrared Scanner (VIRS). The additional instruments are the Lightning Imaging System (LIS) and the Clouds and Earth's Radiant Energy System (CERES). The focus of this study is on physical algorithms producing rain profiles using the TMI and PR instruments, and their combination.

The TMI is a conically scanning nine-channel microwave radiometer measuring five frequencies, 10.7, 19.4, 21.3, 37, and 85.5 GHz, with horizontal–vertical polarization diversity at all but 21.3 GHz (a water vapor channel that requires only single polarization—V-pol on TMI). The TMI measures passive microwave radiation emitted by the earth's surface, water droplets and ice particles in the atmosphere, and the combined effects of gaseous O_2 and H_2O emission, but only after the emitted photons undergo natural scattering and attenuation processes within and throughout the surface and atmospheric media. Such measurements can be used to estimate rain rates because there are physically sound radiative transfer relationships associated with the microphysical properties of rain and the upwelling passive microwave radiation leaving the atmosphere. The different frequencies are important for obtaining information at different vertical levels of the atmosphere, associated with the distinct channel-weighting function properties in the presence of rain. The TMI has a swath of 760 km, the widest scan of the three primary TRMM rain instruments, making it the widest coverage instrument for measuring rain.

The PR instrument is the first-ever spaceborne rain radar. It is based on phased-array slotted waveguide technology, transmitting at a frequency of 13.8 GHz with a swath width of 220 km. The PR measures the attenuated reflectivity of the transmitted signal to rain hydrometeor targets, to the earth's surface, and to the ocean's mirror image. This reflectivity (Z) information can be used to construct three-dimensional renditions of rain rate (R), particularly when the radar is well calibrated and there are representative Z–R relationships on hand. The PR is also important for its ability to measure precipitation over land, where passive microwave imagers traditionally have difficulty because of poor contrast between the generally radiometrically warm precipitation signal and the highly emissive land surface.

Having multiple rain-sensitive instruments on the same satellite platform leads to what may be TRMM's greatest advantage, the possibility for employing combined instrument algorithms. Until TRMM, radar and radiometer techniques have been largely confined to single-instrument methodologies. Using algorithms that combine measurements from more than one instrument sensitive to different microphysical properties allows compensation for some of the individual weaknesses of the single-instrument algorithms, theoretically resulting in more accurate measurements of rainfall.

The organization of this paper is to first describe the four level-2 TRMM facility profile algorithms being intercompared (section 2), then discuss how the intercomparison datasets were created (section 3). The paper then describes the methodology used to compare the algorithms to each other in the context of the underlying hypotheses (section 4), followed by an explanation of the results of the intercomparisons and their interpretation in the context of hypothesis testing (section 5), and finally, conclusions in section 6. The overall finding is that the four level-2 TRMM facility profile algorithms are performing in a manner consistent with the physical limitations of the measurements and the physical assumptions underlying the algorithm designs. This result provides an independent validity check on level-2 TRMM profile algorithms and greater assurance that modeling and diagnostic studies of the earth's climate and general circulation will advance.

2. Description of TRMM facility algorithms

The main dataset used for this study comes from the four level-2 TRMM facility profile algorithms as described in Kummerow et al. (2000). Each algorithm produces vertically distributed rain profiles. Two of the algorithms are single-instrument algorithms, while the two others are combined algorithms.

The nomenclature used to identify the algorithms is the same used by the TRMM science teams and described in Simpson and Kummerow (1996). The nomenclature divides the algorithms into levels and categories according to the type of product generated. This study uses four TRMM facility types of products: products 1b, 2a, 2b, and 2x. Level-1–type-b algorithms produce calibrated radiation quantities at full resolution in earth-located orbit swath format. Level-2–type-a algorithms produce geophysical parameters (e.g., rain rates) at full resolution in earth-located orbit swath format. Level-2–type-b algorithms produce geophysical parameters with some type of averaging process applied, also in earth-located orbit swath format. Finally, level-2–type-x algorithms are advanced experimental versions of 2b algorithms, being readied for replacement of official 2b versions.

a. TMI brightness temperatures (1b11)

The output of the 1b11 algorithm consists of brightness temperatures (T_Bs) for all channels of TMI. This algorithm is not part of the intercomparison per se, but its data are used in stratifying the results from the four level-2 rain-rate algorithms. Specifically, T_Bs from the 10.7- and 85-GHz vertical channels are used in the study.

b. TMI profile algorithm (2a12)

The TMI profile algorithm (2a12) is a single-instrument algorithm. The methodology used in this algorithm is based upon the work of Kummerow et al. (1996), where many realizations of the Goddard Cumulus Ensemble Model (Tao and Simpson 1993) and the University of Wisconsin Nonhydrostatic Modeling System (UWNMS; Tripoli 1992a,b; Panegrossi et al. 1998) are used to establish a set of probability density functions of distinct rainfall profiles. Upwelling T_Bs are obtained from the cloud model profiles by using a forward three-dimensional radiative transfer model. The modeled T_Bs are then compared to the measured T_Bs to determine a metric of agreement. By weighting profiles from the a priori probability density function according to their deviation from the observed T_Bs (Bayesian approach), iterative radiative transfer calculations are avoided, making this technique computationally fast compared to traditional iterative inverse solutions. This technique does not work well when the database of possible cloud structures is not sufficiently populated; then, an iterative optimization-based technique developed by Smith et al. (1994a,b, 1995a) is more appropriate. However, the current TRMM cloud model database is assumed to be sufficiently large to negate the need for iterative methods.

One of the 2a12 algorithm's strengths is that it is an attenuation-type solution, that is, the relationship between radiance (or T_B) and drop size goes according to D^4 instead of a radar-backscatter reflectivity's D^6-type dependence (D representing the diameter of idealized spherical drop). This reduces the sensitivity of the retrieved rain rates to indeterminate vagaries in the drop size distribution (DSD) function. It also has four polarized frequencies, which provide eight attenuation constraints, as well as a ninth constraint provided by the 21.3-GHz V-pol measurement. Another strength is the presence of the 10.7-GHz channels (~2.8 cm), which provide total liquid water path (LWP) estimates since most rain drops are Rayleigh with respect to X-band wavelengths. One of 2a12's biggest weaknesses is its broad weighting functions, which limit vertical resolution. Another potential weakness is 2a12's dependence on the cloud models' renditions of vertical hydrometeor structure. This algorithm also has the lowest spatial resolution, which leads to greater sensitivity to heterogeneous beam filling. Moreover, any purely passive microwave-based algorithm has difficulty obtaining measurements over land because of the aforementioned contrast problem. Finally, the 2a12 algorithm is the most affected by upper cloud ice loading, particularly at higher frequencies (37 and 85 GHz), while the second-lowest frequency 19-GHz measurements are subject to blackbody saturation at higher rain rates.

c. PR profile algorithm (2a25)

The second single-instrument algorithm is 2a25, the PR profile algorithm, a hybrid method similar to the scheme described in Iguchi and Menéghini (1994). This method uses the radar reflectivity (Z) vector to produce a vertical profile of rain. However, in order to get accurate results, it is essential that the path-integrated attenuation (PIA) is known. To do this, a Hitschfeld–Bordan (1954) solution is used in a first step to accumulate attenuation in a top–down sequence, followed by application of the surface reference technique (SRT) to provide correction guidance across the retrieved vertical rain-rate vector according to the difference between the Hitschfeld–Bordan PIA and the SRT-derived PIA [see Menéghini et al. (1983) for a discussion of the SRT method]. The Hitschfeld–Bordan solution requires a consistent pair of Z–R and κ–R functions, that is, $Z = aR^b$ and $\kappa = \alpha R^\beta$, where Z is radar reflectivity factor, κ is attenuation, R is rain rate, and a, b, α, and β are constants. For the SRT, the surface return from a rain-filled path is compared to that from a climatological rain-free path created by compositing all prior cloud-free pixels from the start of the mission up to the present into a background climatology.

The strengths of 2a25 are its use of range-gated returns, which results in highly resolved vertical structure, and its overall sensitivity to rain detection (~17 dBZ). The greatest weakness in using the SRT scheme is the assumption that the differences between two surface returns are caused only by atmospheric path attenuation, and not dissimilar surface boundary conditions. In reality, surface reflectivity depends on the roughness of the surface under control of wind and internal waves over ocean, and seasonally varying biophysical processes over land. The greatest weakness in the measurement itself is that it is a backscatter quantity, proportional to the sixth power of raindrop diameter, and thus sensitive to variant properties of the DSD. Two other potential measurement weaknesses intrinsic to the 2a25 algorithm are its sensitivity to the accuracy of the PR calibration, which may contain residual systematic errors, and severe attenuation at high rain rates, which is not accurately retrieved either from integrating the Hitschfeld–Bordan scheme downward or from application of the SRT scheme.

d. Day-1 combined profile algorithm (2b31)

Algorithm 2b31 is the first of the two TRMM facility combined algorithms and has been described in Smith et al. (1995b) and Haddad et al. (1997). This algorithm is derived from algorithm 2a25, in which the SRT is used to constrain the Hitschfeld–Bordan solution. However, the 2b31 algorithm is implemented in Bayesian form and uses the vertical–horizontal 10.7-GHz T_Bs from the TMI to estimate 13.8-GHz PIA, thus providing a second PIA constraint beyond the SRT constraint. Radiative transfer calculations demonstrate the viability of using 10.7-GHz T_Bs to determine 13.8-GHz PIAs (Smith et al. 1997). Although this algorithm uses measurements from two instruments, and thus is a genuine combined algorithm, the 10.7-GHz-generated 13.8-GHz PIA is used only as a solution constraint provided by a Bayesian equivalent of 2a25 and is thus referred to as a "conservative" combined algorithm.

This algorithm has the same strengths of 2a25, plus an improved PIA constraint derived from the 10.7-GHz T_Bs. Its weaknesses are residual calibration offset and some D^6 sensitivity inherent to the use of backscatter measurements.

e. Tall vector combined profile algorithm (2x31)

The fourth algorithm used in the intercomparison is 2x31, which is a fully combined algorithm referred to as a "tall vector" algorithm. As described in Farrar (1997), a tall vector algorithm is an extension of a passive microwave–based inversion scheme based on cloud modeling. For example, the Smith et al. (1994a) Special Sensor Microwave Imager (SSM/I) algorithm, which used hydrometeor profiles from the UWNMS cloud model as input into a forward radiative transfer model to produce T_B profiles matched to measured T_B profiles, was the kernel algorithm for the tall vector algorithm described in Farrar (1997). The difference for 2x31 is that the TMI measurements are augmented with reflectivities from the PR at range gates throughout the rain column above and below where the weighting functions of the different radiometer channels peak. This concatenation of radar reflectivities with TMI T_Bs creates what is known as a tall vector in an inversion framework, and by giving the radar and passive radiometer measurements equal leverage in an inversion solution, it is theoretically anticipated that this algorithm minimizes the rain-rate retrieval uncertainty. In its TRMM implementation, the 2x31 algorithm has similarities to 2b31 in that it is formulated in Bayesian form, negating the need for computationally intensive radiative transfer iterations (Z. Haddad 1999, personal communication).

This algorithm shares the strengths of both 2a12 and 2a25 with fewer weaknesses. The weaknesses include sensitivity to any residual calibration offset in the PR and physical inconsistencies between the passive radiative transfer equation (RTE) model and the active (radiatively pulsed) RTE model. However, the combined RTE model of Farrar (1997) eliminates the latter problem. It would be anticipated that 2x31 retains some uncertainty due to the PR reflectivity's D^6 sensitivity, but not to the same degree as 2a25 and 2b31 because the attenuation relationships for the entire set of TMI frequencies partially mitigate that sensitivity.

f. Algorithm strengths and weaknesses

The broad weighting functions of 2a12 limit the vertical resolution of this algorithm, while 2a25 benefits from the PR's range-gated estimates, providing narrower "effective" weighting functions and much improved vertical resolution, a strength also shared by 2b31 and 2x31. Algorithm 2a12 also has the lowest spatial resolution, which makes it more susceptible to heteroge-

neous beam filling effects than other algorithms. Moreover, the higher vertical and horizontal resolutions provided by the PR give the 2a25, 2b31, and 2x31 algorithms better rain detection sensitivity than the 2a12 algorithm.

On the other hand, the PR measurements are burdened by D^6 sensitivity in association with variant DSD fluctuations, whereas the TMI's attenuation-type measurements contain only D^4 sensitivity. Since 2a25 and 2b31 are most dependent on the PR measurements, this increased sensitivity to the DSD is expected to cause greater rain-rate uncertainties than with 2a12 and 2x31 at the higher rain intensities. Note that 2x31 is also dependent on PR measurements, but not to the degree of 2a25 and 2b31. Therefore, it is expected to see 2x31 rain-rate uncertainties increase at larger rain rates beyond those of 2a12, but not as great as those of 2a25 or 2b31. Higher rain intensities will also cause saturation at TMI frequencies from 19 GHz on up. This will temporize the constraints offered by those frequencies on the 2a12 and 2x31 algorithms. Since this means 2x31 will become more dependent on the TMI Rayleigh frequency (10.7 GHz), similar in a fashion to algorithm 2b31, 2x31 rain estimates are expected to agree more with the 2b31 estimates at the highest rain rates.

Another problem at higher rain rates is attenuation of the PR signal. Radar rain estimates at attenuating frequencies are very dependent on an accurate estimation of PIA. The 2a25 relies ultimately on the SRT-derived attenuation constraint that is not as precise as 2b31's attenuation constraint derived from TMI 10.7-GHz T_Bs (Smith et al. 1997), meaning that 2a25's rain rates are expected to contain a further source of uncertainty during intense rain. Finally, 2a12 is more dependent on TMI's higher frequencies than the two combined algorithms, which means 2a12 is most sensitive to upper cloud ice loading.

Being able to identify the strengths and weaknesses of physical algorithms in an a priori fashion, as discussed above, is the cornerstone of this study. The strengths and weaknesses are expected to bring about algorithms performing in predictable patterns in relationship to other physical algorithms. If hypotheses can be drawn and tests developed that underscore these patterns, then conclusions can be drawn about a given algorithm's performance with respect to its expected performance. Table 19.1 summarizes the strengths and weakness of the four level-2 profile algorithms as the basis for postulating a set of well-founded hypotheses.

3. Discussion of datasets

In order to compare the algorithms, level-2 retrievals from a number of TRMM overpasses were partitioned into small sectors within major precipitation systems. This section discusses how these data subsets were created. There are three different sizes of datasets used for this analysis. The terms used to describe the three sizes

TABLE 19.1. Known strengths and weaknesses of TRMM facility rain profile algorithms.

Algorithms	Strengths	Weaknesses
2a12	• Attenuation-type measurement (scattering signal in forward diffraction peak) • Four polarized frequencies provide eight attenuation constraints • 10.7-Ghz channel gives total LWP with little D^6 sensitivity	• Broad weighting functions • Ice loading decreases sensitivity to liquid rain • Retrieval of vertical structure tied to cloud model • Saturation at frequencies $\geq$ 19 Ghz • Heterogeneous beam filling
2a25	• Narrow weighting functions (range-gated) • Rain detection sensitivity	• Residual calibration offset • D^6 sensitivity • SRT-derived attenuation constraint • Severe attenuation for large rain rates
2b31	• All advantages of 2a25 • PIA constraint derived from 10.7-Ghz T_B	• Residual calibration offset • D^6 sensitivity
2x31	• All advantages of 2a12 and 2a25	• Residual calibration offset • Physical inconsistencies between passive (steady state) RTE model and active (radiatively pulsed) RTE model • Some D^6 sensitivity • Loss of constraints due to saturation of higher TMI frequencies

are orbit swath, storm region, and storm feature. An orbit swath contains retrievals from one algorithm for one full overpass of the TRMM satellite. A storm region is a subset of an orbit swath. It is made up of all data in a rectangular "box" containing one specific storm of interest. (Note that a box is defined by two lat–long pairs, i.e., minimum and maximum lat–long limits, and contains all orbit swath data from pixels within the limits. Also note that storm regions are intrinsically variable in size.) A storm feature is a subset of a storm region. It is created by all data in a square box containing only a specific storm feature. All storm features are of approximately the same size, and their averaged rain rates make up data points used in much of the intercomparison analysis.

a. Creating datasets

The first step in obtaining the storm data consisted of viewing browse imagery made up of multiple orbit swaths, showing the entire coverage of the TRMM satellite for one full day. When a storm of interest was located, the date and time for the storm were noted with this information used to access specific orbit swath files containing results from the four algorithms being compared. The orbit swath data were first visually quality checked, then minimum–maximum latitude–longitude pairs were selected to contain the individual storm regions and features.

Once the latitude–longitude coordinates for each storm region were assigned, they were used to create five files for each storm of interest, one for each of the level-2 algorithms being compared, and one for the TMI T_B data (1b11). Finally, the same process was used to create storm feature files. Six files for each storm feature were produced, one file for each of the four level-2

algorithms, one for the vertical 10.7-GHz T_Bs from 1b11, and one for the vertical 85-GHz T_Bs from 1b11.

b. Discussion of storm features

The final storm features were all boxes of 0.3° latitude by 0.3° longitude, or approximately 30 × 30 km. The features were made as small as possible to increase resolution but kept large enough to exceed the spatial resolution of the 10.7-GHz measurements. [The instantaneous fields-of-view (FOVs) of the 10.7-GHz TMI channels are ~36-km cross track (CT) × 60-km down track (DT), with effective FOVs of ~9-km CT × 60-km DT.] Only ocean storms were selected, to avoid complex uncertainty problems associated with measuring rainfall over land using passive microwave techniques. From all ocean regions covered by the TRMM satellite during the period 8–31 October 1998, 49 total storms were selected: 12 tropical storm/hurricanes, 14 intertropical convergence zone/South Pacific convergence zone (ITCZ/SPCZ) segments, and 22 extratropical cyclones/fronts (see Tables 19.2, 19.3). When categorizing storms into individual features, care was taken to select areas covering the full range of rain intensity, from lightest to heaviest rain rate. The final number of features was 398: 101 from tropical storms, 119 from large-scale convergence zones, and 178 from extratropical storms.

A frequency distribution of the storm feature rain rates, based on composite rain rates from the four algorithms, is shown in Fig. 19.1. (Note that composite rain rates are obtained by taking the mean value of four separate algorithm results.) The rain-rate histogram peaks at 4 mm h^{-1} with a population of 54. At 10 mm h^{-1}, the populations are approaching 10, and by 15 mm h^{-1}, frequencies are at or below 3.

TABLE 19.2. List of all storms by storm number, type of storm (name given if storm was named), date, and time.

Storm no.	Storm type	Date (1998)	Time (UTC)
1998_281_1352	Hurricane Lisa	8 Oct	1445
1998_281_1524	Cold front	8 Oct	1650
1998_283_1303	Cold front	10 Oct	1350
1998_284_0113	SPCZ	11 Oct	0235
1998_284_1758	Hurricane Zeb	11 Oct	1830
1998_285_0306	Hurricane Zeb	12 Oct	0410
1998_285_1043	ITCZ	12 Oct	1110
1998_288_0540	Tropical Storm 05A	15 Oct	0642
1988_288_0712	ITCZ	15 Oct	0740
1998_288_1317	ITCZ	15 Oct	1420
1998_290_0623	Hurricane Lester	17 Oct	0650
1998_290_0700	Extratropical cyclone	17 Oct	0700
1998_290_0754	Hurricane Madeline	17 Oct	0825
1998_290_0925	ITCZ	17 Oct	0950
1998_290_1050	Extratropical cyclone	17 Oct	1015
1998_291_0512	Cold front	18 Oct	0545
1998_291_1420	Hurricane Babs	18 Oct	1445
1998_292_0000	Hurricane Babs	19 Oct	0030
1998_292_0533	Cold front	19 Oct	0615
1998_292_0600	Hurricane Lester	19 Oct	0600
1998_292_1310	ITCZ	19 Oct	1335
1998_292_1350	Cold front	19 Oct	1350
1998_292_1610	ITCZ	19 Oct	1630
1998_292_1612	Cold front	19 Oct	1650
1998_293_0554	ITCZ	20 Oct	0620
1998_293_1331	Hurricane Babs	20 Oct	1355
1998_293_1634	ITCZ	20 Oct	1735
1998_293_1650	ITCZ	20 Oct	1650
1998_294_1524	Cold front	21 Oct	1605
1998_294_1545	ITCZ	21 Oct	1540
1998_295_0808	ITCZ	22 Oct	0910
1998_295_0940	Cold front	22 Oct	0933
1998_296_0829	Cold front	23 Oct	0925
1998_297_0245	Cold front	24 Oct	0320
1998_297_0547	Cold front	24 Oct	0635
1998_297_0710	Cold front	24 Oct	0710
1998_297_1153	Cold front	24 Oct	1235
1998_297_1324	Cold front	24 Oct	1405
1998_298_1951	Hurricane Babs	25 Oct	2050
1998_299_0156	Extratropical cyclone	26 Oct	0235
1998_300_0046	Cold front	27 Oct	0123
1998_300_0100	Hurricane Mitch	27 Oct	0115
1998_300_1125	Extratropical cyclone	27 Oct	1250
1998_301_0409	Extratropical cyclone	28 Oct	0500
1998_301_1015	Extratropical cyclone	28 Oct	1135
1998_303_0925	Extratropical cyclone	30 Oct	1010
1998_303_1000	ITCZ	30 Oct	0940
1998_304_0512	ITCZ	31 Oct	0615
1998_304_0644	Extratropical cyclone	31 Oct	0720

4. Methodology

Any new rain-measuring algorithm or measurement device is expected to be tested against conventional ground measurements; thus, research along these lines is being done for TRMM as well. However, this study adopts a different philosophy motivated by the fact that long-standing uncertainties in ground verification data translate directly to questions about the validity of the standard verification process. All four of the algorithms being compared are physically based. After designing hypotheses concerning how individual algorithms should perform based on the physical nature of the measurements and the apropos physical assumptions used in the algorithms, we test the hypotheses to see how well the algorithms actually perform against one another in the context of their anticipated performance. Seven hypotheses have been developed and are explained below, along with procedures used to test them. Only pixels with rain rates greater than or equal to 0.01 mm h^{-1} were used in testing the hypotheses, except for hypothesis 3. This was done to eliminate the uncertainty associated with the detection of very low rain rates.

TABLE 19.3. List of all storm numbers, TRMM swath numbers, and locations given by max and min lat and lon.

Storm no.	Swath no.	Max lat	Min lat	Max lon	Min lon
1998_281_1352	981008.4963	31.1°N	26.2°N	33.6°W	42.5°W
1998_281_1524	981108.4964	32.6°S	36.1°S	66.5°E	52.2°E
1998_283_1303	981010.4994	36.0°N	33.8°N	73.0°W	68.1°W
1998_284_0113	981011.5002	21.4°S	26.1°S	140.9°W	147.3°W
1998_284_1758	981011.5013	15.2°N	10.6°N	130.1°E	135.3°E
1998_285_0306	981012.5019	12.9°N	5.9°N	135.0°E	126.6°E
1998_285_1043	981012.5024	13.8°N	7.4°N	123.3°E	131.3°W
1998_288_0540	981015.5068	17.3°N	12.9°N	62.8°E	56.9°E
1998_288_0712	981015.5069	14.5°N	7.0°N	91.7°W	101.4°W
1998_288_1317	981015.5073	12.0°N	3.3°N	40.3°W	51.8°W
1998_290_0623	981017.5100	15.8°N	10.3°N	92.9°W	100.0°W
1998_290_0700	981017.5100	33.5°N	29.0°N	53.3°W	62.8°W
1998_290_0754	981017.5101	22.2°N	18.2°N	105.6°W	110.5°W
1998_290_0925	981017.5102	11.5°N	7.0°N	145.7°W	150.9°W
1998_290_1050	981017.5102	34.7°N	31.5°N	52.3°W	59.2°W
1998_291_0512	981018.5115	32.4°N	26.3°N	47.5°W	60.6°W
1998_291_1420	981018.5121	12.7°N	7.4°N	132.7°E	126.8°E
1998_292_0000	981019.5127	12.6°N	6.0°N	136.3°E	127.9°E
1998_292_0533	981019.5131	36.1°N	34.0°N	35.7°W	40.0°W
1998_292_0600	981019.5131	15.1°N	10.5°N	97.0°W	102.2°W
1998_292_1310	981019.5136	13.3°N	7.2°N	143.8°E	136.6°E
1998_292_1350	981019.5136	35.8°N	33.2°N	157.2°W	168.9°W
1998_292_1610	981019.5138	0.9°S	7.1°S	78.0°E	71.0°E
1998_292_1612	981019.5138	34.7°N	31.4°N	146.7°E	139.6°E
1998_293_0554	981020.5147	14.4°N	8.3°N	111.1°W	118.5°W
1998_293_1331	981020.5152	14.3°N	8.5°N	132.2°E	125.1°E
1998_293_1634	981020.5154	12.3°N	5.8°N	132.7°W	141.6°W
1998_293_1650	981020.5154	1.5°N	8.9°S	68.1°E	55.4°E
1998_294_1524	981021.5169	35.1°N	32.4°N	146.0°E	141.2°E
1998_294_1545	981021.5169	1.1°N	7.5°S	77.7°E	67.6°E
1998_295_0808	981022.5180	8.8°N	3.4°N	16.2°W	23.0°W
1998_295_0940	981022.5180	31.7°S	34.8°S	55.5°E	49.9°E
1998_296_0829	981023.5196	27.8°N	24.6°N	64.1°W	67.6°W
1998_297_0245	981024.5208	32.5°N	28.9°N	56.2°W	62.2°W
1998_297_0547	981024.5210	34.7°N	32.0°N	53.5°W	58.2°W
1998_297_0710	981024.5210	28.9°S	32.3°S	64.4°E	59.7°E
1998_297_1153	981024.5214	36.1°N	33.9°N	169.8°W	176.3°W
1998_297_1324	981024.5215	34.7°N	31.1°N	151.2°E	142.7°E
1998_298_1951	981025.5235	22.8°N	18.3°N	118.9°E	113.0°E
1998_299_0156	981026.5239	34.2°N	29.7°N	51.6°W	62.6°W
1998_300_0046	981027.5254	32.5°N	29.5°N	49.0°W	52.9°W
1998_300_0100	981027.5254	20.9°N	15.0°N	76.1°W	84.8°W
1998_300_1125	981027.5261	30.2°S	33.6°S	39.4°W	45.4°W
1998_301_0409	981028.5272	31.3°N	27.2°N	42.9°W	49.9°W
1998_301_1015	981028.5276	29.0°S	32.4°S	34.4°W	39.7°W
1998_303_0925	981030.5307	36.1°N	33.8°N	179.8°E	172.7°E
1998_303_1000	981030.5307	6.4°S	11.8°S	89.2°E	83.1°E
1998_304_0512	981031.5320	9.3°N	1.4°N	38.7°W	48.9°W
1998_304_0644	981031.5321	35.1°N	29.3°N	158.7°W	176.0°W

a. Hypothesis 1: Effect of ice loading

The first hypothesis is based on the fact that the 2a12 algorithm solution uses measurements at the higher TMI frequencies, which are in turn most influenced by ice loading. Because of this, surface rain-rate differences between 2a12 and the other PR-based algorithms are expected to increase as ice loading increases. To test the hypothesis, the averaged vertical 85-GHz T_Bs (85 GHz being the most sensitive TMI frequency with respect to ice loading) and the averaged surface rain rates from each algorithm for each individual storm feature were compared by graphing the rain rates against the T_Bs (Fig.

19.2). Because of this, as the T_Bs decrease, the 2a12 surface rain rates should diverge from the rain-rate graphs of the other algorithms.

Before graphing, the data were averaged in 10°C bins with bin values assigned to the middles of intervals. For example, the first interval is from 280 to 270 K with an assigned bin value of 275 K. Also, the intervals were overlapped to provide smoothing. For example, the second interval is from 275 to 285 K with the bin value assigned to 280 K. Besides smoothing, data averaging ensures that the data points are spaced evenly and that there is only one value assigned to each bin.

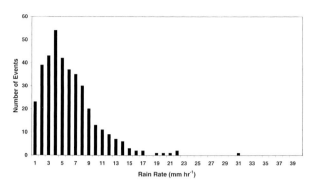

FIG. 19.1. Frequency distribution of feature-averaged composite rain rates.

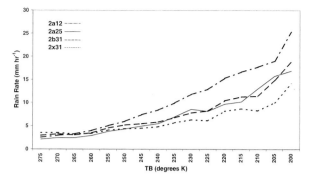

FIG. 19.2. Algorithm-averaged rain rates vs averaged vertical 85-GHz brightness temperatures.

b. Hypothesis 2: Radar D⁶ effect

The amount of power backscattered to a radar by a radar beam intersecting with rain particles is given as $P = (CD^6)/r^2$, where P is power, D is diameter of particles summed over a unit volume, r is range, and C is a constant derived from various measurable characteristics of the radar and of the dielectric constant of the rain particles (Harrold et al. 1973). This demonstrates why the received power at a radar is dependent upon the size of the raindrops, varying by D^6. This is known as the D^6 effect. Because of the D^6 effect, radar measurements are affected more by drop size than passive microwave radiometer measurements. Recall that the latter is an attenuation-type measurement proportional to D^4 and thus less sensitive to indeterminate fluctuations in the underlying DSD functions. Therefore, as storm rain rates increase, signifying larger raindrop sizes, the algorithms most dependent on the PR, that is, 2a25 and 2b31, should exhibit the largest deviations in rain rate relative to the others. Since the 2x31 algorithm is also dependent on PR input, it should also be affected by the D^6 effect and deviate from the remaining algorithms (i.e., specifically 2a12) at higher rain rates. However, 2x31 should not deviate as much as 2a25 or 2b31 (particularly at intermediate rain rates), because it has a greater degree of attenuation constraint than 2b31, as offered by the tall vector approach.

To test this hypothesis, the averaged surface rain rates from each algorithm for each storm feature were obtained, as were the composite surface rain rates for each storm feature. The results were then placed in order of their composite surface rain-rate value and binned using the same procedure described in section 19.4.1, but at 2-mm-h^{-1} intervals with the overlap between intervals set at 1 mm h^{-1}. The results were then graphed with the algorithm-averaged rain rates being a function of the composite rain rates (diagram not shown). As the composite rain rates increased, the graphs of the 2a25 and 2b31 algorithms were expected to deviate from the graphs of the other two algorithms due to the increased number and size distribution dispersion of raindrops. However, the results of this test were inconclusive be-

cause too few samples of high-intensity rain rates were used, which is most important for a satisfactory resolution of this test.

In order to get a conclusive result for hypothesis 2, the data population of the high-intensity rain rates had to be increased. To do so, individual pixel values were used. Using the pixel data in the storm features' files, the first step was to obtain 2a12 surface rain rates pixel by pixel. (The 2a12 algorithm was used as a starting point because it has the lowest resolution and, thus, fewer pixels per unit area.) If a surface rain rate was greater than 5 mm h^{-1}, the nearest 2a25 pixel was located. Once the closest 2a25 pixel was found, it was tested to ensure that the combined difference between the latitudes and longitudes of the two pixels were within 0.03°. If the pixel passed the test, the same latitude–longitude pair was used to select the 2b31 and 2x31 pixels. After finding all pixels obeying the search criteria, the composite algorithm value was calculated. If a composite rain rate was above 8 mm h^{-1}, the pixel value was added to the averaged storm feature value distribution obtained earlier. This increased the number of values to over 1000, all at the higher rain rates where they were needed.

Once the larger dataset was obtained, the individual algorithm rain rates were again plotted against the composite rain rates (Fig. 19.3), but using a discrete bining procedure, not an overlapping one. From 0 to 16 mm

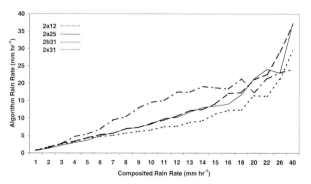

FIG. 19.3. Algorithm-averaged surface rain rates vs composite rain rates.

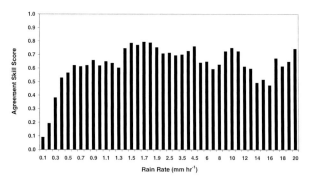

FIG. 19.4. Histogram of 2a12/2a25 rain detection agreement skill score binned by averaged rain rate.

h^{-1}, the interval was 1 mm h^{-1}; from 16 to 22 mm h^{-1}, 2 mm h^{-1}; and finally above that, the interval was changed to maintain an adequate population of data in a given interval.

c. Hypothesis 3: TMI and PR rain detection agreement

As rain rate increases, the detection agreement skill score between 2a12 and 2a25 would be expected to increase, since 2a12 suffers greatest from light rain detection capability. Since the 2a25 algorithm outputs approximately three times the number of pixels per feature as does the 2a12 algorithm, because of its higher resolution, the first step in testing this assumption was to make sure that pixels being tested were from the same location. The 2a12 pixel latitudes and longitudes were obtained first. Next, a 2a12 storm feature was searched for any 2a25 pixel that was within 0.01° latitude and 0.05° longitude. These nonidentical values were used because the TMI footprint is larger across the path of a swath than along its path (related to its conical scan pattern), while a PR footprint's dimensions are nearly equivalent in both directions except at extreme scan angles. This allowed for more exact latitude measurements. The 0.05° longitude difference allowed room for location uncertainty but still kept the 2a25 pixels close to the center of the 2a12 pixel.

Once the collocated pixels were selected, they were both tested for rain/no rain. The results were placed in contingency table bins, three bins for each 0.01 mm h^{-1} interval up to 20 mm h^{-1}. The three bins, B1, B2, and B3, represent yes–no, no–yes, and yes–yes. The bin no–no was not allowed because the pixel selection was made on the basis that one of the algorithms had to include rain. The agreement skill score for each rain rate was then calculated by dividing the number of times both algorithms detected rain, B3, by the total number of times either algorithm detected rain, that is, B3/(B1 + B2 + B3). The agreement skill scores were then graphed as a function of rain rate (Fig. 19.4).

d. Hypothesis 4: Vertical structure differences

The vertical structure differences between the 2a12 algorithm and the other algorithms are expected to be larger than the differences between the other algorithms and themselves, due to the TMI's lack of vertical resolution. The testing of this hypothesis begins by calculating the average rain rate from each algorithm for every level of each feature, as well as the composite surface rain rate for each feature. Considering the results of the distribution of the composite surface rain rates, three categories of rainfall with approximately the same population were established. The categories are light (0 mm h^{-1} < R ≤ 4 mm h^{-1}), medium (4 mm h^{-1} < R ≤ 9 mm h^{-1}), and heavy (9 mm h^{-1} < R). The feature-averaged rain rates are then grouped into the three rain categories and the mean rain rates for each algorithm for each height category are calculated. Finally, these height-averaged rain rates are graphed against the composite surface rain rate (diagram not shown).

The 2a12 algorithm would be expected to indicate the largest difference between itself and the other algorithms. However, for this first test the results are inconclusive. The 2a12 values are indeed larger at every level, but the version-4 2a12 rain-rate estimates are known to be approximately 20% larger on a systematic basis than the conservative level-3 reference algorithm 3a11, related to a deficiency in how the 10.7-GHz channels were employed (C. Kummerow 1999, personal communication). (This deficiency has been corrected in the version-5 2a12 estimates for which reprocessed rain rates were completed in mid-2000.) Otherwise, the diagram shows no significant indication of vertical structure differences besides magnitude. Therefore, a different test was needed for validating or invalidating the hypothesis.

In reexamining the hypothesis, the issue of how TMI's lack of vertical resolution affects 2a12 results was considered in more detail. The peak values of any TMI channel-weighting function extend over a much greater depth of atmosphere than does any individual PR range-gated return (see Mugnai et al. 1990). Also, the TMI has fewer channels than the PR has range gates. These two characteristics should lead to a relative vertical smoothing effect in the 2a12 retrievals, resulting in smaller differences between the vertical levels in the 2a12 algorithm than between vertical levels in the other algorithms using range-gated measurements.

To test the refined hypothesis, the algorithm-averaged rain rates for each level and each storm feature were calculated. Then the coefficient of variation for each algorithm's rain rate with height was calculated. (The coefficient of variation, which is standard deviation divided by sample mean, eliminates the effect of magnitude bias among the algorithms.) Once the coefficient of variation from each algorithm for each storm feature was obtained, a composite coefficient of variation was calculated using the mean of the four individual algo-

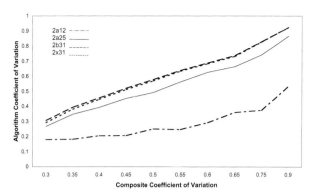

FIG. 19.5. Algorithm coefficient of variation along vertical axis vs composite coefficient of variation for first six algorithm reporting levels.

rithm coefficients of variation. Finally, the algorithm coefficients of variation with height were graphed as functions of the composite coefficient of variation (Fig. 19.5). According to the hypothesis, as the composite coefficient of variation increases, the 2a12 algorithm should exhibit the smallest increase.

e. Hypothesis 5: Attenuation effect

Because of liquid water attenuation, which influences radar estimates more than TMI estimates when 10.7-GHz information is used, the differences between the 2a25 algorithm and the other algorithms are expected to increase at a given rain-layer depth as the 10.7-GHz T_B increases (assuming heavy ice loading is not present). To test this hypothesis, storm features were first limited

to those having an 85-GHz T_B of greater than 240 K to ensure no heavy ice loading (the cutoff was determined from results given in Fig. 19.2). The averaged rain rates from each algorithm for each layer of the feature and the averaged vertical 10.7-GHz T_B for each feature were then calculated. Finally, graphs for each layer with the averaged rain rate being a function of the 10.7-GHz T_B were created (Fig. 19.6).

The 10.7-GHz T_Bs are proportional to the path-integrated liquid water content of a column because the 10.7-GHz frequency is taken as nearly Rayleigh for the characteristic DSD at even heavy rain rates. As the 10.7-GHz T_B increases and the distance the radar beam must travel increases (the lower the range gate, the farther the distance), then the differences between the 2a25 algorithm and the other algorithms would be expected to increase.

f. Hypothesis 6: Saturation of higher TMI frequencies

As rain rate increases, the higher TMI frequencies would be expected to saturate insofar as emission is concerned, making less contribution to the solution of the 2x31 algorithm. This would cause 2x31 to become mostly dependent on the 10.7-GHz frequency. Since 2b31 is only dependent on the 10.7-GHz measurement from TMI, the agreement between the rain-rate estimations of the two combined algorithms would be expected to increase as rain rate increases. To test this hypothesis, the averaged surface rain rates of the two algorithms were graphed as a function of a "restricted" composite surface rain rate as seen in Fig. 19.7. The

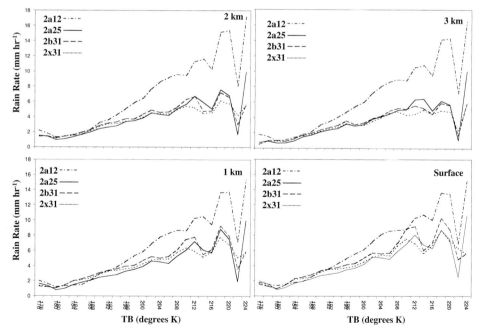

FIG. 19.6. Algorithm-averaged rain rates vs averaged 10.7-GHz brightness temperatures at four rain heights.

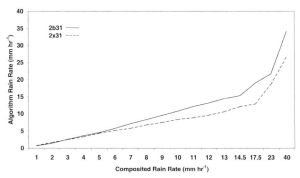

FIG. 19.7. Algorithm-averaged rain rates (2b31 and 2x31 only) vs composite rain rates.

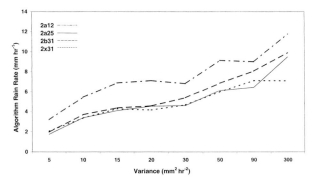

FIG. 19.8. Algorithm-averaged rain rates vs variance of 2a25 storm feature rain rates.

restricted composite was created by calculating the mean of just the 2b31-and 2x31-averaged surface rain rates. If hypothesis 6 is valid, the surface rain rates of the 2b31 and 2x31 algorithms will converge as the restricted composite surface rain rate increases. The difficulty here is that the effect of hypothesis 2 is also at work on the relative 2b31–2x31 behavior, thus requiring special attention (see section 19.5.6 below).

g. Hypothesis 7: Beam filling effects

Because the TMI is a diffraction-limited measurement, its various channels have broader beamwidths than the PR. Thus, the 2a12 algorithm is expected to be more sensitive to beam filling heterogeneities. Beam filling heterogeneity occurs when the atmospheric volume being sensed is occupied with nonuniform rain rates and possibly subpixel rain-free areas. In this instance, all the different rain rates within the volume of a pixel being sensed are blended together radiometrically, causing inaccuracies in the estimated rain rates, since T_B–rain-rate relationships are inherently nonlinear. Therefore, since the TMI has the greatest beamwidths, there is a greater chance that it will encounter nonuniform precipitation in its measurements.

The storm features used to test this hypothesis were limited to those with 85-GHz T_Bs greater than 240 K to ensure there was low ice loading, just as was done when testing hypothesis 5. The averaged surface rain rates from each algorithm for each feature were then calculated, as were the variances of the 2a25 algorithm surface rain rates across each feature. The data were then averaged, with the value being assigned to the largest whole number in a variance interval. From 0 to 20 mm^2 h^{-2} the interval was 5 mm^2 h^{-2}, while above that point the intervals were selected to maintain an adequate sampling in each interval. After averaging, the results were graphed with the averaged surface rain rates being a function of the 2a25 variance (Fig. 19.8). As variance increases, rain rates for the 2a12 algorithm are expected to diverge from those of the other algorithms.

5. Results and interpretation

a. Test of hypothesis 1

Hypothesis 1 states that the 2a12 algorithm would be most affected by ice loading, since it is the algorithm most dependent on the higher TMI frequencies, which are most sensitive to the volume extinction of ice. Since 85-GHz T_Bs represent the highest TMI frequency, they are necessarily the most sensitive measurements with respect to ice loading. Therefore, to test hypothesis 1, the averaged surface rain rate from each algorithm was graphed as a function of the averaged 85-GHz V-pol T_B as shown in Fig. 19.2.

At the higher 85-GHz T_Bs (low ice loading) all four algorithms show good agreement, with largest differences of about 1.6 mm h^{-1} at 250 K. Around 250 K is also where the four algorithms begin to separate. The 2a12 algorithm shows the largest increase, growing by 217% between 250 and 205 K. The 2x31 algorithm shows the smallest increase, growing by only 127% over the same T_B interval. The other two algorithms, 2a25 and 2b31, maintain good agreement between each other throughout the range of the graph, increasing by ~188%. Therefore, the 2a12 algorithm does show a greater change in rain rate with a decrease in 85 GHz T_B than any of the other three algorithms, confirming that hypothesis 1 is valid.

b. Test of hypothesis 2

Because of the D^6 effect in radar measurements, the algorithms most dependent on PR data will show the largest increases in rain-rate difference as storm intensity and concomitant rain rate increase. This was the concept expressed in hypothesis 2. To test the hypothesis, the algorithm-averaged surface rain rates were graphed as functions of composite surface rain rate (Fig. 19.3). There is not as much agreement evident in this diagram as there was in Fig. 19.2. The plot of the 2a12 algorithm separates from the others at 3 mm h^{-1} after only three data bins. The 2x31 plot begins to diverge from the other two algorithms three bins later at 6 mm h^{-1}. The 2a25 and 2b31 algorithms show good agreement throughout the rain-rate range. Between 3 and 14 mm h^{-1}, the 2a12 graph exhibits the largest increase, growing by 660%, whereas 2x31 exhibits the smallest increase at 268%, with 2a25 and 2b31 in between at a growth rate of 420%. However, the 14 mm h^{-1} data point marks the end of 2a12's rapid growth. After 14 mm h^{-1}, the 2a25 and 2b31 algorithms indicate the largest rate of change, growing by 295% between 14 mm h^{-1} near the end of the abscissa, while 2x31 indicates the next highest rate of change at 220%, and 2a12 indicates the smallest rate of change growing by only 26%.

The 2a12 algorithm indicates the largest growth rate for this test, but this occurs during the first half of the graph. Once into the higher rain rates, 2a12's rate of growth decreases dramatically and becomes the algorithm with the lowest rate of growth. Algorithms 2a25 and 2b31, the two algorithms most dependent on radar measurements, indicate the largest rate of change for the higher rain rates, while 2x31, which also uses radar measurements, indicates the next largest rate of change. This is consistent with the expectations for the D^6 effect; the algorithms most dependent on the radar (2a25 and 2b31) exhibit the largest growth rate increase at large rain rates, while the tall vector combined algorithm (2x31), which is less dependent on radar measurements, exhibits a solid increase but not quite as large as for the two former algorithms. Therefore, the results of this test indicate that hypothesis 2 is valid.

c. Test of hypothesis 3

Figure 19.4 is a histogram of the 2a12/2a25 rain detection agreement skill score as a function of averaged surface rain rate. According to hypothesis 3, the agreement skill score between the two algorithms should increase as the surface rain rate increases. The histogram shows that the agreement between the two algorithms at very low levels is indeed poor. For rain rates below 0.2 mm h^{-1}, the agreement is less than 20%. However, the agreement skill score climbs quickly. Once rain rates exceed 0.4 mm h^{-1}, the agreement skill score goes above 50%, where it stays along the remainder of the abscissa, occasionally reaching skill scores above 80%. There are a few skill scores dropping near 50% at rain rates of 14–16 mm h^{-1}, but these are believed to be due to limited sampling of intense rain rates, more so than a deterioration of agreement skill scores at the higher rain rates. In any case, and first noting that our interpretation of these results has to be treated cautiously because of inherent resolution differences between TMI-based and PR-based algorithms, the results bear out the validity of hypothesis 3.

d. Test of hypothesis 4

Hypothesis 4 states that the 2a12 algorithm will have larger vertical differences between it and the other al-

gorithms than the other algorithms will have between themselves. This is due to the lack of vertical resolution in the TMI measurements. The first attempt to validate this hypothesis was inconclusive as noted above. Another attempt at validating hypothesis 4 resulted in Fig. 19.5, a diagram of the algorithm coefficient of variation with height graphed as a function of the composite coefficient of variation with height. Since the 2a12 algorithm's vertical resolution is limited by the TMI's broad weighting functions and fewer measurements along the vertical axis, it is expected that this algorithm would exhibit smaller variability along the vertical axis, or correspondingly, a smaller coefficient of variation.

The Fig. 19.5 diagram indicates that 2a12 does indeed have the smallest coefficient of variation. The 2a12 plot increases by less than 0.2 for a 0.45 change in the composite value. The 2a25 plot is next with an increase of 0.47 across the same interval. Finally, the plots for 2b31 and 2x31 are nearly identical, indicating the largest increases at ~0.54, indicating that hypothesis 4 is valid.

e. Test of hypothesis 5

Hypothesis 5 states that 2a25 would differ more from the other algorithms as attenuation increases. The results of the test for this hypothesis are shown in Fig. 19.6, with diagrams of the algorithm rain rates for the first four vertical levels graphed as functions of 10.7-GHz T_B. At first glance the test seems to have failed, with 2a12 indicating the greatest differences. While this is true behavior, 2a12's differences are not caused by attenuation alone. The major difference between 2a12 and the other algorithms is the result of the overestimated systematic differences between the version-4 form of the 2a12 algorithm discussed above. As noted, the version-4 2a12 rain-rate estimates are about 20% greater than the version-4 estimates of the other algorithms.

If only the other three algorithms are considered, the results adhere to the hypothesis. At all vertical levels the algorithms are in agreement, but with a change in the 2a25 algorithm's position relative to the other algorithms as the level decreases and the 10.7-GHz T_B increases. At all four levels the graphs of the 2a25 algorithm indicate the lowest rain rates at the lowest T_Bs, but at upper levels the 2a25 graphs increase to the highest rain rates at the higher T_Bs. At level 4, the increase begins around 198 K while at level 3 it begins around 206 K. By the time the second level is reached, the 2a25 graph does not surpass the 2x31 graph until 210 K, and it surpasses the 2b31 graph only briefly between 213 and 215 K. At level 1, the 2a25 graph again surpasses the 2x31 graph at about 210 K, but sinks below it again at 216 K. They are nearly identical from 217 to 220 K, then 2a25 sinks below 2x31 again. The 2a25 graph also surpasses 2b31, but only from 213 to 215 K. As for the 2b31 and 2x31 graphs, they maintain the same relative position to each other for all four vertical levels.

In summary, the 2a25 graph has the same value relative to the other two algorithms at low T_Bs at all four vertical levels but transitions from a greater relative value to a lesser value for the higher T_Bs as the level decreases. Meanwhile, 2b31 and 2x31 maintain their values relative to each other. This means that 2a25 is most influenced by attenuation, indicating hypothesis 5 is valid.

While describing the algorithms in section 19.2, it was mentioned that 2b31 was derived from 2a25. If this is so, why is 2x31 not affected as much by attenuation as is algorithm 2a25? The answer is the difference in how each of the algorithms estimate PIA. The accuracy of a radar using attenuating frequencies is highly dependent on the accuracy of PIA estimates (Hitschfeld and Bordan 1954). Algorithm 2a25 uses the SRT as the final arbitrator for estimating PIA. This means that the PIA estimate is referenced to clear-sky conditions based on a cloud-free surface reflectivity (σ_o) climatology (i.e., cloud-free σ_os), not a σ_o surface capped by heavy rainfall, which may disturb the surface and diminish the reflectivity from a more mirror surface. By the same token, algorithm 2b31 uses 10.7-GHz T_Bs to estimate PIA over and above the SRT constraint (see Haddad et al. 1997). The result for 2b31 is an improved estimate of PIA and thus a better estimate of rain rate with less attenuation error.

f. Test of hypothesis 6

Hypothesis 6 states that saturation of the higher TMI frequencies at higher rain rates would lead to better agreement between the two combined algorithms, 2b31 and 2x31. Figure 19.7 illustrates the algorithm-averaged surface rain rates for 2b31 and 2x31, graphed as a function of restricted composite surface rain rate (calculated from 2b31 and 2x31 alone). This diagram does not indicate convergence of the 2b31 and 2x31 graphs at the highest rain rates. However, a comparison of the graph slopes is revealing. From 5 mm h^{-1} (where the graphs begin to diverge) to 14.5 mm h^{-1}, the slope of 2b31 is 40% larger than the slope of 2x31 (1.14 and 0.82, respectively). From 14.5 mm h^{-1}, where 2b31 begins its next slope change, to the end of the graph, the difference has dropped to 30%, with 2b31 indicating a slope of 0.74 and 2x31 a slope of 0.57. Finally, from 17.5 mm h^{-1}, where 2x31's slope changes to the end of the graph, the difference has dropped to only 10%, with 2b31's slope at 0.67 and 2x31's slope at 0.61. This reduction in the difference of slopes should be viewed as a partial success for the hypothesis 6.

The reason hypothesis 6 cannot be considered a complete success stems from the results of the hypothesis-2 test. As noted, the 2b31 algorithm is more dependent on PR measurements than is the 2x31 algorithm (or perhaps less influenced by the TMI measurements as an alternate way to describe the basic difference). In that context, 2b31 is more influenced by the radar's D^6 effect. While the effect of cloud blackbody saturation in hy-

pothesis 6 is expected to lead to better agreement between 2b31 and 2x31 at higher rain rates, the greater influence of the D^6 effect on 2b31 stemming from hypothesis 2 is reducing the expected agreement at higher rain rates under hypothesis 6. It is important to note that at this juncture there is no straightforward way to quantify the competing effects.

g. Test of hypothesis 7

Figure 19.8, a diagram of the algorithm-averaged surface rain rates graphed as a function of variance of the 2a25 surface rain rate for each storm feature, conveys the test of hypothesis 7. This states that 2a12 should be most sensitive to beam filling heterogeneity. The diagram indicates that the 2a12 algorithm graph systematically exceeds those of the other algorithms. It climbs rapidly from 5 to 10 mm^2 h^{-2}, levels off between 15 and 30 mm^2 h^{-2}, then continues climbing along the remainder of the abscissa. However, the fluctuations of the 2a25 graph are nearly identical to those of the 2a12 graph, the 2a25 graph being slightly flatter between 10 and 30 mm^2 h^{-2}.

Clearly, the magnitudes of the 2a25 graph are smaller than those for 2a12. The graphs for the 2b31 and 2x31 algorithms do not track the 2a12 graph as well as 2a25, but the same overall patterns emerge. Thus, the most salient feature of this diagram is the systematically larger rain rates for algorithm 2a12. But these differences are anticipated and are related to the 2a12 version-1 algorithm's systematic overestimation of ~20% as discussed previously. Therefore, putting aside the 2a12 differences in magnitude, this diagram does not exhibit any significant difference between any of the individual algorithm graphs. This means that the test results for hypothesis 7 must be considered inconclusive.

6. Conclusions

The purpose of this study was to seek verification of some existing methods of satellite precipitation estimation, as well as some newer combined methods that are possible for the first time, using radiometer and radar measurements from the recently launched TRMM satellite. The four level-2 TRMM facility rain profile algorithms under investigation, 2a12, 2a25, 2b31, and 2x31, are all physically based. Given the characteristics of the TMI and PR measurements and the governing assumptions used in designing the algorithms, seven well-founded hypotheses were formed concerning the expected performance of the algorithms relative to one another. After testing, it was found that 1) the 2a12 algorithm is the most influenced by ice loading, as hypothesized; 2) 2a25 and 2b31, the algorithms most dependent on the influence of PR measurements, are also most affected by the underlying radar D^6 effect (with 2x31 affected to a lesser degree), as hypothesized; 3) rain detection agreement skill scores between 2a12 and

2a25 increase as storm rain intensity increases, as hypothesized; 4) vertical structure differences are greater between 2a12 and the other algorithms than between the other algorithms and themselves, as hypothesized; and 5) 2a25 is the algorithm most affected by path-integrated attenuation, as hypothesized.

Testing of hypothesis 6 concerning TMI's blackbody saturation effects, led to partial success. Although the results did not indicate 2b31 and 2x31 fully converging at higher rain rates as postulated, they did show strong decreases in the difference between the slopes of the 2b31 and 2x31 rain-rate graphs. Since the saturation effect of hypothesis 6 is offset by 2b31's increased sensitivity to the D^6 effect from hypothesis 2, the decrease in the difference between the slopes of the rain-rate graphs can be viewed as a conditional success for hypothesis 6.

Finally, the test results for hypothesis 7, concerning 2a12 beam filling uncertainty, were inconclusive. While 2a12 exhibits the greatest differences between itself and the other algorithms, there is not sufficient evidence to determine whether the differences are caused by beam filling effects or by the 20% systematic overestimation known to exist in version-1 of the 2a12 rain retrievals.

Notwithstanding the results for hypothesis 7, the results summarized above suggest that the four level-2 TRMM facility rain profile algorithms are performing within the limitations of the TMI and PR measurements and according to the underlying physical assumptions used in the algorithm designs. Whereas some of the above interpretation of the hypothesis testing could be modified, our interpretation is consistent with the results of the tests as dictated by the hypotheses themselves, nothing more. It remains debatable whether our posed hypotheses are sufficiently complete and formulated precisely enough to fully defend the reliability of the TRMM level-2 facility algorithms.

In the area of precipitation retrieval, as in a number of other remote sensing areas, competition between algorithm producers runs high. Our emphasis here is to show that when the hierarchy of TRMM level-2 physical algorithms stemming from single instrument to combined instrument are run together with controlled inputs, expected performance metrics can be quantitatively tested. In this respect, the results of our study offer solid evidence that the TRMM rain estimates are reliable within the context of the physics employed.

There is much more to be done in fully evaluating the accuracy and precision of the TRMM algorithms. Likely, this can never be accomplished to the satisfaction of everyone using conventional ground truth schemes, and it remains a severe challenge to verify TRMM rain retrievals based on a precision "error model" that can serve as a calibration bench, simply because developing such an error model is so overwhelming a project. A true error model would have to be capable of 1) simulating near-exact three-dimensional, non-steady-state, and nonelastic radiative transport in complex and fluctuating gaseous–hydrometeor media (unconstrained by simplifying assumptions concerning hydrometeor shape, size distribution, orientation, and complex dielectric properties); 2) fully reproducing from first principles the emittance–reflectance properties of all varieties of natural earth surfaces over which rain falls; and 3) characterizing near-exact renditions of all realistic precipitating clouds with nonhydrostatic three-dimensional cloud models using fully explicit microphysical process models and capable of getting all the rest of the dynamics, thermodynamics, and hydrological processes of the atmosphere near perfect. That is a serious and possibly unachievable challenge. Nevertheless, the error model approach remains as the final arbitrator and final goal for satellite rain verification since TRMM scientists have explored the most readily available accuracy and precision metrics attainable with established in situ measuring systems, but without actually producing final quantitative conclusions concerning the TRMM algorithm validation problem.

Therefore, we view these results as a "path" between ground truth comparison and classic error modeling. The results also have a broader meaning. While the testing methods used in this study cannot be used alone to quantify TRMM precipitation uncertainty in the conventional terms of accuracy and precision, they offer an independent means for corroborating other verification analyses. This is important considering that direct comparison methods have uncertainty problems of their own.

Acknowledgments. We wish to acknowledge the support of various individuals on this project: Dr. Harry Cooper, Dr. James Lamm, and Mr. Jim Merritt, all of Florida State University, and especially Dr. Song Yang of NASA/Goddard Space Flight Center/UMBC. Our appreciation is also extended to the staff at TSDIS/GSFC, who quickly and professionally handled all our requests for data transfer and consultation. The first author also extends his appreciation to the U.S. Air Force for supporting his graduate education program at Florida State University. This project was supported by NASA TRMM Grant NAG5-4752.

REFERENCES

Arkin, P. A., and B. Meisner, 1987: The relationship between large-scale convective rainfall and cold cloud over the Western Hemisphere during 1982–1984. *Mon. Wea. Rev.,* **115,** 51–74.

Barrett, E. C., and D. W. Martin, 1981: *The Use of Satellite Data in Rainfall Monitoring.* Academic Press, 340 pp.

Bell, T. L., A. Abdullah, R. L. Martin, and G. L. North, 1990: Sampling errors for satellite derived tropical rainfall: Monte Carlo study using a space-time stochastic model. *J. Geophys. Res.,* **95D,** 2195–2205.

Bellon, A., and G. L. Austin, 1986: On the relative accuracy of satellite and raingage rainfall measurements over middle latitudes during daylight hours. *J. Climate. Appl. Meteor.,* **25,** 1712–1724.

Damant, C., G. L. Austin, A. Bellon, and R. S. Broughton, 1983:

Errors in the Thiessen technique for estimating areal rain amounts using weather radar data. *J. Hydrol.,* **62,** 81–94.

Dodge, J. C., and H. M. Goodman, 1994: The WetNet project. *Remote Sens. Rev.,* **11,** 5–21.

Farrar, M. R., 1997: Combined radar-radiometer rainfall retrieval for TRMM using structure function-based optimization. Ph.D. dissertation, Florida State University, 185 pp.

Gadgil, S., and S. Sajani, 1998: Monsoon precipitation in the AMIP runs. *Climate Dyn.,* **14,** 659–689.

Haddad, Z. S., E. A. Smith, C. D. Kummerow, T. Iguchi, M. R. Farrar, S. L. Durden, M. Alves, and W. S. Olson, 1997: The TRMM 'Day-1' radar/radiometer combined rain-profiling algorithm. *J. Meteor. Soc. Japan,* **75,** 799–809.

Harrold, T. W., E. J. English, and C. A. Nicholass, 1973: The Dee weather radar project: The measurement of area precipitation using radar. *Weather,* **28,** 332–338.

Hitschfeld, W., and J. Bordan, 1954: Errors inherent in the radar measurement of rainfall at attenuating wavelengths. *J. Meteor.,* **11,** 58–67.

Hou, A. Y., S. Zhang, A. da Silva, and R. Rood, 1999: Improving assimilated global data sets using SSM/I-derived precipitation and columnar moisture observations. *GEWEX News,* **9,** 4–5.

——, D. Ledvina, A. da Silva, S. Zhang, J. Joiner, R. Atlas, G. Huffman, and C. Kummerow, 2000: Assimilation of SSM/I-derived surface rainfall and total precipitable water for improving the GEOS analysis for climate studies. *Mon. Wea. Rev.,* **128,** 509–537.

Iguchi, T., and R. Meneghini, 1994: Intercomparison of single-frequency methods for retrieving a vertical rain profile from airborne or spaceborne radar data. *J. Atmos. Oceanic Technol.,* **11,** 1507–1511.

Kummerow, C., W. S. Olson, and L. Giglio, 1996: A simplified scheme for obtaining precipitation and vertical hydrometeor profiles from passive microwave sensors. *IEEE Trans. Geosci. Remote Sens.,* **34,** 1213–1232.

——, W. Barnes, T. Kozu, J. Shiue, and J. Simpson, 1998: The Tropical Rainfall Measuring Mission (TRMM) sensor package. *J. Atmos. Oceanic Technol.,* **15,** 809–817.

——, and Coauthors, 2000: The status of the Tropical Rainfall Measuring Mission (TRMM) after two years in orbit. *J. Appl. Meteor.,* **39,** 1965–1982.

Meneghini, R. J., J. Eckerman, and D. Atlas, 1983: Determination of rain rate from a spaceborne radar technique. *IEEE Trans. Geosci. Remote Sens.,* **21,** 34–43.

Morrissey, M. L., 1991: Using sparse raingages to test satellite-based rainfall algorithms. *J. Geophys. Res.,* **96,** 18 561–18 571.

Mugnai, A., H. J. Cooper, E. A. Smith, and G. J. Tripoli, 1990: Simulation of microwave brightness temperatures of an evolving hail storm at the SSM/I frequencies. *Bull. Amer. Meteor. Soc.,* **71,** 2–13.

Panegrossi, G., S. Dietrich, F. S. Marzano, A. Mugnai, E. A. Smith, X. Xiang, G. J. Tripoli, P. K. Wang, and J. P. V. Poiares Baptista, 1998: Use of cloud model microphysics for passive microwave-based precipitation retrieval: Significance of consistency between model and measurement manifolds. *J. Atmos. Sci.,* **55,** 1644–1673.

Rodriguez-Iturbe, I., and J. M. Mejia, 1974: The design of rainfall networks in time and space. *Water Resour. Res.,* **10,** 713–728.

Seed, A., and G. L. Austin, 1990: Variability of summer Florida rainfall and its significance for the estimation of rainfall by gages, radar, and satellite. *J. Geophys. Res.,* **95D,** 2207–2215.

Shin, K.-S., and G. R. North, 1988: Sampling error study for rainfall

estimate by satellite using a stochastic model. *J. Appl. Meteor.,* **27,** 1218–1231.

Simpson, J., and C. Kummerow, 1996: TRMM Science Operations Plan. [Available online at http://www.trmm.gsfc.nasa.gov.]

——, R. F. Adler, and G. R. North, 1988: A proposed Tropical Rainfall Measuring Mission (TRMM) satellite. *Bull. Amer. Meteor. Soc.,* **69,** 278–295.

——, C. Kummerow, W.-K. Tao, and R. F. Adler, 1996: On the Tropical Rainfall Measuring Mission (TRMM). *Meteor. Atmos. Phys.,* **60,** 19–36.

Smith, E. A., X. Xiang, A. Mugnai, and G. Tripoli, 1992: A cloud radiation model for spaceborne precipitation retrieval. *Extended Abstract, Int. TRMM Workshop on the Processing and Utilization of the Rainfall Data Measured from Space,* Tokyo, Japan, Communications Research Laboratory, 273–283.

——, ——, ——, and ——, 1994a: Design of an inversion-based precipitation profile retrieval algorithm using an explicit cloud model for initial guess microphysics. *Meteor. Atmos. Phys.,* **54,** 53–78.

——, C. Kummerow, and A. Mugnai, 1994b: The emergence of inversion-type profile algorithms for estimation of precipitation from satellite passive microwave measurements. *Remote Sens. Rev.,* **11,** 211–242.

——, A. Mugnai, and G. Tripoli, 1995a: Theoretical foundations and verification of a multispectral, inversion-type microwave precipitation profile retrieval algorithm. *Passive Microwave Remote Sensing of Land–Atmosphere Interactions,* B. J. Choudhury et al., Eds., VSP Press, 599–621.

——, Z. Haddad, and C. Kummerow, 1995b: Overview on TRMM combined algorithm development. *TRMM Science Operations Plan,* J. Simpson and C. Kummerow, Eds., Florida State University, 65 pp.

——, F. J. Turk, M. R. Farrar, A. Mugnai, and X. Xiang, 1997: Estimating 13.8-GHz path-integrated attenuation from 10.7-GHz brightness temperatures for the TRMM combined PR-TMI precipitation algorithm. *J. Appl. Meteor.,* **36,** 365–388.

——, and Coauthors, 1998: Results of WetNet PIP-2 project. *J. Atmos. Sci.,* **55,** 1483–1536.

Tao, W.-K., and J. Simpson, 1993: Goddard cumulus ensemble model. Part I: Model description. *Terr. Atmos. Oceanic Sci.,* **4,** 35–72.

——, S. Long, J. Simpson, and R. Adler, 1993a: Retrieval algorithms for estimating the vertical profiles of latent heat release: Their applications for TRMM. *J. Meteor. Soc. Japan,* **71,** 685–700.

——, J. Simpson, C.-H. Sui, B. Ferrier, S. Lang, J. Scala, M.-D. Chou, and K. Pickering, 1993b: Heating, moisture, and water budgets of tropical and midlatitude squall lines: Comparison and sensitivity to longwave radiation. *J. Atmos. Sci.,* **50,** 673–690.

Tripoli, G. J., 1992a: A nonhydrostatic model designed to simulate scale interaction. *Mon. Wea. Rev.,* **120,** 1342–1359.

——, 1992b: An explicit three-dimensional nonhydrostatic numerical simulation of a tropical cyclone. *Meteor. Atmos. Phys.,* **49,** 229–254.

Wilheit, T., and Coauthors, 1994: Algorithms for the retrieval of rainfall from passive microwave measurements. *Remote Sens. Rev.,* **11,** 163–194.

Yang, S., and E. A. Smith, 1999a: Moisture budget analysis of TOGA COARE area using SSM/I-retrieved latent heating and large-scale Q2 estimates. *J. Atmos. Oceanic Technol.,* **16,** 633–655.

——, and ——, 1999b: Four-dimensional structure of monthly latent heating derived from SSM/I satellite measurements. *J. Climate,* **12,** 1016–1037.

——, and ——, 2000: Vertical structure and transient behavior of convective-stratiform heating in TOGA COARE from combined satellite-sounding analysis. *J. Appl. Meteor.,* **39,** 1491–1513.

Chapter 20

Status of TRMM Monthly Estimates of Tropical Precipitation

ROBERT F. ADLER

Laboratory for Atmospheres, NASA/Goddard Space Flight Center, Greenbelt, Maryland

CHRISTIAN KUMMEROW

Department of Atmospheric Sciences, Colorado State University, Fort Collins, Colorado

DAVID BOLVIN

Science Systems and Applications, Inc., Lanham, Maryland

SCOTT CURTIS

Joint Center for Earth Systems Technology, Baltimore, Maryland

CHRIS KIDD

University of Birmingham, Birmingham, United Kingdom

ABSTRACT

Three years of Tropical Rainfall Measuring Mission (TRMM) monthly estimates of tropical surface rainfall are analyzed to document and understand the differences among the TRMM-based estimates and how these differences relate to the pre-TRMM estimates and current operational analyses. Variation among the TRMM estimates is shown to be considerably smaller than among a pre-TRMM collection of passive microwave-based products. Use of both passive and active microwave techniques in TRMM should lead to increased confidence in converged estimates.

Current TRMM estimates are shown to have a range of about 20% for the tropical ocean as a whole, with variations in heavily raining ocean areas of the Intertropical Convergence Zone (ITCZ) and South Pacific Convergence Zone (SPCZ) having differences over 30%. In midlatitude ocean areas the differences are smaller. Over land there is a distinct difference between the Tropics and midlatitude with a reversal between some of the products as to which tends to be relatively high or low. Comparisons of TRMM estimates with ocean atoll and land rain gauge information point to products that might have significant regional biases. The bias of the radar-based product is significantly low compared with atoll rain gauge data, while the passive microwave product is significantly high compared to rain gauge data in the deep Tropics.

The evolution of rainfall patterns during the recent change from intense El Niño to a long period of La Niña and then a gradual return to near neutral conditions is described using TRMM. The time history of integrated rainfall over the tropical oceans (and land) during this period differs among the passive and active microwave TRMM estimates.

1. Introduction

The Tropical Rainfall Measuring Mission (TRMM), a joint satellite mission of the United States and Japan, was launched in late November 1997 and is providing a wealth of information related to precipitation in the Tropics. A description of the mission and the satellite instruments, along with a summary of initial results from the mission, is given by Kummerow et al. (2000) and by Simpson et al. (2000). This paper examines the estimates of tropical surface precipitation made by TRMM and compares these results to those of the pre-

TRMM era and the monthly global analyses of the Global Precipitation Climatology Project (GPCP; Huffman et al. 1997).

When the concept of a TRMM-like mission was first proposed in the early to mid-1980s (the first TRMM Science Working Group Meeting was in 1986) there was little quantitative knowledge of the magnitude and geographic and seasonal distribution of rainfall in the Tropics, especially over the oceans. Climatologies based on ship reports of weather (e.g., Jaeger 1976; Legates and Wilmott 1990) described the oceanic intertropical convergence zone (ITCZ) and other features but differed considerably on the magnitude of tropical rainfall and exactly how it was distributed over the tropical oceans, even in terms of a long-term climatology. Satellite-based estimates during the period 1975–85 focused on the use of both infrared (IR) satellite data (especially geosynchronous data) and on data from the early passive microwave instruments. Cloud statistics from geosynchronous IR observations were compared to Global Atmospheric Research Programme (GARP) Atlantic Tropical Experiment (GATE) surface-based radar data (Arkin and Meisner 1987) to produce a simple relation that when applied to geosynchronous data from around the globe gave rainfall estimates that allowed seasonal and interannual changes to be described. The Electronic Scanning Microwave Radiometer (ESMR), flying on the *Nimbus-5* polar-orbiting satellite launched in 1974, measured upwelling radiation at 19 GHz, enabling estimates of precipitation to be made over the ocean (Wilheit et al. 1977). A later Nimbus instrument, the Scanning Multichannel Microwave Radiometer (SMMR) provided multifrequency observations up to 37 GHz and was also used to estimate precipitation. However, over-water precipitation measurements from both of these instruments were limited in accuracy because of data quality, instrument calibration, and sampling issues.

In mid-1987 the first in a series of Special Sensor Microwave Imager (SSM/I) instruments (Hollinger et al. 1990) was launched on board a U.S. Department of Defense polar orbiting satellite. This well-calibrated, conically scanning instrument observing frequencies between 19 and 86 GHz provided operational agencies and researchers the observational basis for developing and applying passive microwave precipitation algorithms.

TRMM built on the hardware and science experience provided by results from these SSM/I instruments. TRMM combined a SSM/I-like instrument, with an additional frequency at 10 GHz [TRMM Microwave Imager (TMI)], with an active microwave sensor at 14 GHz [Japan's Precipitation Radar (PR)]. These two instruments, combined with a lower orbit altitude for higher spatial resolution and a precessing orbit to observe the diurnal cycle, provide the most complete precipitation-observing complement of instruments ever sent into orbit. Two additional instruments complete TRMM's precipitation package. The Visible and Infrared Scanner (VIRS) provides the connection from precipitation observations to cloud information available from high–time resolution (hourly) geosynchronous observations. The Lightning Imaging Sensor (LIS) provides lightning occurrence information critical in determining lightning–precipitation relations and microphysical insights. The TRMM instruments are described in Kummerow et al. (1998).

TRMM was launched late in November 1997 and more than three years of data have been recorded and archived. This paper will assess these three years of TRMM surface precipitation estimates, how they com-

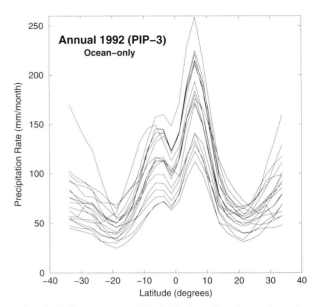

Fig. 20.1. Zonally averaged, latitudinal profiles of oceanic precipitation for 1992 for 18 algorithms using SSM/I passive microwave data submitted for the PIP-3 algorithm intercomparison activity.

pare with the pre-TRMM state of precipitation estimation, and how they compare with a standard research analysis of monthly precipitation.

2. Tropical precipitation estimates at the time of TRMM launch

At the time of TRMM's launch in late 1997 there was still considerable variation in the estimation of rainfall over the tropical oceans. Figure 20.1 shows a collection of zonally averaged rainfall over ocean for the year 1992 based on SSM/I microwave data and various algorithms submitted for an algorithm intercomparison exercise in 1996 called the Third Precipitation Intercomparison Project (PIP-3; see Adler et al. 2001). Obviously, the estimated values vary considerably. At the peak at 8°N the values range from 120 to 260 mm month^{-1}, a factor of more than 2. In the subtropic minima the range of values remains approximately a factor of 2 and increases to over a factor of 3 at 40° latitude in both hemispheres. An even larger range of values results when all the PIP-3 observational products (including IR-based and others) are included.

The collection of passive microwave estimates in Fig. 20.1 is a good representation of the broad state of knowledge of tropical oceanic rainfall in the middle 1990s. In the following section the TRMM-based estimates resulting from both the TMI (passive microwave) and PR (active microwave) instruments are compared to the pre-TRMM results of Fig. 20.1.

3. Climatological rainfall from TRMM

A summary of TRMM rain products discussed in this paper is presented in Table 20.1. The levels (2 or 3)

TABLE 20.1. TRMM satellite products.

Name	Ref. no.	Purpose
Level-2 data		
TMI profiles	2A-12	Surface rainfall and 3D structure of hydrometeor and heating over TMI swath (Kummerow et al. 2001).
PR profiles	2A-25	Surface rainfall and 3D structure of hydrometeors (Iguchi et al. 1998).
PR TMI combined	2B-31	Surface rainfall and 3D structure of hydrometeors derived from TMI and PR simultaneously (Haddad et al. 1997)
Level-3 data		
TMI monthly rain	3A-11	Monthly rainfall maps, ocean only (Chang et al. 1999).
TRMM and other satellites	3B-42	Geostationary precipitation data calibrated by TRMM daily, 1° resolution (Adler et al. 2000).
TRMM and other data	3B-43	TRMM, calibrated IR, and gauge products; data merged into single rain product, monthly, 1° res. (Adler et al. 2000).

follow the standard National Aeronautics and Space Administration (NASA) nomenclature. Level 2 consists of the retrieved geophysical parameters at the satellite footprint level, while level-3 products represent either space- or time-averaged geophysical parameters. All rainfall products discussed here are version 5, introduced on 1 October 1999. Details on the algorithms can be found in the references in Table 20.1.

Maps of TRMM climatologies from January 1998 to December 2000 are shown in Fig. 20.2 for each of the satellite products resulting from application of the algorithms (version 5) at a spatial resolution of 2.5° latitude and longitude, except for the TMI statistical product, which is only available at 5° resolution over ocean. The patterns are very similar, with the differences in magnitude to be discussed shortly. Portions of the ITCZ are evident in both hemispheres over the Atlantic and Pacific Oceans, along with land maxima in Africa, South America, and over the maritime continent. Midlatitude maxima are evident across and to the east of Japan and the United States.

An intercomparison of zonal mean rainfall accumulations for 1998–2000 for the five major rainfall algorithms (version 5) is presented in Fig. 20.3. Version 5 of the algorithms represents the initial improvements of the algorithms during the first two years after launch, beyond corrections made to eliminate software errors. As can be seen from Fig. 20.3, the zonal averages for this near-3-yr period have a wide range between the TMI profiling algorithm and that of the PR. Table 20.2 gives the ocean, land, and total precipitation in the 37.5°N–37.5°S band for each of the TRMM products and also includes the estimate based on the GPCP monthly analyses. The tropical mean estimates (ocean only) vary from 92 mm month^{-1} for the TMI (2A-12) to 75 mm month^{-1} for the PR (2A-25) estimate, a range of 17 mm month^{-1} or 20%. This is the same approximate range of values shown by Kummerow et al. (2000), using a period of 1 yr (1998). The land values also have a similar value of range (20%). The TRMM version-5

results do indicate a significant narrowing of the differences among the algorithms as compared to the earlier version 4 (Kummerow et al. 2000). The GPCP value over the oceans is the same as that from TRMM 3A-11, which is not surprising because a very similar algorithm applied to SSM/I data is the driver for the combination of data over the oceans for the operational GPCP analyses. The GPCP analyses include rain gauge information over land and therefore gives a first indication of possible biases of the TRMM satellite products over land.

As a means to estimate how this variation among the TRMM-based estimates relates to the pre-TRMM spread, the standard deviation of the 18 SSM/I-based estimates over ocean from PIP-3 in Fig. 20.1 is compared to the same statistic from the collection of four TRMM estimates, also over the ocean. The product that includes the geostationary observations was not included in this exercise in order to keep it an all-microwave comparison. The result in Fig. 20.4 shows that the variation among the TRMM microwave products is significantly smaller than from the earlier collection of SSM/I-based, passive microwave products. These results can be interpreted as TRMM making a significant improvement in the estimation of total rainfall in the Tropics as compared to the wide range of estimates available before TRMM, although it should be noted that some of the estimates in Fig. 20.1 were considered experimental. Perhaps more importantly it should also be remembered that the two TRMM algorithms that use the PR data had no previous application to satellite data. They are, therefore, at an earlier stage of testing with space data as compared to the passive microwave algorithms, which have had extensive testing with SSM/I data.

The fact that at this point in the analysis of TRMM data there is still a fairly significant difference in magnitude among the TRMM products is not that surprising considering the variability of the pre-TRMM products and the "youth" of the PR-related algorithms. Because the passive microwave and radar estimates depend in

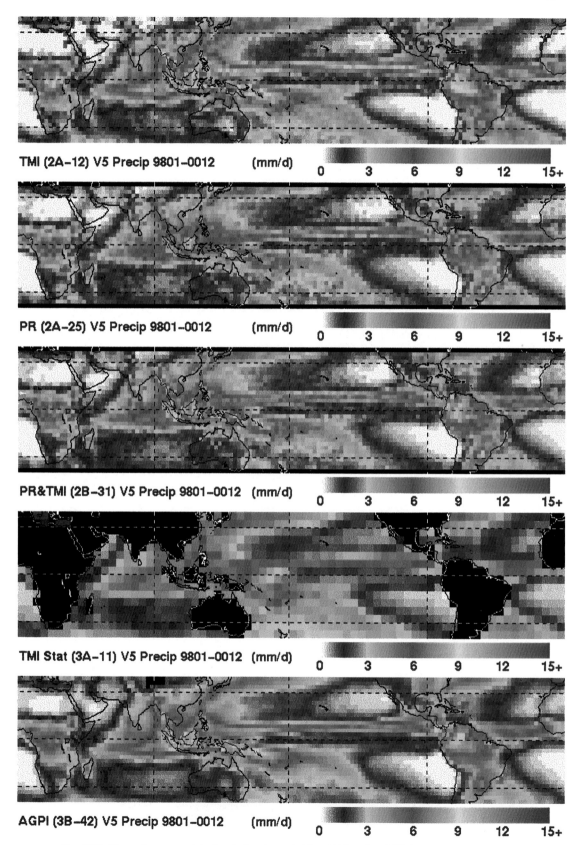

FIG. 20.2. Maps of mean precipitation during the period Jan 1998 to Dec 2000 from five TRMM products.

TRMM V5 Zonal Mean Rainfall

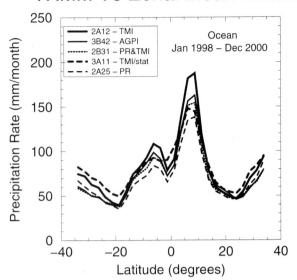

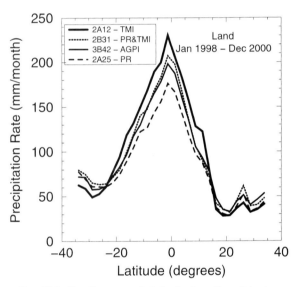

FIG. 20.3. Zonally averaged, latitudinal profiles of (top) oceanic and (bottom) land precipitation from five TRMM products (four over land) for the period Jan 1998 to Dec 2000.

Standard Deviations

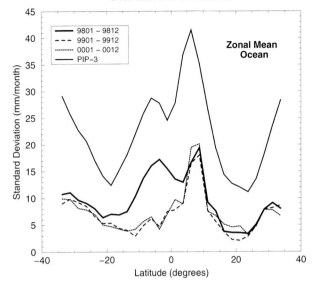

FIG. 20.4. Standard deviation as a function of latitude among monthly rainfall estimates. Three of the curves are for the standard deviation during the three years of TRMM among the four TRMM ocean estimates. The PIP-3 curve is for the standard deviation among the 18 estimates seen in Fig. 20.1.

very different ways on the microphysics and structure of the rainfall, their eventual convergence should strengthen our confidence in the resulting estimates.

4. Time evolution of TRMM estimates

During the first three years of TRMM, the patterns in tropical precipitation underwent a significant change. A rapid transition from El Niño to La Niña rainfall patterns occurred during the first half of 1998 (Adler et al. 2000), followed by a long period of La Niña pattern into the middle of 2000, and then a return to a near-neutral ENSO status. This evolution during the three year TRMM dataset can be seen in Fig. 20.5, where anomalies from monthly precipitation climatologies are shown. The TRMM 3B-43 product is used for this figure, although each of the products would show qualitatively similar evolutions. The 3B-43 product is a blend

TABLE 20.2. Tropical rainfall totals.

Method	Ocean	Land	Total
TMI-profiling			
2A-12	92 mm month^{-1}	100 mm month^{-1}	94 mm month^{-1}
PR			
2A-25	76	81	78
PR-TMI			
2B-31	80	93	83
TMI-Stat			
3A-11	88		
TRMM/other satellites			
3B-42	83	94	86
GPCP	89	80	86

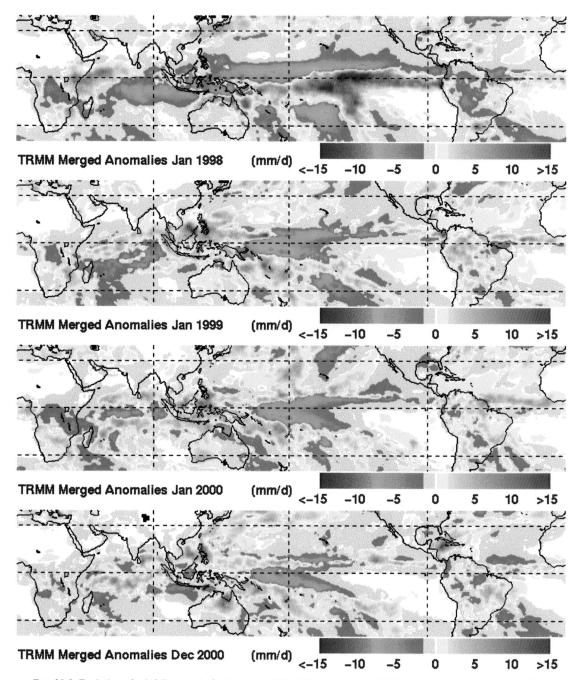

FIG. 20.5. Evolution of rainfall anomaly fields during 1998–2000 based on the TRMM merged analysis product (3B-43).

of the 3B-42 satellite estimate and monthly gauge estimates over land. Product 3B-42 currently uses the combined TMI/PR product (2B-31) to calibrate the Geostationary Precipitation Index (GPI; Arkin and Meisner 1987) IR-based estimates to create an adjusted GPI, or AGPI. The monthly climatologies used are based on the GPCP 20-yr climatology, adjusted to match the magnitude of the 3B-43 product over the 3-yr period. The top panel in Fig. 20.5 shows the anomalies in January 1998, near the beginning of the TRMM mission and

when the El Niño was still very strong. A very large excess in rainfall is obvious in the eastern Pacific Ocean along with a rainfall deficit extending from the Indian Ocean, through the Maritime Continent, and into the western North Pacific Ocean. A significant area of above average rain is evident in east Africa and also a general deficit of rain over the Amazon. The negative anomaly along the north side of the Pacific Ocean maximum is the beginning of the negative anomalies that come to dominate this area in the coming La Niña.

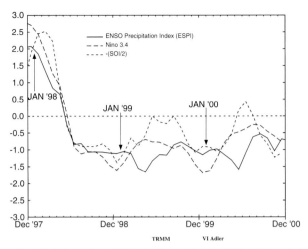

FIG. 20.6. Evolution of ENSO indices during the 1998–2000 period. The three indices are the ESPI of Curtis and Adler (2000), the Niño-3.4 SST anomaly, and the SOI.

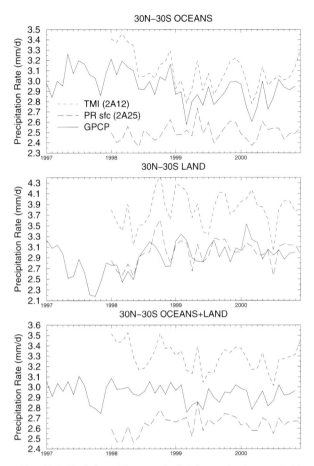

FIG. 20.7. Evolution of integrated rainfall over (top) ocean, (middle) land, and (bottom) total area during the period 1998–2000 from two TRMM estimates and the GPCP analysis.

The second panel of Fig. 20.5 shows the situation a year later, in January 1999. The anomaly pattern is strikingly different at this point with above average rainfall over the Maritime Continent and a rainfall deficit over the central Pacific Ocean. The transition from El Niño to La Niña occurred very rapidly in early 1998 as can be seen in Fig. 20.6, which shows the evolution of the ENSO Precipitation Index (ESPI; Curtis and Adler 2000) and the Niño-3.4 SST anomalies and Southern Oscillation index (SOI) during the TRMM mission. The ESPI is a measure of the strength of the anomalous Walker circulation based on gradients of the precipitation anomalies over the Maritime Continent and the central Pacific Ocean. The ESPI, and the other indices, show a rapid change from strong positive (El Niño) to strong negative (La Niña) during the first half of 1998. Thus the first year of TRMM encompassed the end of a major El Niño and the first part of a La Niña.

The La Niña continued throughout 1999 (see Fig. 20.6) and the anomaly map for January 2000 (third panel in Fig. 20.5) shows a pattern very similar to that of January 1999. By July 2000 (bottom panel of Fig. 20.5) the La Niña had weakened and the anomaly pattern had smaller-scale features over the Indian and Pacific Oceans. The indices in Fig. 20.6 were nearing neutral conditions. Thus TRMM viewed La Niña conditions for about two years between mid-1998 and mid-2000.

Although all the TRMM products show the same pattern of anomaly fields during the 1998–2000 period, there is a difference in the evolution of the tropical total rainfall during the 1998–2000 period. Figure 20.7 shows the time change of the TRMM estimates integrated over water and over ocean plus land. The top panel (Fig. 20.7a) is an extension in time of a figure in Adler et al. (2000), here using version-5 data. The results confirm that, over ocean, the TRMM passive microwave product (2A-12) shows a decrease from the El Niño still in progress in early 1998 into the period of the extensive La

Niña through 1999 and into 2000. However, the two products that use the TRMM radar data have nearly constant ocean totals over the three years (only 2A-25 is shown). When land is included to produce total tropical rainfall, the decrease from 1998 to 1999 in the passive product is muted. This difference in trends among the TRMM products may be related to regional differences in the accuracy of the different TRMM estimates and the shift of the location of the rainfall maxima in the transition from El Niño to La Niña during the first year of TRMM. Figure 20.8 displays the matched probability density function (pdf) values of one passive microwave algorithm (2A-12) versus the radar algorithm (2A-25) for monthly values for the 1998 and 1999 months of January, February, and March. One can see that in both years the passive-based estimates exceed the radar estimates above very low values. However, there is a distinct difference in the magnitude with 1998, the El Niño year, having much higher 2A-12 relative values. The difference between the years is small at values below about 150 mm month^{-1} but increases drastically at higher values. Thus the difference between these two estimates as a function of time in Fig. 20.7a

TRMM JFM PDF (30N – 30S)

Ocean–only

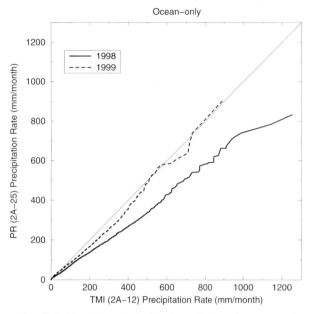

FIG. 20.8. Plots of matched pdf values of rain estimates over the tropical ocean using the TMI profiling (2A-12) and radar (2A-25) algorithms for Jan, Feb, and Mar of 1998 (El Niño) and 1999 (La Niña).

seems to be related to differences in the estimates in the heaviest raining areas. Even though this analysis is based on monthly values, high rainfall amounts on a monthly scale are typically related to occurrences of large convective systems with large instantaneous rainfall rates and large stratiform regions. These types of systems need to be investigated to determine how the algorithms perform in these situations.

5. Regional variations among TRMM products

To examine the variations among the TRMM algorithms on a regional basis, a mean of the four TRMM-alone products for the 2.5 years of data was made and maps of the deviation from that mean were done for each product (Fig. 20.9). Remember that there are only three products over land. In the deep Tropics over ocean, in the areas of rainfall maxima (e.g., ITCZ and SPCZ), the algorithms give relatively similar results. The passive profiling algorithm (2A-12) is highest with the radar-only algorithm (2A-25) and the passive statical (3A-11) having the lowest values and the combined radar-radiometer algorithm (2B-31) in between. The passive-statistical product (3A-11) has narrow bands of less than the mean values exactly in the peak areas along the ITCZ, with above-average values just outside the maximum rain areas. This pattern is due partially to the 5° latitude–longitude areas in which the product is computed, but may also reflect a difference in areas of very heavy rain between estimating rain on a pixel basis, as in the profiling algorithm, and on a distribution-fitting

basis as in the statistical technique. The ratio of values between 2A-12 and 2A-25 is over 1.3 in these areas of heavy climatological rainfall over the ocean.

In the rain maxima of midlatitudes over oceans the differences among the estimates is somewhat muted. The passive profiling algorithm is generally still the largest, but the difference between it and the radar product is smaller ratio-wise, about 1.1 to 1.2. The combined product is somewhat lower than the radar-only product in these areas. However, there are some interesting regional differences in the relative magnitudes among the products above 30° latitude. In the Northern Hemisphere the radar product (2A-25) is nearly the same as the profiling product in the North Atlantic, unlike the situation east of Japan in the North Pacific. In the Southern Hemisphere similar differences are noted with the most dramatic being off the southeast coast of Australia where 2A-25 is significantly larger than 2A-12.

In the subtropical minima over the oceans things are somewhat reversed, with the radar product (2A-25) being slightly higher than the passive profiling product (2A-12). The passive statistical (3A-11) estimates are the highest and this translates to this product being highest in the latitudinal profiles of Fig. 20.2, both in the subtropics and at midlatitudes. Thus, over oceans, there are some generalities as to the relative magnitudes of the four products, but there is significant variation regionally.

Over land there are also large differences among the algorithms as is clear in Figs. 20.3 and 20.9. Over the tropical land maxima of Africa and the Amazon (Fig. 20.9), the TMI profiling algorithm is the highest, the radar-based estimate the lowest, and the combined algorithm has intermediate values. The difference between 2A-12 and 2A-25 is about 30% over the Amazon and nearly 50% over Africa. The combined product (2B-31) is above the three product mean in the high rain areas of the Amazon but lower than the mean over similar areas of Africa. In midlatitudes over land, for example, in southeast China or the southern United States, the combined product (2B-31) is highest, the TMI product the lowest, and the radar-only product is intermediate in value. The differences in midlatitude are relatively small, however, about 10%–15%. Therefore, there are significant regional differences in the relative biases of the TRMM products, with the primary variation related to latitude.

6. Comparison with ground-based estimates

Comparison of monthly surface rainfall estimated from TRMM with ground-based estimates from gauges and from radar–gauge merged datasets can be valuable to help diagnose the large-scale and regional differences among the TRMM estimates and possibly point to algorithm improvement strategies. However, because the ground-based datasets do not cover all regions (especially over the ocean) and have their own measurement

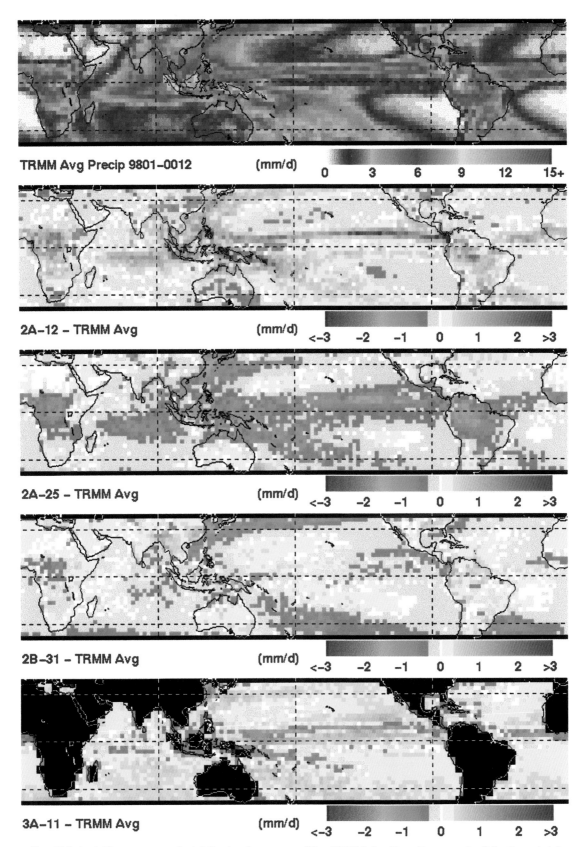

TRMM Avg Precip 9801–0012   (mm/d)   0   3   6   9   12   15+

2A–12 – TRMM Avg   (mm/d)   <–3   –2   –1   0   1   2   >3

2A–25 – TRMM Avg   (mm/d)   <–3   –2   –1   0   1   2   >3

2B–31 – TRMM Avg   (mm/d)   <–3   –2   –1   0   1   2   >3

3A–11 – TRMM Avg   (mm/d)   <–3   –2   –1   0   1   2   >3

FIG. 20.9. (top) The mean map of rainfall using the average of four TRMM algorithms (three over land) for the period Jan 1998 and Dec 2000 and the difference from that mean for each of the four algorithms (the lower four panels).

TRMM vs. Pacific Atoll Gauges

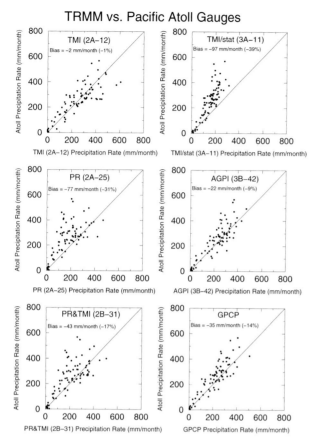

FIG. 20.10. Comparison of TRMM (and GPCP) monthly estimates with Pacific atoll gauges. The analysis is done for 2.5° lat–lon boxes with at least one gauge.

TRMM vs. GPCC Gauges (15N–15S)

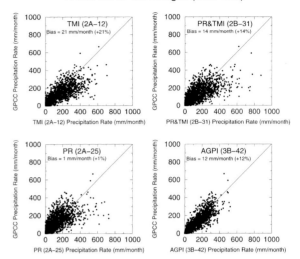

FIG. 20.11. Comparison of TRMM monthly estimates with rain gauge analyses over the lat band 15°N–15°S. The analysis is done for 2.5° lat–lon boxes with at least two gauges.

the passive profiling and combination products are closer to the mean values estimated from gauges and that the radar and passive-statistical products appear low compared to the gauge-based estimates.

Comparisons of the TRMM estimates with a gauge-based analysis (Rudolf et al. 1994) (2.5° lat–lon boxes) was used to diagnose some of the regional differences seen in the algorithms over land. Figures 20.11 and 20.12 show results of the satellite–gauge comparison for locations inside and outside of 15° latitude. The

errors, it is not always easy to draw concrete conclusions. Figure 20.10 shows results of comparing the TRMM products and the GPCP analysis for the last three years with the monthly estimates from the western Pacific Ocean atoll rain gauge dataset (Morrissey and Green 1991). Although the scatter of points is large, due to both the sampling errors of TRMM and those of the sparse gauge coverage, the results indicate that the monthly estimates have a wide range in the calculated bias between the satellite and the gauge estimate. The TMI (2A-12) algorithm has an overall small negative bias (−1%) (Fig. 20.10a). The monthly estimates based on the PR (2A-25) algorithm (Fig. 20.10b) show a much larger bias (−31%). Surprisingly, the TMI-statistical product (3A-11) has a large, negative bias, especially in the high rain areas. The similar algorithm applied to SSM/I data does not have the same large bias. The GPCP plot (Fig. 20.10f) gives an indication of that since that algorithm drives the combination product for the GPCP analysis. The combined radar–radiometer product (2B-31) and the product that uses 2B-31 to adjust the geosynchronous data (3B-42) have intermediate, negative biases. These comparisons indicate that, at least in the heavily raining area of the western Pacific Ocean,

TRMM vs. GPCC Gauges (>15NS)

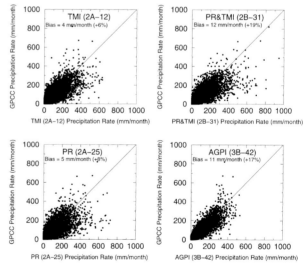

FIG. 20.12. Comparison of TRMM monthly estimates with rain gauge analyses over the lat band poleward of 15° latitude in the Tropics. The analysis is done for 2.5° lat–lon boxes with at least two gauges.

results indicate that, in the deep Tropics over land, the TMI profiling algorithm has a much larger positive bias ($+29\%$) than the PR algorithm ($+3\%$; Fig. 20.11). Outside of $15°$ latitude, the biases are the same and moderately positive ($+11\%$; Fig. 20.12). These results point to the need for evaluation of the passive microwave algorithm over land to understand the regional variation.

In summary, comparison of TRMM results with existing gauge analyses over land and water indicate that, over water, the more mature TMI profile product compares well with atoll-based rain gauges, while the more experimental PR algorithm produces estimates significantly lower than the atoll gauges in the western Pacific Ocean, as does the TMI-statistical product. Over land, comparison of TRMM products with gauge analyses produces reasonable results with relatively small biases outside of $15°$ latitude, but the TMI-based product has large positive biases relative to the gauges in the deep Tropics.

7. Conclusions

With nearly three full years of data, the Tropical Rainfall Measuring Mission (TRMM) is making a considerable contribution to our knowledge of climatological rainfall over the Tropics. The large range of possibilities with regard to absolute values that existed in the pre-TRMM era has been reduced, even though the TRMM estimates include those incorporating spaceborne radar data for the first time. Because the TRMM passive microwave and radar estimates depend in different ways on the microphysics and structure of the rainfall, their eventual convergence should strengthen our confidence in the resulting estimates. However, at this relatively early stage of the TRMM research effort the TRMM estimates still differ among themselves and those differences vary regionally.

With three years of data the TRMM estimates are shown to have a range of about 20% for the tropical ocean as a whole. Similar differences exist over land. The TRMM estimates vary around an ensemble mean of 84 mm month^{-1} (2.8 mm day^{-1}) over the tropical ocean, 97 mm month^{-1} (3.2 mm day^{-1}) over the land, and 88 mm month^{-1} (2.9 mm day^{-1}) combined. Regional variations among the algorithms are noted in maps and zonal averages with differences in heavily raining ocean areas of the ITCZ and SPCZ having differences over 30%. In midlatitude ocean areas the differences are smaller. Over land there is a distinct difference between the Tropics and midlatitude with a reversal between some of the products as to which tends to be relatively high or low. Surface-based comparison data indicate that in the deep Tropics the radar algorithm may be underestimating over the ocean, whereas the passive product may be overestimating over land.

TRMM began its flight during an intense El Niño and there was a rapid transition to a La Niña during 1998. The evolution of this ENSO event can be followed with the TRMM data in terms of movement of precipitation anomalies as is typically seen with these events. However, when the precipitation is integrated over the tropical oceans (and land) the time variation of these quantities is not the same depending on whether one is using the active or passive microwave products from TRMM. This difference in trend is, at least, partially related to the divergence of estimates in the very heavy raining areas.

The differing TRMM estimates of surface rainfall noted in this paper should converge when the physical basis for the algorithm differences are understood. Resolving these differences among the TRMM estimates and producing improved, converged estimates of tropical climatological rainfall remain among TRMM's highest scientific priorities.

REFERENCES

Adler, R. F., G. J. Huffman, D. T. Bolvin, S. Curtis, and E. J. Nelkin, 2000: Tropical rainfall distributions determined using TRMM combined with other satellite and raingauge information. *J. Appl. Meteor.,* **39,** 2007–2023.

——, C. Kidd, G. Petty, M. Morrissey, and H. Goodman, 2001: Intercomparison of global precipitation products: The third Precipitation Intercomparison Project (PIP-3). *Bull. Amer Met. Soc.,* **82,** 1377–1396.

Arkin, P. A., and B. N. Meisner, 1987: The relationship between large-scale convective rainfall and cold cloud over the Western Hemisphere during 1982–1984. *Mon. Wea. Rev.,* **115,** 51–74.

Chang, A. T. C., L. S. Chiu, C. Kummerow, and J. Meng, 1999: First results of the TRMM Microwave Imager (TMI) monthly oceanic rain rate: Comparison with SSM/I. *Geophys. Res. Lett.,* **26,** 2379–2382.

Curtis, S., and R. Adler, 2000: ENSO indices based on patterns of satellite-derived precipitation. *J. Climate,* **13,** 2786–2793.

Haddad, Z. S., E. A. Smith, C. D. Kummerow, T. Iguchi, M. R. Farrar, S. L. Durden, M. Alves, and W. S. Olson, 1997: The TRMM "Day-1" radar/radiometer combined rain-profiling algorithm. *J. Meteor. Soc. Japan,* **75** (4), 799–809.

Hollinger, J. P., J. L. Pierce, and G. A. Poe, 1990: SSM/I instrument evaluation. *Trans. IEEE Geosci. Remote Sens.,* **4,** 781–790.

Huffman, G. J., and Coauthors, 1997: The Global Precipitation Climatology Project (GPCP) version 1 dataset. *Bull. Amer. Meteor. Soc.,* **78,** 5–20.

Iguchi, T., T. Kozu, R. Meneghini, J. Awaka, and K. Okamoto, 1998: Preliminary results of rain profiling with the TRMM precipitation radar. *Proc. 8th Int. Union of Radio Science (URSI) Com F Triennial Open Symp. on Wave Propagation and Remote Sensing,* Aveiro, Portugal, URSI, 147–150.

Jaeger, L., 1976: Monatskarten des Niederschlags fur die ganze Erde. *Ber. Dtsch. Wetterdienstes,* **139,** 33.

Kummerow, C., W. Barnes, T. Kozu, J. Shiue, and J. Simpson, 1998: The Tropical Rainfall Measuring Mission (TRMM) sensor package. *J. Atmos. Oceanic Technol.,* **15,** 808–816.

——, and Coauthors, 2000: The status of the Tropical Rainfall Measuring Mission (TRMM) after two years in orbit. *J. Appl. Meteor.,* **39,** 1965–1982.

——, Y. Hong, W. Olson, S. Yang, R. Adler, J. McCollum, R. Ferraro, G. Petty, and T. Wilheit, 2001: The evolution of the Goddard profiling algorithm (GPROF) for rainfall estimation from passive microwave sensors. *J. Appl. Meteor.,* **40,** 1801–1820.

Legates, D., and C. J. Wilmott, 1990: Mean seasonal and spatial variability in gauge-corrected, global precipitation. *Int. J. Climatol.,* **10,** 111–127.

Morrissey, M. L., and J. S. Green, 1991: *The Pacific Atoll Raingauge*

Data Set. Planetary Geoscience Division Contribution 648, The University of Hawaii, 45 pp.

Rudolf, B., H. Hauschild, W. Rueth, and U. Schneider, 1994: Terrestrial precipitation analysis: Operational method and required density of point measurements. *Global Precipitations and Climate Change,* Vol. 26, M. Desbois and F. Desalmond Eds., *NATO ASI Series I,* Springer-Verlag, 173–186.

Simpson, J., and Coauthors, 2000: The Tropical Rainfall Measuring Mission (TRMM) progress report (in Russian). *Earth Obs. Remote Sens., 4,* 71–90.

Wilheit, T. T., T. C. Chang, M. S. V. Rao, E. B. Rodgers, and J. S. Theon, 1977: A satellite technique for quantitatively mapping rainfall rates over the oceans. *J. Appl. Meteor., 16,* 551–560.